AF597448

ACS SYMPOSIUM SERIES 379

Biotechnology for Crop Protection

Paul A. Hedin, EDITOR
U.S. Department of Agriculture

Julius J. Menn, EDITOR
U.S. Department of Agriculture

Robert M. Hollingworth, EDITOR
Michigan State University

Developed from a symposium sponsored
by the Division of Agrochemicals
at the Third Special Conference
of the Division of Agrochemicals
of the American Chemical Society,
Snowbird, Utah,
June 28–July 3, 1987

American Chemical Society, Washington, DC 1988

Library of Congress Cataloging-in-Publication Data

Biotechnology for crop protection.
Paul A. Hedin, editor, Julius J. Menn, editor, Robert M. Hollingworth, editor

"Developed from a symposium sponsored by the Division of Agrochemicals at the Third Special Conference of the Division of Agrochemicals of the American Chemical Society, Snowbird, Utah, June 28–July 3, 1987."

p. cm.—(ACS Symposium Series, 0097–6156; 379).

Includes index.

ISBN 0–8412–1545–6
1. Plants, Protection of—Congresses. 2. Plant biotechnology—Congresses.

I. Hedin, Paul A (Paul Arthur), 1926– . II. Menn, Julius J. III. Hollingworth, Robert M., 1939– . IV. American Chemical Society. Division of Agrochemicals. V. American Chemical Society. Division of Agrochemicals. Special Conference (3rd: 1987: Snowbird, Utah). VI. Series.

SB950.A2B58 1988
632—dc19 88–28125
CIP

PRINTED IN THE UNITED STATES OF AMERICA

ACS Symposium Series

M. Joan Comstock, *Series Editor*

Foreword

The ACS SYMPOSIUM SERIES was founded in 1974 to provide a medium for publishing symposia quickly in book form. The format of the Series parallels that of the continuing ADVANCES IN CHEMISTRY SERIES except that, in order to save time, the papers are not typeset but are reproduced as they are submitted by the authors in camera-ready form. Papers are reviewed under the supervision of the Editors with the assistance of the Series Advisory Board and are selected to maintain the integrity of the symposia; however, verbatim reproductions of previously published papers are not accepted. Both reviews and reports of research are acceptable, because symposia may embrace both types of presentation.

Contents

CONTROL OF INSECTS

Preface

BIOTECHNOLOGY HAS BEEN DEFINED as the use of biological means to develop processes and products by studying organisms and their components. Biological means include bioreactors, bioreactions, immunolocalization, biologically active cells and enzymes, cell tissue and organ cultures, genetic engineering, gene transfer, recombinant DNA technology, and hybridoma techniques. Although attaining results has generally been a slow process, genetics and breeding are successful ways to solve the pest problems of crop plants, and the development of new biocontrol agents may now be expedited by the use of genetic engineering and other manipulative techniques. These new methods offer approaches and precision never before available to scientists.

The emergence of biotechnology research and products has had a broad impact on science, the economy, and the public. As we begin to implement biotechnology projects, the public has become involved with the ethical implications of transferring DNA between different species. This has led to questions about regulation of new products and the need to examine these exciting opportunities and their potential consequences.

The objectives of the Division of Agrochemicals are to consider all scientific aspects of chemistry relevant to the control of pests of agricultural significance and other methods for enhancing or modifying agricultural productivity. We therefore brought together scientists who work in biotechnology to encourage continued research and to disseminate information to other scientists about the status of and opportunities in this field. This book is a collection of chapters and abstracts based on presentations given at the conference.

- Ganesh Kishore organized the section on the control of weeds. Chapters cover biorational synergists for oxidant-generating herbicides; the molecular biology of resistance to sulfonylurea herbicides; genetic engineering of atrazine resistance; and increasing EPSP–synthase activity in crop plants to increase tolerance to glyphosate.

- Horace Cutler organized the section on controlling plant diseases. These chapters discuss trichothecenes and their role in the expression of plant disease; the role of phytoalexins in resistance of peanuts to aflatoxin contamination; preinfection changes in germlings of rust fungus induced to differentiate by host contact; and studies of structure and biogenesis, toxic effects, and metabolites of fungal pathogens.

- Herbert Oberlander coordinated the section that features chapters on sex-specific selection using chimeric genes; potential applications of neuroendocrine research to insect control; insect cuticle structure and metabolism; molecular aspects of immune mechanisms in insects; molecular genetics of nerve insensitivity resistance to insecticides; and inhibition of juvenile hormone esterase by transition-state analogs.

- Jerome Miksche, who developed the section on plant gene expression, reports on biotechnology research in the Agricultural Research Service. Also presented are chapters on the regulation of gene expression in the sunflower embryo; biosynthesis; post-translational acylation and degradation of the chloroplast 32kDa Photosystem II reaction center protein; and applications of restriction fragment length polymorphisms to crop plants.

- Bruce Carlton organized the section that summarizes recent progress in the development of genetically engineered microbial pesticides. Many *Bacillus thuringiensis* strains contain plasmid-encoded genes for insecticidal proteins that can be further enhanced by gene-splitting techniques. Chapters about the transfer of *B. thuringiensis* genes into *E. coli* and tobacco and their toxic effects on insects are included. Progress in the development of plant-colonizing microbes as pesticidal protein-delivery systems is also reported.

- Bruce Hammock developed the section on new immunochemical techniques. These chapters describe investigation of disease resistance in plants; neuronal development in insect embryos; pesticide residue analysis for plant diagnostics and quarantine; and the development of a biosensor for applying monoclonal antibodies and microelectronics for environmental analysis.

- Philip Kearney coordinated the section on monitoring the use of biotechnology and genetic engineering products released into the environment. It includes a chapter on monitoring *B. thuringiensis* in the environment with ELISA; DNA probe technology to track engineered microorganisms; using DNA detection cassettes based on unique DNA sequences to evaluate plasmid gene transfer to the environment; and using ribosomal RNA to establish a normal sequence base and then to determine whether changes may occur because of releases.

- Frederick Betz organized the section on regulatory considerations for biotechnology products which includes chapters from a member of a public interest group, the biotechnology industry, the U.S. Department of Agriculture, a state experiment station, and the U.S. Environmental Protection Agency. The consensus of these authors is that biotechnology will develop products for the market at an ever-increasing rate, but caution must be exercised to prevent environmental damage and concomitant damage to the health of the science industry.

We hope that this book will contribute to the understanding and subsequent adoption of additional concepts and research strategies for the control of pests. We believe that biotechnology, with all its ramifications, will contribute to the common good of public health, economic vigor, and the overall quality of life. We also hope that this book will serve as a tool to identify unifying themes by which biotechnology can be applied to control the pests of crop plants.

We thank Willa Garner for her excellent administrative management of the symposium on which this book is based. We are also grateful to all of the participants for their contributions, which add materially to this book. Finally, we thank the Agricultural Research Service of the U.S. Department of Agriculture for a financial grant that aided in the organization of the symposium.

PAUL A. HEDIN
Agricultural Research Service
U.S. Department of Agriculture
Crop Science Research Laboratory
P.O. Box 5367
Mississippi State, MS 39762–5367

JULIUS J. MENN
Agricultural Research Service
U.S. Department of Agriculture
Plant Sciences Institute
BARC–West, Building 003
Beltsville, MD 20705

ROBERT M. HOLLINGWORTH
Michigan State University
Pesticide Research Center
East Lansing, MI 48824

May 27, 1988

CONTROL OF WEEDS

New Horizons in Weed Control

Weed control is basic to efficient and profitable agriculture. Specific damages caused by weeds include, lower crop yields, higher costs to control pests, poorer quality products, and higher consumption of nutrients. It is therefore not surprising that weed control has been practiced since ancient times. Until recently, weed control was primarily achieved by mechanical methods. Following the discovery of organic chemicals which kill weeds, the use of herbicides for weed control has gained wide spread acceptance.

One of the limitations of current day herbicides is selectivity. Although herbicides are known which kill every major weed in every major crop, most herbicides do not show selectivity between the crop and weed. Until recently, selectivity was addressed mostly by chemical synthesis of analogs of the active molecule. This approach has been successful with herbicides such as sulfonylureas, phenoxyacetic acids, amides etc. However, frequently it is not possible to design analogs of herbicides which maintain the weed spectrum and yet display crop selectivity. In these instances, attempts have been made to identify "safeners" which can selectively safen the crop towards the herbicide. This approach has only been successful with a few herbicides. Biologically, attempts have been made to identify genotypes which show higher levels of tolerance to a herbicide and breed those traits into commercial cultivars. With this approach, it is essential to identify genotypes with resistance to the herbicide, and these genotypes may have other undesirable characteristics not eliminated by breeding. Mechanical methods to specifically deliver herbicides to weeds have also been explored with limited success.

More recently, Gressel and his co-workers have suggested identifying compounds which selectively enhance the toxicity of the herbicide. For example, herbicides such as paraquat can be synergized by copper chelators which primarily inhibit metabolic detoxification of the oxygen radicals essential for herbicidal activity. Other compounds may serve to inhibit enzymes such as glutathione S-transferase which metabolically inactivate the herbicide. A more detailed discussion of this approach can be found in a later chapter in this book.

Recent developments in plant genetic engineering have provided a novel means for introduction of herbicide resistance into crop plants. With this approach, it is possible to use genes from diverse sources including bacteria, plants and animals for conferring herbicide resistance to crop plants. Herbicide resistance can be engineered into crop plants by several mechanisms. These include overproduction of a target enzyme/protein, mutation of the target protein so as to reduce herbicide sensitivity, metabolic inactivation of the herbicide and inhibition of herbicide uptake/transport. Resistance to Roundup® and sulfonylurea has been engineered by either overproduction or mutation of the target protein as discussed in the following chapters. Metabolic inactivation has proved useful for engineering tolerance to Basta®, bromoxynil and atrazine. The advantage of the metabolic inactivation approach is that it addresses both primary and secondary effects of the herbicide. It is, however, essential that the metabolite of the herbicide has environmentally desirable characteristics and, is non-toxic to birds, animals and humans. There are no published reports of engineering resistance to plants based on alterations in herbicide uptake/translocation. This is primarily because of lack of model plant/bacterial systems with resistance to herbicides (of interest) based on transport.

From an agronomic perspective, availability of genetically engineered herbicide tolerant crop plants should significantly impact our current day practices. In particular, postemergent herbicides such as Roundup® (and Basta®) provide efficient, broad spectrum, weed control and can be used when the weed problem is encountered instead of on a scheduled basis. This should lower the extent of use as well as promote the use of environmentally safe, non-toxic herbicides in agriculture. Since the herbicides currently being used in different crops do not fully address the complete weed spectrum, the ability to use broad spectrum herbicides should enhance the weed control capability of the farmer. In this context, it is interesting to note that on an annual basis, in U.S. alone, approximately 17% of the soybean yield (1.9 billion dollars) is lost due to the presence of weeds. With herbicides such as sulfonylureas and 2 imidazolinones which have soil activity, availability of herbicide tolerant crops will permit crop rotation with sensitive crops before the herbicide is inactivated in the soil. This will also ensure maximal use of the weed control capability of these herbicides.

In summary, plant genetic engineering promises to add a new dimension to weed control technology by permitting full use of the herbicide potential of the current day, environmentally safe, nontoxic chemicals.

G. Kishore
Monsanto Company
700 Chesterfield Village Parkway
St. Louis, MO 63198

RECEIVED June 3, 1988

Chapter 1

Biorational Herbicide Synergists

J. Gressel and Y. Shaaltiel

Department of Plant Genetics, The Weizmann Institute of Science, Rehovot, IL–76100, Israel

Compounds added to herbicides in order to rationally suppress the plants' tolerance mechanisms can synergize the herbicides. Tridiphane suppresses glutathione-*S*-transferase and thus prevents detoxification of chloro-*s*-triazine herbicides as well as some other pesticides, that are degraded by glutathione conjugation. Monoxygenase inhibitors can prevent many oxidative herbicide detoxifications. Our own work on biochemical, physiological and genetic studies of paraquat resistance led to a rationale for synergists. Paraquat rapidly, but transiently, inhibited photosynthesis of a resistant weed while permanently inhibiting the wild type. Chloroplasts of the resistant biotype dominantly and pleiotropically inherited constitutively elevated levels of superoxide dismutase, ascorbate peroxidase and glutathione reductase, enzymes engaged in detoxifying the active oxygen species generated with paraquat. Inhibition of these enzymes could lower the required threshold for phytotoxicity, synergizing the herbicides. The first two enzymes contain copper and the first also zinc. Copper and zinc chelators inhibited the enzymes, synergizing paraquat and other active oxygen generating herbicides, in all weed species tested.

We are blessed, in the past number of years, with a better and better understanding of the modes of action and the modes of resistance to herbicides. This is especially true of the photosystem II inhibiting herbicides *(1)* but also of the dinitroanilines such as trifluralin *(2)* and particularly with the herbicides affecting amino acid

0097–6156/88/0379–0004$06.25/0

biosyntheses. The latter includes the EPSP-synthase (enolpyruvylshikimate phosphate synthase) inhibiting herbicides such as glyphosate *(3)*, the sulfonylureas and imidazolinones inhibiting acetolactate synthase *(4,5)* and bialophos and glufosinate inhibiting glutamine synthase *(6,7)*. The understandings of the biochemical underpinnings of modes of action and modes of resistance have allowed the rapid advances in biotechnology of engineering herbicide resistant crops *(8, 9)*, the subject of the following three chapters. The greatest advance in this area, the engineering of a gene for an enzyme degrading bialophos and glufosinate *occurred too late to be included in this meeting. This is unfortunate as it is a prime example of a simple enzyme, with a readily available herbicide conjugating substrate (acetyl co-enzyme A) that confers full resistance to agricultural rates of herbicide with a large margin of error at very low levels of gene expression (7)*. The advantages of such low expression levels of this type of herbicide degradases in having the least possible deleterious effects on crop yields has been discussed at length *(8, 9)*.

Much of our new understanding of modes of herbicide action and resistance emanate from industry and industry-academia collaboration. The gaining of this information is not always altruistic; most of the information on modes of crop resistance to herbicides comes from the metabolic studies required for registration. There is a huge variation in company policies on how much of this information is available to the scientific community. One area where our information is far too limited is in the mode of resistance to herbicides in weeds that are normally tolerant to the herbicide in question. These are the weeds that are normally not controlled by a given selective herbicide, at the time it was first tested; not weeds that evolved resistance. As there is no need to have metabolism and toxicology studies on such weeds, this information is to a large extent awaiting elucidation. As we shall see below, such information is of critical importance for the biorational choice or design of herbicide synergists. We know much more about the modes of evolved resistance to herbicides, and much of our first intuitions on synergies derived from these studies.

Synergists are any combination which has a greater effect together than the sum of the components. Various methods of proving synergies between compounds have been proposed (10-13) and are needed when both compounds are active by themselves. From an industrial registration point of view synergists can be divided (sometimes arbitrarily, as we shall see) into two groups: combinations of active herbicides which are synergistic with each other; herbicide -adjuvant combinations where one of the components is not phytotoxic.

One can obtain pesticide synergists by random screening of additives using sublethal rates of the pesticide. This is not the subject of this chapter. We term "biorational" as any choice of possible synergists that is based on our understanding of the mode of resistance of the pest. Adding a compound to the herbicide which we can presume on the basis of biochemical knowledge, will suppress resistance, is "biorational". The other side of the coin is the biorational protection of crops against herbicides (Table I): e.g. the use of an auxin conjugation inhibitor 2,6-dihydroxyacetophenone *(14)* to protect against glyphosate based on the researchers' understanding of the multitude of glyphosate actions in plants *(15)*, is a case of such biorational protection.

Biochemical Limitations of Synergy. The duty of the biorationally chosen synergist is to prevent the degradation of a herbicide, or the phytotoxic products produced when herbicide is present. When a weed performs no such degradation there is nothing to synergize. The classic selective herbicide of the previous generation (e.g. atrazine) is not overly phytotoxic to corn at 5 kg/ha; it kills some weed species at 30 g/ha, some at 100 g/ha, or 300 g/ha, but the full 1-2 kg/ha are used in the field because there are some pernicious weeds (mainly grasses in this case), that degrade atrazine (far more slowly than maize, but still degrade it) and the high rates are needed to control these species. A synergist blocking atrazine degradation probably would not lower the 30 g/ha required to kill the species usually controlled by that rate. The fully active synergists should be expected to bring all species down from the 1-2 kg/ha rate towards the much lower rates.

This lowering of rates should work equally well with the newer generation of herbicides (e.g. sulfonylureas). There is far too much toxic carryover from a wheat field treated with 30 g/ha chlorsulfuron to allow cultivation of many crops, and many weeds remain suppressed. If all the weeds in wheat could be controlled by the carryover level remaining, then there would be no carryover the following year. This requires ascertaining why some weeds are more tolerant to the herbicide than others; information not fully available. It is known that wheat, a resistant species degrades chlorsulfuron *(16, 17)* so it can be expected that some weeds do the same, but at a lower rate.

The main biorational problems will be in finding selective synergists. If a weed and crop detoxify the herbicide or its toxic products by the same means, finding a synergist will indeed be daunting. For this reason it is envisaged that many synergists will be used when the crop is not present (no-till, certain pre-emergence or directed sprays) because of this biochemical limitation.

Advantages of synergists. The main advantage of synergists is that they should allow the farmer to substantially lower the herbicide rates used. The foremost interest of the farmer is cost-effective weed-control. If the synergist is less or equally expensive in a combination using far less herbicide, giving the same quality of weed control, then that criterion is met. In many cases it is expected that the spectrum of weeds controlled will be increased by addition of a synergist. There are many other short and long range advantages. With lower herbicide rates there should be less carryover, allowing a larger choice of crops for the following year. The lower the herbicide rate, the less the chance of the herbicide reaching ground water before being metabolized by soil micro-organisms. If one were to use a fifth of the present herbicide rates, the chance of a herbicide appearing in ground waters is probably less than a fifth, as the capacity to degrade would not be oversaturated.

The prospects of synergists hurting industry are not as gloomy as some have assumed. Synergist sales will replace some of the decreased herbicide sales; synergists will make less market-competitive herbicides more competitive because of lower total price; the lessened carryover will interest the farmers; industry will have less ecological combats with agencies over residue and toxicology problems. From this it is clear that there need not be a massive tonnage reduction in production.... and some reduction is better than being forced to remove a herbicide from the market-place.

Synergists and the Evolution of Herbicide Resistance. Selection pressure is the most important factor controlling the rate at which herbicide resistant populations will evolve in a susceptible species *(18)*. Selection pressure is a more or less direct function of herbicide rate. The first and most wide-spread weeds to evolve atrazine resistance were *Senecio* sp. *Amaranthus* spp. and *Chenopodium* spp. These are controlled by exceedingly low atrazine doses, 30-100 g/ha, i.e. the selection pressure exerted by 1-2 kg/ha is very great. The last species to evolve resistance were the grasses *(19)*, that are controlled by only the highest levels, i.e. the selection pressure for grasses was lower than that for broadleaf species. Had synergists such as tridiphane (see below) been used earlier to lower atrazine rates (and thus lower the selection pressure), it would be expected that the evolution of the resistant weeds would be considerably delayed. One could make the same predictions for synergized sulfonylurea herbicides.

Herbicide-Herbicide Synergistic Combinations.

A computer survey of the patent literature made a year ago came up with 515 recent patents claiming herbicidal synergies. A perusal of the abstracts of 45 of them, chosen as a sample, showed that the patent community uses a broader definition of synergy than used here; they believe that there is a synergism when two herbicides control more weed species than each separately. This would better be termed "complementarity". The overlap of control range allows a lowering of herbicidal rates, which may or may not be due to a metabolic synergy.

Some synergies appeared fortuitously (atrazine-tridiphane) and not by rational discovery. From what we now know, tridiphane is a metabolic class inhibitor and may have other rational future uses with herbicides that are detoxified by the same mechanism.

Tridiphane-atrazine. Tridiphane was initially developed as a grass controlling herbicide, and it was later found that atrazine and tridiphane synergistically control grasses in combination *(20)*. From indirect evidence showing the stoppage of atrazine catabolism it was initially thought that tridiphane acts by inhibiting the specific glutathione-*S*-transferase for atrazine *(21)*. It was later found that a glutathione-*S*-transferase conjugates tridiphane with glutathione *(22)*:

$$Cl_3C{-}CH_2{-}C(O)CH_2 \xrightarrow[\text{Glutathione-}\underline{S}\text{-transferase}]{\text{reduced glutathione}} Cl_3C{-}CH_2{-}C(OH){-}CH_2{-}SG$$

tridiphane (3,5-dichlorophenyl, epoxide) → <u>S</u>-(tridiphane)GSH

This is similar to the first step in atrazine degradation:

$(CH_3)_2$-CH-NH / CH_3-CH_2-NH triazine-Cl (atrazine) —reduced glutathione, Glutathione-S-transferase→ $(CH_3)_2$-CH-NH / CH_3-CH_2-NH triazine-S-CH_2-CH(C(=O)-NH-CH_2-COOH)-NH-C(=O)-CH_2-CH_2-CH(NH_2)COOH

atrazine

(NON-TOXIC)

The tridiphane-glutathione conjugate is a far more potent inhibitor of the glutathione-*S*-transferase measured than tridiphane alone (Fig.1A). Tridiphane synergizes atrazine, killing weeds in maize in the field, yet the maize is unaffected. This is surprising as maize utilizes a glutathione-*S*-transferase to degrade the atrazine and this glutathione-*S*-transferase is inhibited by the tridiphane-glutathione conjugate. This inhibition of glutathione-*S*-transferase activity is less strong than with weeds (Fig.1A). A kinetic analysis showed that the tridiphane-glutathione conjugate is then a competitive inhibitor of glutathione binding to glutathione-*S*-transferase with a 4 times higher affinity for the enzyme from weeds vs. maize *(22)*. These data are probably not enough to explain why tridiphane is inactive as a synergist in corn: i.e. why is corn not killed. Young corn leaves have 6 times more glutathione than *Setaria,* which could partially explain the difference.... but older maize and *Setaria* leaves have the same concentration *(22)*. Tridiphane does not kill older maize plants. The answer probably lies in a rapid catabolism of the tridiphane-glutathione complex in maize which does not occur in *Setaria* (Fig. 1B) *(22)*. Thus, we see in balance that maize is able, by virtue of having more glutathione, less tridiphane-glutathione conjugate, with less activity and faster degradation of the conjugate, maize can "save itself" from tridiphane synergism of atrazine.

As stated before, the choice of tridiphane to synergize atrazine was not "biorational". If we use the information to synergize other pesticides, this will be "biorational" by our broad definition. Many pesticides have glutathione conjugation as the first step in their downfall from toxicity. From our understanding above, tridiphane must first be conjugated by the target organism. The glutathione-*S*-transferase must then be inhibited, and the tridiphane-glutathione conjugate must be stable for tridiphane to synergize other pesticides in other pests. Indeed the tridiphane conjugate prevented degradation of a number of pesticides (Table II). Tridiphane was also able to synergize EPTC and alachlor in maize *(20)*. The mode of synergy has not been checked, but both of these herbicides are degraded by glutathione conjugation. Tridiphane was also excellent at synergizing the insecticide diazinon against houseflies. Tridiphane had a twenty fold greater rate of conjugate formation with glutathione than diazinon has, and diazinon does not seem to compete with this step *(23)*. The LD_{50} with normal houseflies was reduced threefold from near 0.09 μg diazinon per fly to about 0.03 μg diazinon per fly - but 20 μg tridiphane per fly was required for this *(23)*. Houseflies have other mechanisms of diazinon degradation and that may be the reason for the small increment of synergy with much tridiphane. Tridiphane may be more useful with roaches, which only degrade diazinon by glutathione conjugation *(24)*. Tridiphane would also be very useful

Table I. Biorationally Protecting Against Glyphosate Action by 2, 6-Dihydroxyacetophenone (DHP) on *Teucrium canadense*

Treatment	Plant height	Free IAA content
	(% of control)	
Water (control)	100	10
0.2mM glyphosate	56	20
0.5mM 2,6-DHP	92	101
glyphosate + DHP	88	74

Source: Calculated from data of Lee and Starratt *(14)*.

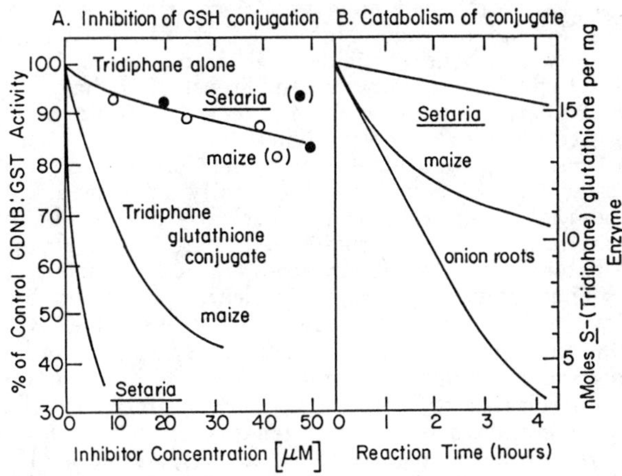

Figure 1(A). Tridiphane blockage of glutathione-*S* transferase activity, and (B) catabolism of the conjugate to an inactive form in various species. Source: Replotted from data in Lamoureux and Rusness *(22)*.

with flies or any other insect that evolves diazinone resistance due to elevated levels of glutathione conjugation.

Glyphosate-Imidazolinone mixtures. An excellent example of an intuitive rational mixture of herbicides for a synergism was the choice of two compounds which each inhibit different amino acid biosyntheses *(25)*. Glyphosate alone at 100 g/ha caused no necrosis, and 250 g/ha of an imidazolinone herbicide caused an average of 25% necrosis on 8 weed species. A mixture of the two at those rates caused 97% necrosis *(25)*. This synergism was found with all imidazolinone herbicides tested *(25)*. and all inhibited the enzyme acetolactate synthase *(26)*. The synergism may be due to either a superior starvation of amino acids or a synergistic buildup of toxic intermediates due to the double blockage.

Synergistic adjuvants

The concept of using specific enzyme inhibitors which, at the concentrations used, do not cause visible signs of toxicity is most beckoning. The first such inhibitors used had high mammalian toxicity; their biorationale came from knowledge of mammalian systems. Later compounds come from our knowledge of plant systems, and the mammalian toxicity dropped by orders of magnitude.

Monoxygenase inhibitors. Monoxygenase (=mixed function oxidases= cytochrome P_{450}s, etc.), can have two diametric effects vis a vis pesticides. They can degrade some and they can activate, others. In the case of activation, the pesticide itself is not toxic; it must be oxidized to a toxic form. Fedtke and Trebst *(27)* have rationally taken these diametric effects to formulate a "unifying concept" of herbicide protection and synergy. There are many known herbicide protectants (=safeners=antidotes) and their mode(s) of protection are not at all clear *(28)*. To validate the "unifying concept", Fedtke *(29)* demonstrated that compounds that are clearly and probably specifically monoxygenase inhibitors can protect plants against a group of herbicides (Table III). The commercial protectants safen the same herbicides. The "unifying concept" is clearly supported by the fact that a variety of workers have shown that these same two compounds; piperonyl butoxide (a widely use insecticide synergist) and aminobenzotriazole (a suicide inhibitor of mammalian monoxygenases) *(30)* also synergize many herbicides *(31)*. Aminobenzotriazole has been shown to inhibit plant monoxygenases as well *(32)*. These synergies work for herbicides that lose activity because of monoxydation. An example of such a synergy is shown in Fig. 2, and many examples are summarized in Table IV. Like most "unifying concepts" in biology, this hypothesis is limited. It can only include those compounds that act as monoxygenase inhibitors and cannot encompass all the synergists discussed in earlier and later sections. Still, it may be valid (upon further experimentation) to say, as a "unifying concept" that monoxygenase inhibitors will synergize some herbicides and safen others. In some respects the "unifying concept" can be broadened to other pests. Certain herbicide protectants that act as monoxygenase inhibitors were able to synergize the insecticide propoxur in a resistant strain of houseflies *(38)*. A herbicide metolachlor, which must be oxidized to be activated in plants, also synergized propoxur in flies, probably by competing for the monoxygenase against propoxur, and thus sparing the propoxur *(39)*.

Table II. Inhibition of Glutathione Conjugation of Pesticides by the Tridiphane Glutathione Conjugate

Pesticide	glutathione-*S*-transferases from pea epicotyl	horse liver	house fly	ref.
	(I_{50} - μM tridiphane)			
PCNB	3	4		*(22)*
Propachlor	1	11		*(22)*
Fluorodifen	9	2		*(22)*
Diazinon			16	*(24)*

Table III. Protecting Against Herbicide Action with Monooxygenase Inhibitors

Herbicide	Control I_{50} (μM)	Piperonyl-butoxide I_{50} (μM)	Piperonyl-butoxide factor	aminobenzotriazole I_{50} (μM)	aminobenzotriazole factor
alachlor	25	3980	159	631	25
metolachlor	79	501	6	40	0.5
propachlor	63	100	1.6	631	10
diallate	631	5010	8	631	10
mefenacet	0.1	1	10	4	40

The effects of piperonyl-butoxide (148μM) and aminobenzotriazole (373μM) were measured in an oat rooting test in the dark.
Source: Collated and calculated from Fedtke *(29)*.

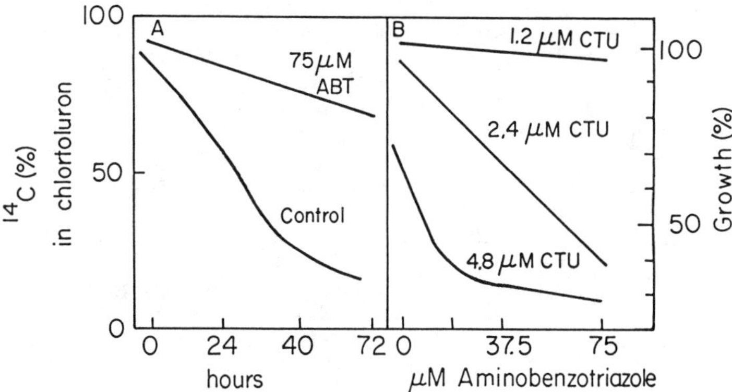

Figure 2. Synergy of chlortoluron (CTU) by aminobenzotriazole (ABT) in wheat.(A) Inhibition of chlortoluron degradation, (B) Synergism of inhibition of wheat growth. Source: Replotted from data in Cabanne et al. *(33)*.

The similarities of the plant and insect monoxygenases was also verified by showing that plants could enzymatically oxidize aldrin to dieldrin *(38)* in the same way as insects.

Aminobenzotriazole is a specific monoxygenase synergist. Unlike tridiphane it could not synergize metolachlor, which is degraded by a glutathione-*S*-transferase *(36)*. For unclear reasons another monoxygenase inhibitor 2,4-DP was able to prevent metolachlor degradation in cotton but not in maize *(36)* even though the degradation products in cotton were glutathione derivatives.

There is some evidence that a previously known synergy may be due to inhibiting an oxidase. Bromoxynil and ioxynil synergize amitrole; it was later found that their degradation products (their respective acids) had the same effect *(39)*. It was proposed that this effect was due to the release of oxydase inhibiting bromine and iodine from the herbicides *(39)*.

Fungicide research has fortuitously come up with new herbicide synergists. Industry wide research has been centered on finding inhibitors of sterol biosynthesis which are fungicidal but not phytotoxic. These azole containing compounds mainly effect late monoxygenase steps leading to ergosterol which is necessary for the growth of many fungi *(40)*, an oxidative C-14 demethylation of 24-methylenedihydroanosterol. With an only partially understood structural-activity relationship some of these were found to be useful plant growth retardants, due to a blockage in gibberellic acid biosynthesis or other plant sterol biosyntheses *(40)*. In the case of tetcyclasis this was traced to blockage of kaurene-oxidase, a plant monoxygenase converting kaurene to kaurenol and then to kaurenoic acid *(41)*. Tetcyclasis also inhibits at least two plant monoxygenases, as well *(42)*. On the possibility that it might inhibit herbicide degrading monoxygenases, tetcylacis was tested for its ability to prevent chlortoluron degradation in maize and cotton (Fig. 3). It was a hundred times more active than aminobenzotriazole in these systems *(36)*. Clearly industry should be screening all the potential sterol biosynthesis inhibitor derivatives from their fungicide programs for herbicide synergies at the levels of monoxygenases. As fungicides and chlortoluron are both used in wheat, compatibilities of azole fungicides with chlortoluron must be checked.

Almost all of the monoxygenase synergies have been shown in crops (Table IV). It is the weeds we wish to kill, not wheat. Too little is about herbicide metabolism in weeds to know if the monoxygenase inhibitors will be useful with herbicides in agricultural situations. Indirect suggestions, by analogies with insecticides, suggest a new and important use for monoxygenase inhibitors as synergists. Some of the multi-resistances to insecticides of totally different modes of action and chemistries have been due to the evolution of elevated levels of monoxygenases which non-specifically degrade many insecticides *(43)*. As of a few years ago, no such multi-resistances were found in weeds. The herbicide resistances that did evolve were mainly target-protein modifications *(1, 2, 19)*. Recently, there are reports of multiple cross resistances that are quite worriesome to the farmer. These include widespread co-evolution of *Lolium rigidum* throughout Australia with resistances to diclofop-methyl, and chlorsulfuron, among others *(44)*. Chlortoluron resistant *Alopecurus*

Table IV. Herbicide Synergisms by Monooxygenase Inhibitors

Herbicide	presumed monooxygenase inhibitor[a]	test species[e]	ref.
EPTC	SKF 525A[b], PBO[c]	corn	*(34)*
tebuthiuron	EPTC/butylate with dichlormid	corn	*(35)*
chlortoluron	PBO, metopyroine, aminobenzotriazole	wheat	*(31,36)*
		wheat,cotton	*(31)*
		corn	*(31)*
	2,4-DP[d]	cotton,corn	*(31)*
	tetcyclasis	cotton,corn	*(36)*
metolachlor	2,4-DP	cotton	*(36)*
	PBO	sorghum	*(37)*
MCPA	aminobenzo-triazole	potato	*(38)*

[a]Inhibitor is defined loosely here as any compound that prevents loss of herbicidal effect. The compound may inhibit the monooxygenases or replace the herbicide as a substrate; [b] SKF 525A = diethyaminoethyl-2-2-diphenylvalurate HCl; [c] PBO piperonylbutoxide; [d] 2,4-DP (2,4 dichlorophenoxy)-1-propyne; [e] not all synergies are in whole plant systems.

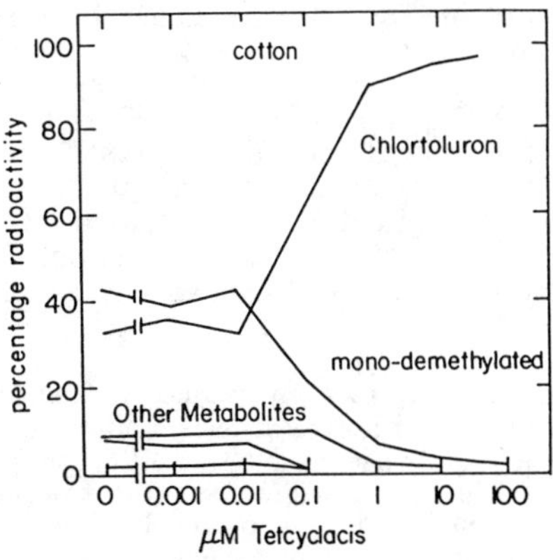

Figure 3. Tetcyclasis inhibition of chlortoluron catabolism in cotton. Note the much greater level of activity than that of aminobenzotriazole (Figure 2). Source: Redrawn from data in Cole and Owen *(36)*.

myosuroides evolved in England with cross resistances to isoproturon, methabenzthiazuron, pendimethalin, terbutryne, diclofop-methyl and chlorsulfuron *(45)*. The researchers "rationally" decided to test monoxygenase inhibitors to suppress resistance, as chlortoluron (Fig. 3,4) is among the herbicides. Indeed, aminobenzotriazole quite effectively reduced the resistance, i.e. synergized the herbicides *(46)*.

Aryl-acylamidase inhibitors. Rice is resistant to the herbicide propanil due to a rice specific aryl-acylamidase which degrades the herbicide. Various carbamate and organophosphorothioate insecticides were found to be incompatible with propanil because they prevented propanil degradation. This was found to be due to a direct inhibition of this enzyme *(47, 48)*. Such insecticides or their derivatives can be "rational" synergists when this enzyme is found to be active in degrading herbicides in weeds, and must be suppressed.

Inhibitors of the oxygen detoxification pathway. A few major groups of herbicides kill plants by photogeneration of active oxygen species. The triazines, uracils, phenylureas, and some of the pyridazinones block electron transport at the reducing side of photosystem II of photosynthesis just before plastoquinone reduction *(1)*. Some phenolic herbicides act at a nearby site on photosystem II. The bipyridilliums drain electrons from photosystem I, probably from ferredoxin *(49)*. A large group of nitro-diphenylethers require light to generate active oxygen species although it is clear from the diverse findings in the literature that either; no bulk pigment acts as the photoacceptor in all systems; or that different pigments act in different systems *(50)*. It is also not clear which active oxygen species is generated first. As soon as there is more active oxygen generated than the endogenous detoxification system can cope with (Fig. 4), there is membrane lipoxidation. This results in water loss, a general breakdown in the electron transport systems, and the release of the chemically transformed solar energy as a variety of active oxygen species which can be confused with the first one generated. Singlet oxygen can be produced, and so can chlorophyll radical. This self-amplifying chain reaction of generation of active oxygen, membrane lipoxidation and water leakage leads to the rapid desiccation caused by these herbicides. The stronger the light, the faster this chain reaction. Certain environmental xenobiotics: SO_2, O_3, NO_x, and some fungal toxins such as cercosporin can have the same effects.

Organisms have endogenous oxygen radical detoxification systems, that probably co-evolved with photosynthesis and aerobic respiration. They include superoxide dismutase and the ability to produce and recycle oxy-radical quenching agents such as glutathione and ascorbate. Plants also produce carotenes and α-tocopherol which quench these radicals. This detoxification system is especially necessary to cope with energy "leakages" from the photosynthetic electron chain. For a herbicide to be toxic, it must produce more radicals than can be quenched by this native system before the herbicide is dissipated. Chloro-*s*-triazine herbicides inhibit photosystem II in isolated thylakoids at the same concentration in all species. Despite, this, high rates are required in the field to kill grasses, due to the glutathione-*S*-transferase degradation described in an earlier section. This leads to a balance between the amount of herbicide remaining with the ability of the plant to detoxify radicals until the herbicide is degraded. As much of the "action" of producing oxygen radicals is in the chloroplast both with and without herbicide, and most active species such as singlet oxygen and superoxide have very short diffusion distances, it is natural to assume that the chloroplast has its own concerted anti-oxidant defence mechanisms. This includes the slowly recyclable carotenes, as well as a pathway for oxygen detoxification which recycles its components. This pathway has been elucidated by many, and put in context especially by Halliwell *(51)* and Asada *(52)*. The first enzyme is superoxide dismutase (Fig.4) which dismutates the superoxide to peroxide. The peroxide is less energetic and can diffuse, but is still dangerous; it can react with ferrous iron and form hydroxyl radicals (the Fenton reaction) which are extremely energetic. If the peroxide is slowly formed from superoxide, the excess superoxide can react with the ferric ions formed by the Fenton reaction, re-reducing them and making them available to form more hydroxyl ions. The Fenton reaction coupled with the iron recycling reaction is termed the Haber-Weiss reactions *(51)*. It is thus imperative for the chloroplast to rapidly detoxify both the superoxide and the peroxide but alas, chloroplasts contain no catalase. They do contain ascorbate peroxidase and ascorbate, which can adequately compete with the Fenton reaction to remove the potentially dangerous peroxide. The dehydroascorbate formed is reduced back to ascorbate by glutathione, either spontaneously or by a dehydroascorbate reductase. The oxidized glutathione is recycled back to glutathione by glutathione reductase, utilizing NADPH *(51, 52)*. Some electrons must pass through the photosynthetic electron chain to produce NADPH for this reaction, or the system will break down. This system will recycle both glutathione and ascorbate used quench singlet oxygen and other radical reactions.

Our interest in the system began when it was intimated that superoxide dismutase levels might be responsible for different levels of paraquat tolerance *(53-55)*. As it is, some weeds such as *Amaranthus* spp. are killed by 30 g/ha paraquat and others at a variety of rates up to the 1-2 kg/ha used in weed control. The paraquat resistant *Conyza* spp. that evolved were only controlled by >15 kg/ha paraquat *(56, 57)*. In the meantime, a second theory was promulgated to explain this paraquat resistance; that paraquat was sequestered before it could reach green tissue. This was mainly based on radioautographic evidence, obtained 4 h after paraquat was treated through cut petioles *(57, 58)*. By this time sensitive plants have usually whithered under strong light. We believe that the radioautographic evidence shows that resistant plants can *actively* sequester paraquat or its metabolites, but that paraquat does get to the chloroplast, and that elevated levels of the enzymes of the Halliwell-Asada pathway keep the plant alive until paraquat is actively sequestered. The evidence that we obtained to support this view allowed us to rationally choose compounds that synergize paraquat and other active oxygen generating herbicides.

The lines of research that led us to believe that elevated levels of the Halliwell-Asada pathway have primary responsibility for increased tolerance to active oxygen are as follows:

(a) Kinetic evidence showed that paraquat sprayed on leaves of the paraquat resistant biotype rapidly but transiently inhibited photosynthesis (Fig. 5A). This shows that paraquat did reach the chloroplasts, affected them, but the plant remained alive while the paraquat was dissipated. In contrast, photosynthesis was inhibited just as rapidly in the sensitive wild type, but irreversibly so (Fig. 5A).

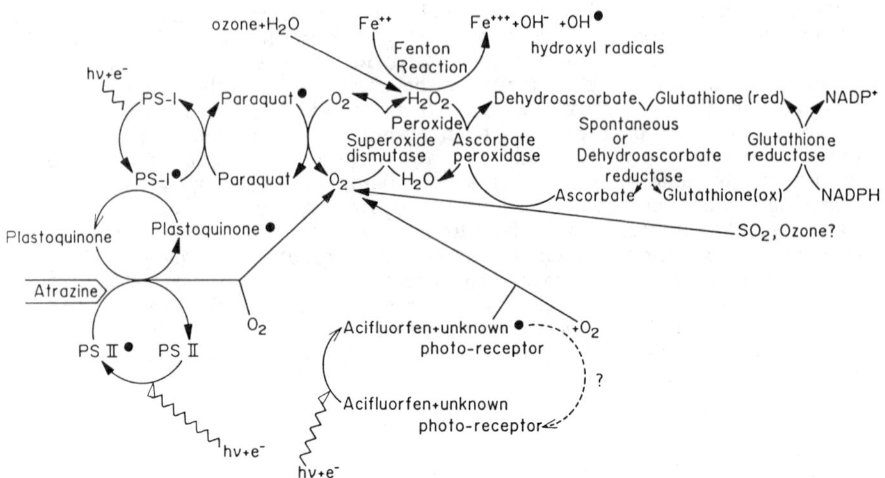

Figure 4. Scheme of interactions between active oxygen generating xenobiotics and the oxygen detoxification system. Light intensity, the rate constants of photogeneration of active oxygen, dissipation of the xenobiotics and the levels and rate constants of the enzymes interact to determine whether a plant will be spared or killed.

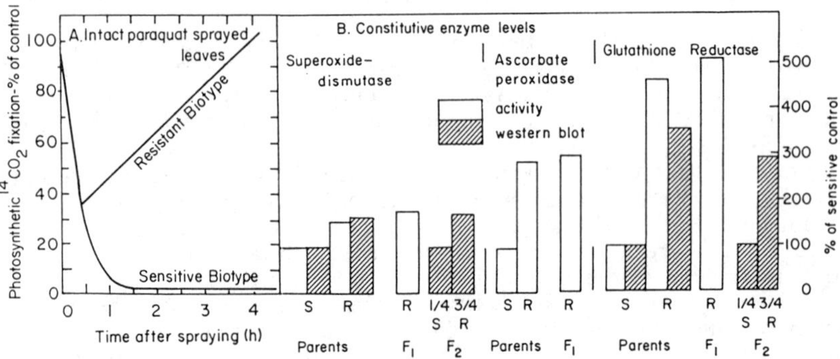

Figure 5 (A) Evidence that paraquat transiently affects chloroplasts of resistant *Conyza* and (B) that there are constitutively elevated levels of the Halliwell-Asada active oxygen detoxification pathway in the chloroplasts. A. Resistant and susceptible plants of *Conyza bonariensis* were sprayed to runoff with 0.1 mM paraquat and whole leaves were removed for measurement of photosynthesis at times thereafter as an estimation of paraquat arriving at, and affecting chloroplasts. Simultaneous measurements of stomatal aperture were made to ascertain that the stomates remained open. Source: Data redrawn from *(60)*. B. Normal enzyme levels (*without* paraquat treatment) in resistant and susceptible *Conyza bonariensis*. Source: Collated and drawn from *(57, 61)*.

(b) Isolated intact (Class A) chloroplasts of the paraquat sensitive *Conyza bonariensis* biotype evolved 3 times more ethane than the resistant type under high light (59). This is a sign of resistance to membrane damage due to oxidant "leakage" from photosynthesis. When paraquat was added, ethane evolution from the sensitive biotype tripled; it also increased in the resistant biotype, but to less than the normal level of the sensitive biotype. These data indicate that the primary mechanism for tolerating paraquat is in the chloroplast.

(c) We measured the levels of the three major enzymes in the pathway in isolated intact chloroplasts and found tham to be constitutively elevated in the resistant *Conyza* biotype (Fig. 5B). Resistance was dominantly inherited under the control of a singe gene *(60)*. The levels of the three enzymes were elevated in the F_1 generation. The critical correlation was in the F_2 generation where we immunologically determined the levels of two of the enzymes in single plants that were tested for resistance / susceptibility to paraquat. All the resistant plants had elevated enzyme levels and all the sensitive plants had normal enzyme levels (Fig. 5B). This lack of F_2 segreation is a very strong correlative phenomenon. These data are supported by others' findings that induced paraquat tolerance can be correlated with increased chloroplast superoxide dismutase *and glutathione reductase (62)*. Interspecific comparisons have correlated acifluorfen tolerance to increased levels of ascorbate and α-tocopherol *(63)*, and to increased superoxide dismutase *(64)*.

(d) If high levels of the Halliwell-Asada pathway enzymes confer resistance to paraquat, they should confer increased tolerance to other herbicides and oxidant stresses. Indeed we and others have shown such correlations (Table V) in a variety of species. The level of resistance to paraquat is usually higher than to other herbicides, probably because the other herbicides remain active in the plant for longer durations.

The above data supply considerable evidence for the involvement of the Halliwell-Asada pathway and its products in tolerance to oxidant generating herbicides. It only remains to be shown that suppression of the pathway suppresses tolerance. This would be further evidence for the involvement of the pathway. Any compound suppressing the pathway would have considerable possibilities as a synergist. The plastid superoxide dismutase and ascorbate peroxidase are both copper containing enzymes *(52)*. The former contains zinc and the latter tightly bound iron, and thiol groups. The later enzymes of the pathway contain thiol groups as well *(52)*. Thiol binding reagents and compounds which complex copper and/or zinc should inhibit the pathway. A suppression by thiol biding reagents at the plant level would hardly be good evidence as the plant has many thiol containing enzymes. Such compounds usually have high mammalian toxicity and would thus have little potential as synergists. There are few copper containing enzymes and fewer yet zinc containing enzymes in plants. Thus compounds which bind them should serve as somewhat specific synergists. We tested a variety of relatively specific copper and zinc chelators in *in vitro*, cellular, tissue and whole plant systems (Table VI). It is clear that they all severely affected the plants when used with active oxygen generating herbicides. None of them had any obvious effects on plants at the concentrations used *(68)*. Their effects were probably due to the chelation of copper

Table V. Herbicide Cross Tolerances to Oxidative Stresses and Relations with the Halliwell-Asada pathway

species	primary tolerance	cross tolerances	enzymatic correlations[a]	ref.
Conyza bonariensis	paraquat	atrazine SO_2 acifluorfen	high SOD,GR,AP	(65)
		photoinhibition		(b)
Lolium perenne	paraquat	SO_2	elevated SOD/GR	*(65)*
Lolium perenne	SO_2	paraquat	elevated SOD/GR	*(65)*
Nicotiana tabacum cv Florida	O_3	paraquat	elevated SOD/GR	*(65)*
cotton	drought	paraquat	elevated GR	*(63)*

[a]SOD, superoxide dismutase; GR, glutathione reductase; AP, ascorbate peroxidase.
[b] Jansen, Canaani, Shaaltiel, Gressel (unpub. results).

0.75% DDC alone	1μM paraquat + DDC	10μM paraquat alone	1μM paraquat alone

Figure 6. Synergism of paraquat by diethydithiocarbamate, a copper chelator. Paraquat sensitive *Conyza bonariensis* plants of slightly different sizes were sprayed to run-off 24h before photographing. Source: From *(66)*.

Table VI

Synergism of active-oxygen generating herbicides by copper and zinc chelators

Chelator	matal chelated	conc. (mM)	Herbicide synergized	System	I_{50} shift down
diethyldithiocarbamate	Cu	0.9	Paraquat	*Asparagus* cells	4
		9		bean leaf discs	>180
		22		R *Conyza*	> 10
		22		S *Conyza*	>100
		22		*Gallium*	> 3
		22		*Sinapis*	10
		22		*Stellaria*	3
		22	Acifluorfen	R *Conyza*	4
		22		S *Conyza*	5
		0.9		*Asparagus* cells	8
		22	Atrazine	R *Conyza*	10
		22		S *Conyza*	12
oxime of 4-octyl-2 -acetyl phenol	Cu	0.6	Paraquat	*Asparagus* cells	13
		1.2	Acifluorfen	bean leaf discs	31
pyridine-2,6-dicarboxylic acid	Zn	0.4	Paraquat	*Asparagus* cells	3
di-2-ethyl-hexyl-phosphoric acid	Zn	0.5	Paraquat	*Asparagus* cells	8
		0.5		bean leaf discs	35
		0.5	Acifluorfen	bean discs	22
tetraethylene-pentamine	Cu	1.3	Paraquat	*Asparagus* cells	>4
triethylene-tetramine	Cu	0.5	Paraquat	*Asparagus* cells	>2.5

Source: Data collated from *(68, 69)*.

and zinc for the following reasons; related compounds such as Mn or zinc bisdiethyldithiocarbamates (maneb and zineb) were not synergistic as their chelation capacity was blocked. Iron chelating compounds protected cell cultures *(68, 69)*. This is probably due to a blockage of the Haber-Weiss reactions, preventing hydroxyl radical formation. The iron of ascorbate peroxidase is tightly bound and, unlike the copper and zinc, inaccessible to chelators. A much lower concentration of the chelators was required to synergize these herbicides in the isolated cells, or leaf discs than was needed with whole plants (Table VI), while the herbicide rates remained essentially the same *(69)*. This suggests that the chelators do not efficiently penetrate the leaf cuticle. Once formulations of these chelators which have better penetration are found, or better penetrating chelators are discovered, we may have a new means of effective weed control. It is envisaged that the first use of such synergists will be in no-till situations or wherever a general herbicide is needed. Chelating herbicide synergists will only be useful in foliage contact or postemergence modes; they will probably bind to soils, and chelate the metals there before entering plants. Such compounds may well broaden the spectrum of weeds controlled by some herbicides.

Chelators have been shown to synergize the insecticide *trans* -permethrim by preventing its hydrolysis in the soybean looper *(70)*. The enzyme is clearly not a typical esterase as it contains a metal group. The authors did not add excesses of various metal ions to show that they could suppress the synergism; this would show that the compounds actually acted as chelators in their system.

Conclusions

The examples shown above suggest that the more we know about herbicide biochemistry in weeds and crops, the more we can rationally choose compounds and find leads for herbicide synergists. There is still a long way from this step to arriving at compounds for field use... but this rationale can also save many of the initial steps from random screening. Synergists are one logical way to lower the xenobiotic load on the environment while still providing for adequate weed control.

Acknowledgments. The authors wish to thank their colleagues who have provided unpublished and in press material for this review. The authors research on the mode of paraquat resistance was supported by a grant form the Israel Academy for Sciences and Humanities and their work on synergists by the Yeda Fund for Applied Research. J. G. is the Gilbert de Botton Professor of Plant Sciences.

Literature Cited

1. Gressel, J., *In "Herbicide Physiology"*, Duke, S.O. Ed., C.R.C. Press: Boca Raton, 1985, Vol. 2. pp. 159-189.

2. Vaughan, K.C.; Marks, M.D.; Weeks, D.P., *Plant Physiol.*, 1987, 87, 956-961.

3. Coggins, J.R., *In "Herbicides and Plant Metabolism"*,Dodge, A.D., Ed. Cambridge Univ. Press: Cambridge, 1987, In press.

4. Ray, T.B., *Trends in Biochem Sci. (TIBS)*, 1986, 11, 180-183.

5. Shaner, D.L.; Anderson, P.C.; Stidham, M.A., *Plant Physiol.*, 1984, 76, 545-546.

6. Leason, M.; Cunliffe, D.; Parkin, D.; Lea, P.J.; Mifflin, B.J., *Phytochem.*, 1982, 21, 855-857.

7. DeBlock, M.; Botterman, J.; Vandewiele, M.; Dockx, J; Thoen, C.; Gossele, V.; Movva, R.; Thompson, C.; VanMontagu, M.; Leemans, J. *EMBO J.*, 1987, In press.

8. Comai, L.; Stalker, D., *Oxford Surv. Plant Molec. Cell Biol.*, 1986, 3, 166-195.

9. Gressel, J., *In "Herbicides and Plant Metabolism"*, Dodge, A.D., Ed., Cambridge Univ. Press: Cambridge, 1987, in press.

10. Hewlett, P.S., *Biometrics*, 1969, 25, 477-487.

11. Akubundu, I.O.; Sweet, R.D.; Duke, W.B., *Weed Sci.*, 1975, 23, 20-25.

12. Morse, P.M., *Weed Sci.*, 1978, 26, 58-71.

13. Nash, R.G., *Weed Sci.*, 1980, 29, 147-155.

14. Lee, T.T.; Starratt, A.N., *In "Proc. 1st Intl. Symp. on Àdjuvants for Agrochemicals"*, 1987, In press.

15. Lee, T.T.; Starratt, A.N., *Phytochem.*, 1986, 2457-2461.

16. Sweetser, P.B.; Schow, G.S.; Hutchinson, J.M., *Pestic. Biochem. Physiol.*, 1982, 17, 18-23.

17. Hutchinson, J.M.; Shapiro, R.; Sweetser, P.B., *Pestic. Biochem. Physiol.*, 1984, 22, 243-247.

18. Gressel, J.; Segel, L.A., *J. Theor. Biol.*, 1978, 75, 349-371.

19. LeBaron, H.M.; Gressel, J., Eds., *"Herbicide Resistance in Plants"*, Wiley: New York, 1982.

20. Ezra, G.; Dekker, J.H.; Stephenson, G.R., *Weed Sci.*, 1985, 33, 287-290.

21. Zorner, P.; Stafford, L.; McCall, P., *Proc. North Central Weed Control Conf.*, 1983, p. 109.

22. Lamoureaux, G.L.; Rusness, D.G., *Pestic. Biochem. Physiol.*, 1986, 26, 323-342.

23. Lamoureux, G.L.; Rusness, D.G., *Pestic. Biochem. Physiol.*, 1987, 27, 318-329.

24. Shishido, K.; Usui, M.; Fukami, J.-I. *Pestic. Biochem. Physiol.*, 1972, 2, 51-00.

25. Bocion, P.F., *Eur. Patent Application*, 1987, 87101712.5.

26. Hagmann, M.-L.; Bocion, P.F., 1987, personal communication.

27. Fedtke, C.; Trebst, A., *In "Pesticide Science and Biotechnology"*, Greenhalgh, R.; Roberts, T.R., Eds., Blackwell: Oxford, 1987, pp. 000-000.

28. Hatzios, K.K.; Hoagland, R.E., Eds., *"Herbicide Safeners Development, Uses and Mechanism of Action"*, Academic Press: New York, 1987, in press.

29. Fedtke, C., *Weed Res.*, 1987, 27 (in press).

30. Ortiz de Montellano, P.R.; Kunze, K.L., *J. Biol. Chem.*, 1980, 255, 5578-5585.

31. Gaillardon, P.; Cabanne, F.; Scalla, R.; Durst, F., *Weed Res.*, 1985, 25, 397-402.

32. Reichert, D.; Simon, A.; Durst, F.; Mathews, J.M.; Ortiz de Montellano, P.R., *Arch. Biochem. Biophys.*, 1982, 216, 522-529.

33. Cabanne, P.; Gaillardon, P.; Scalla, R.; Durst, F., *Brit. Crop Protection Conf.-Weeds*, 1985, pp. 1163-1170.

34. Komives, T.; Dutka, F., *Cereal Res. Comm.*, 1980, 8, 627-633.

35. Hatzios, K.K., *Weed Sci.*, 1981, 29, 601-604.

36. Cole, D.J.; Owen, W.J. *Plant Sci.*, 1987, (in press).

37. Hatzios, K.K., *Weed Sci.*, 1983, 31, 280-284.

38. Ketchersid, M.L.; Plapp, F.W.; Merkle, M.G., *Weed Sci.*, 1985, 33, 774-448.

39. Balayannis, P.G., *Ann.. Appl. Biol.*, 1969, 109, 573-580.

40. Köller, W., *Pestic. Sci.*, 1987, 129-147.

41. Grossmann, K.; Weiler, E.W.; Jung, J., *Planta*, 1985, 164-370-375.

42. Fritsch, H.; Kochs, G.; Hagmann, M.; Heller, W.; Jung, J.; Grisebach, H., *Abstracts 6th Intl. Cong. Pestic. Chem.*, IUPAC: Ottawa, 1986, p. 3B-16.

43. Georghiou, G.P.; Saito, T., Eds. *"Pest Resistance to Pesticides"* Plenum: New York, 1983.

44. Heap, I.; Knight, R., *Aust. J. Agric. Res.* 1986, 37, 149-156.

45. Moss, S.R.; Cussans, G.W., *In "Biological and Chemical Approaches to Combating Resistance to Xenobiotics"*, Ford, M.; Hollomon, D.; Khambay, B.; Sawicki, R., Eds., Soc. Chem. Ind.: London, 1987 (in press).

46. Cussans, G.W.; Moss, S.R., (personal communication)

47. Lamoureux, G.L.; Frear, D.S. *In "Xenobiotic Metabolism, In vitro Methods"*, Paulson, G.D.; Frear, D.S.; Marks, E.P.; Eds., Amer. Chem. Soc.: Washington, D.C., 1979, p. 95.

48. Pelsey, F.; Leroux, P.; Heslot, H., *Pestic. Biochem. Physiol.*, 1987, 27, 182-188.

49. Summers, L.A., Ed., *"The Bipyridilium Herbicides"*, Academic Press: New York, 1980.

50. Gaba, V.; Cohen, N.; Shaaltiel, Y.; Ben-Amotz, A.; Gressel, J., 1987 (submitted).

51. Halliwell, B.; Gutteridge, J., Eds., *"Free Radicals in Biology and Medicine"*, Oxford Sci. Publ.: Oxford, 1985.

52. Asada, K.; Nakano, Y.; Hossain, M.A., *In: "Oxidative Damage and Related Enzymes"*, Rotilio, G.; Bannister, J.V., Eds., Harwood:London, 1984, pp. 342-347.

53. Harvey, B.M.R.; Harper, D.B., *In "Herbicide Resistance in Plants"*, Le Baron, H.M.; Gressel, J., Eds., Wiley: New York, 1982, pp. 215-233.

54. Youngman, R.J.; Dodge, A.D., *In "Proc. 5th Intl. Cong. Photosynthesis*, Balaban: Philadelphia, 1981, pp. 537-544.

55. Watanabe, Y.; Honma, T.; Ito, K.; Miyahara, M., *Weed Res.* (Japan), 1982, 27, 49-54.

56. Shaaltiel, Y.; Gressel, J., *Pestic. Biochem. Physiol.*, 1986, 26, 22-28.

57. Fuerst, E.P.; Nakatani, H.Y.; Dodge, A.D.; Penner, D.; Arntzen, C.J., *Plant Physiol.*, 1985, 77, 984-989.

58. Vaughn, K.C.; Fuerst, E.P., *Pestic. Biochem. Physiol.*, 1985, 24, 86-94.

59. Shaaltiel, Y.; Gressel, J., 1987 (submitted).

60. Shaaltiel, Y.; Chua, N.-H.; Gepstein, S.; Gressel, J., 1987 (in manuscript).

61. Lewinsohn, E.; Gressel, J., *Plant Physiol.* 1984, 76, 125-130.

62. Burke, J.J.; Gamble, P.E.; Hatfield, J.L.; Quisenberry, J., *Plant Physiol.,* 1985, 415-419.

63. Finck, B.F .; Kunert, K.J., *J. Agric. Food Chem.,* 1985, 33, 574-577.

64. Hagmann, M.-L.; Shaaltiel, Y.; Bocion, P.F.; Gressel, J., 1987 (unpublished results).

65. Shaaltiel, Y.; Glazer, A.; Gressel, J. (in manuscript), 1987.

66. Jansen, M.; Canaani, O.; Shaaltiel, Y.; Gressel, J., 1986 (unpublished results).

67. Shaaltiel, Y., *Ph.D. Dissertation,* The Weizmann Institute of Science: Rehovot, 1987.

68. Shaaltiel, Y.; Gressel, J., *In "Pesticide Science and Biotechnology",* Greenhalgh, R.; Roberts, T.R., Eds., Blackwell: Oxford, 1987, pp. 183-186.

69. Shaaltiel, Y.; Warshawsky, A.; Gressel, J., (in manuscript).

70. Dowd, P.F.; Sparks, T.C., *Pestic. Biochem. Physiol.,* 1986, 25, 73-81.

RECEIVED May 4, 1988

Chapter 2

Molecular Biology of Resistance to Sulfonylurea Herbicides

Julie K. Smith[1], C. Jeffry Mauvais, Susan Knowlton, and Barbara J. Mazur

Agricultural Biotechnology Division, Experimental Station, E. I. du Pont de Nemours and Company, Wilmington, DE 19898

The sulfonylureas, an extremely potent class of herbicides, act by inhibiting acetolactate synthase (ALS), which is the first common enzyme in the biosynthetic pathways leading to the branched chain amino acids. Two other unrelated classes of herbicides also act by interfering with this enzyme. We have cloned and characterized the genes encoding ALS from several higher plants. The ALS genes isolated from herbicide sensitive and herbicide resistant plants have been compared, and several mutations which confer the herbicide resistant phenotype have been identified. Cloned herbicide resistant ALS genes have been used to transform both homologous and heterologous plant species. ALS genes can be modified *in vitro* in order to achieve selective resistance toward broad or narrow classes of inhibitors. The modified genes can be introduced into a variety of commercial crops.

The sulfonylurea herbicides are a relatively new class of crop protection chemicals which are notable for their low application rates (typically grams/hectare) and low mammalian toxicity. In addition many are selective, having the ability to kill weeds without injuring the target crop. Bacteria, fungi, and plants have all been shown to be naturally sensitive to these compounds. The mode of action of the sulfonylureas was initially discovered in a microbial system. Certain strains of bacteria were able to tolerate the sulfonylureas when grown on rich media but not when grown on minimal media (1). By studying the response of various bacteria to sulfonylureas in media supplemented with various nutrients, it was determined that these herbicides act by interfering with the branched chain amino acid biosynthetic pathways. The target enzyme was found to be acetolactate synthase (ALS, EC 4.1.3.18) which is the first common enzyme in the pathways leading to leucine,

[1]Current address: Biology Department, Philadelphia College of Pharmacy and Science, Philadelphia, PA 19104

0097–6156/88/0379–0025$06.00/0

isoleucine, and valine. ALS was subsequently shown to be the target of the sulfonylureas in yeast (2) and in higher plants (3, 4). Animals do not utilize this pathway, and must ingest the branched chain amino acids in their diets. This presumably contributes to the low mammalian toxicity of the sulfonylureas.

ALS is also inhibited by a number of compounds which are structurally unrelated to the sulfonylureas. These include two other classes of herbicides: the imidazolinones (5) and the triazolopyrimidines (Hawkes, T.R.; Howard, J.L.; Pontin, S.E. In *Herbicides and Plant Metabolism*, in press). LaRossa *et al.* have speculated on why ALS is such an effective target for so many inhibitors (6). Blocking ALS leads to the buildup of the toxic substrate a-ketobutyrate. The elevated levels of this metabolite combined with the reduced levels of the branched chain amino acids appear to make the inhibition of ALS a particularly lethal event.

Tolerance toward the sulfonylureas is known to occur naturally due either to the presence of a form of ALS that is insensitive to the inhibitors (7) or to a mechanism for detoxification of the inhibitors (8). Another mechanism that could in principle lead to tolerance is the overproduction of the target (ALS) enzyme. We are interested in engineering herbicide tolerance in crop plants in order to increase the margin of safety for the application of existing selective chemicals, to achieve selectivity in crops where selective chemicals do not currently exist, and to reduce damage in rotated crops which is due to the presence of herbicide residues. As a first step toward this goal we have characterized the ALS genes from several higher plants, including *Arabidopsis thaliana* and *Nicotiana tabacum* (tobacco). ALS genes have been cloned from both wild type plants and from lines which were selected to be herbicide resistant. The cloned ALS genes are now being engineered *in vitro* and are being reintroduced into several crop species.

Cloning and Characterization of Plant ALS Genes

The genes encoding each of the three known isozymes of ALS in *E. coli* have been cloned and sequenced (9-15). The single ALS gene in the yeast *Saccharomyces cerevisiae* has also been cloned (2), and its sequence has been determined (16). The deduced amino acid sequences of yeast and bacterial ALS proteins have been compared (16). There are three blocks of high sequence homology interspersed with blocks of sequence that are not well conserved. The degree of sequence conservation in the ALS enzymes from these diverse organisms suggested that the ALS genes from higher plants could be isolated using the method of heterologous hybridization. The yeast ALS gene was used to probe genomic libraries from the blue-green alga *Anabaena* and from the higher plants *Arabidopsis* and tobacco. Hybridizations were carried out under conditions of reduced stringency, and phage containing putative ALS genes were isolated in each case (17). The genes from the *Arabidopsis* and tobacco libraries were subcloned in plasmid vectors, mapped and sequenced, and their deduced amino acid sequences were determined (Mazur, B.J.; Chui, C.-F.; Smith, J.K. *Plant Physiol.*, in press).

Neither plant gene has introns. The genes encode proteins of 667 (tobacco) or 670 (*Arabidopsis*) amino acids, with predicted molecular weights of approximately 73,000 daltons. Figure 1 shows a

```
          10                  30                  50                  70
MAAAA..PSPSSS.AFSKTLSPSSSTSSTLLPRSTFPFPHHPHKTTPPPLHLTHTHIHIHSQRRRFTISNVISTNQKVSQ
MAAATTTTTTSSSISFSTKPSPSSSKSPLPISRFSLPFSLNPNKSSSSSRRRGIKSSSPSSISAVLNTTTNVTTTPSPTK
     MIRQSTLKNFAIKRCFQHIAYRNTPAMRSVALAQRFYSSSSRYYSASPLPASKRPEPAPSFNVDPLEQPAEPSKL

          90                 110                 130                 150
TEKTETFVSRFAPDEPRKGSDVLVEALEREGVTDVFAYPGGASMEIHQALTRSSIIRNVLPRHEQGGVFAAEGYARATGF
PTKPETFISRFAPDQPRKGADILVEALERQGVETVFAYPGGASMEIHQALTRSSSIRNVLPRHEQGGVFAAEGYARSSGK
AKKLRAEPDMDTSFVGLTGGQIFNEMMSRQNVDTVFGYPGGAILPVYDAIHNSDKFNFVLPKHEQGAGHMAEGYARASGK
    MASSGTTSTRKRFTGAEFIVHFLEQQGIKIVTGIPGGSILPVYDALSQSTQIRHILARHEQGAGFIAQGMARTDGK

         170                 190                 210                 230
PGVCIATSGPGATNLVSGLADALLDSVPIVAITGQVPRRMIGTDAFQETPIVEVTRSITKHNYLVMDVEDIPRVVREAFF
PGICIATSGPGATNLVSGLADALLDSVPLVAITGQVPRRMIGTDAFQETPIVEVTRSITKHNYLVMDVEDIPRIIEEAFF
PGVVLVTSGPGATNVVTPMADAFADGIPMVVFTGQVPTSAIGTDAFQEADVVGISRSCTKWNVMVKSVEELPLRINEAFE
PAVCMACSGPGATNLVTAIADARLDSIPLICITGQVPASMIGTDAFQEVDTYGISIPITKHNYLVRHIEELPQVMSDAFR

         250                 270                 290                 310
LARSGRPGPILIDVPKDIQQQLVIPDWDQPMR.....LPGYMSRLPKLPNEMLLEQIVRLISESKKPVLYVGGGCSQSSE
LATSGRPGPVLVDVPKDIQQQLAIPNWEQAMR.....LPGYMSRMPKPPEDSHLEQIVRLISESKKPVLYVGGGCLNSSD
IATSGRPGPVLVDLPKDVTAAILRNPIPTKTTLPSNALNQLTSRAQDEFVMQSINKAADLINLAKKPVLYVGAGILNHAD
IAQSGRPGPVWIDIPKDVQTAVFEIETQ.........PAMAEKAAAPAFSEESIRDAAAMINAAKRPVLYLGGG...VIN

         330                 350                 370                 390
DLRRFVEL...TGIPVASTLMGLGAFPTGDELSLSMLGMHGTVYANYAVDSSDLLLAFGVRFDDRVTGKLEAFASRAKIV
ELGRFVEL...TGIPVASTLMGLGSYPCDDELSLHMLGMHGTVYANYAVEHSDLLLAFGVRFDDRVTGKLEAFASRAKIV
GPRLLKELSDRAQIPVTTTLQGLGSFDQEDPKSLDMLGMHGCATANLAVQNADLIIAVGARFDDRVTGNISKFAPEARRA
APARVRELAEKAQLPTTMTLMALGMLPKAHPLSLGMLGMHGVRSTNYILQEADLLIVLGARFDDRAIGKTEQFCPNAKII

         410                 430                 450                 470
HIDIDSAEIGKNKQPHVSICADIKLALQGLNSILESKEGKLKLDFSAWRQELTEQKVKHPLNFKTF.........GDAIP
HIDIDSAEIGKNKTPHVSVCGDVKLALQGMNKVLENRAEELKLDFGVWRNELNVQKQKFPLSFKTF.........GEAIP
AAEGRGGIIHFEVSPKNINKVVQTQIAVEGDATTNLGKMMSKIFPVKERSEWFAQINKWKKEYPYAYMEETPGSKIKPQT
HVDIDRAQLGKIKQPHVAIQADVDDVLAQLIPLVEAQPRAEWHQLVADLQREFPCPIPKA..............CDPLS

         490                 510                 530                 550
PQYAIQVLDELTNGNAIISTGVGQHQMWAAQYYKYRKPRQWLTSGGLGAMGFGLPAAIGAAVGRPDEVVVDIDGDGSFIM
PQYAIKVLDELTDGKAIISTGVGQHQMWAAQFYNYKKPRQWLSSGGLGAMGFGLPAAIGASVANPDAIVVDIDGDGSFIM
VIKKLSKVANDTGRHVIVTTGVGQHQMWAAQHWTWRNPHTFITSGGLGTMGYGLPAAIGAQVAKPESLVIDIDGDASFNM
HYGLINAVAACVDDNAIITTDVGQHQMWTAQAYPLNRPRQWLTSGGLGTMGFGLPAAIGAALANPDRKVLCFSGDGSLMM

         570                 590                 610                 630
NVQELATIKVENLPVKIMLLNNQHLGMVVQWEDRFYKANRAHTYLGNPSNEAEIFPNMLKFAEACGVPAARVTHRDDLRA
NVQELATIRVENLPVKVLLLNNQHLGMVMQWEDRFYKANRAHTFLGDPAQEDEIFPNMLLFAAACGIPAARVTKKADLRE
TLTELSSAVQAGTPVKILILNNEEQGMVTQWQSLFYEHRYSHTHQL........NPDFIKLAEAMGLKGLRVKKQEELDA
NIQEMATASENQLDVKIILMNNEALGLVHQQQSLFYEQGAFVATYP.......GKINFMQIAAGFGLETCDLNNEADPQA

         650                 670
AIQKMLDTPGPYLLDVIVPHQEHVLPMIPSGGAFKDVITEGDGRSSY                    TOBACCO ALS
AIQTMLDTPGPYLLDVICPHQEHVLPMIPSGGTFNDVITEGDGRIKY                ARABIDOPSIS ALS
KLKEFVSTKGPVLLEVEVDKKVPVLPMVAGGSGLDEFINFDPEVERQQTELRHKRTGGKH         YEAST ALS
SLQEIINRPGPALIHVRIDAEEKVYPMVPPGAANTEMVGE                            E. COLI ALS
```

Figure 1. ALS sequences. Deduced amino acid sequences of ALS enzymes from plants, yeast (16), and E. coli (large subunit of ALS I, 9, 15). Vertical lines indicate identical residues.

comparison of the deduced amino acid sequences of ALS proteins from tobacco, Arabidopsis, yeast, and bacteria. This diagram illustrates the presence of blocks of high sequence conservation which are separated by regions where the various proteins have diverged. The presence of the strongly conserved regions explains why the plant genes could be cloned using the yeast gene as a hybridization probe. The two plant proteins are highly conserved with respect to each other, even in regions where the yeast and E. coli proteins have diverged. Approximately 75% of the nucleotides and 85% of the amino acids are identical between the ALS genes and proteins from the the two plants. This level of conservation suggested that ALS genes isolated from one plant species would likely be functional when introduced into a heterologous species.

The only region where the two plant proteins are not well conserved is in the N-terminal region, which encodes the chloroplast transit peptides. In this region only about 23% of the amino acid residues and 40% of the nucleotides are identical. In plants ALS is encoded in the nucleus yet localized in chloroplasts (18-20). The transit peptide is thought to direct the nascent protein post-translationally into the chloroplasts. The transit peptide is then cleaved to yield the mature ALS protein. This process can be studied in a model system containing isolated chloroplasts and ALS protein which has been transcribed and translated in vitro (Bascomb, N.; Gutteridge, S.; Leto, K.; Smith, J.K., submitted for publication). The putative ALS transit peptides of tobacco and Arabidopsis show little homology when compared with each other or with yeast, which has a transit peptide that directs ALS into the mitochondria (21, 22). The two plant ALS transit sequences also show little homology with the transit sequences which have been determined for other chloroplast-localized proteins (23). Structural similarities can be seen, however, when the hydropathy profiles of the tobacco and Arabidopsis ALS transit peptides are compared (not shown). This suggests that a functional transit sequence depends more on secondary or higher order structural constraints than on primary sequence information. The in vitro uptake system described above can be used to further investigate the transit peptide domain.

The cloned plant ALS genes have been used to study the organization and regulation of ALS in plants. By hybridization analysis, it was shown that Arabidopsis has only a single ALS gene (Mazur, B.J.; Chui, C.-F.; Smith, J.K. Plant Physiol., in press). Tobacco was shown to have two ALS genes, consistent with it being an allotetraploid (24), and with earlier genetic data suggesting two ALS loci (18).

At the RNA level, each of the two tobacco genes is transcribed, although one is consistently expressed at higher levels than the other (Martin, S.; Mazur, B.J.; Smith, J.K., manuscript in preparation). The highest levels of ALS message are seen in flowers and in the youngest leaves, while expression is barely detectable in roots and in the older tissues.

These cloned Arabidopsis and tobacco ALS genes have been used as hybridization probes to isolate ALS genes from other crop species and to isolate ALS genes from plants selected for resistance to sulfonylurea herbicides.

Cloning and Characterization of Mutant ALS Genes

Two selection strategies have been used to obtain plants resistant to sulfonylurea herbicides.

Tobacco Mutants. Chaleff and Ray (18) plated callus cultures from haploid *Nicotiana tabacum* on media containing 2 ppb (approximately 6 nM) chlorsulfuron or sulfometuron methyl. Resistant cell lines were regenerated and diploid plants were characterized genetically. The herbicide resistance mutations fell into two classes, represented by mutant lines C3 and S4, which defined two semidominant nuclear loci, *SuRA* and *SuRB*. Homozygous mutant plants of the S4 type were able to tolerate at least 100-fold more herbicide than were their wild type progenitors. Callus initiated from homozygous S4 plants was subsequently exposed to even higher levels of herbicide (200 ppb), and resistant lines were regenerated into plants (25). One such line, designated Hra, was able to tolerate concentrations of herbicide 1000-fold greater than that required to inhibit wild type lines. The Hra mutation was shown to be genetically linked to the S4 mutation. These three mutant tobacco lines (C3, S4, and Hra) all had an altered form of ALS which was less sensitive to inhibition by the sulfonylurea herbicides than was enzyme extracted from wild type plants. This suggested that cloned mutant ALS genes could be used to transform other plants to herbicide resistance. Although the tobacco ALS gene described above was isolated from an S4 genomic library, transformation of this gene into tobacco cells showed that it encoded a herbicide-sensitive form of ALS. This gene was therefore used as a hybridization probe to isolate the ALS genes from genomic libraries made from the C3 and the Hra mutant lines (Lee, K.Y.; Townsend, J.; Tepperman, J.; Black, M.; Chui, C.-F.; Dunsmuir, P.; Mazur, B.J.; Bedbrook, J., submitted for publication). By independently introducing each cloned gene into sensitive tobacco cells and then assaying transformants for resistance, mutant and wild type genes were distinguished. Each gene was sequenced and the mutations responsible for the herbicide resistant phenotype were identified. The C3 gene has a single mutation while the Hra gene has alterations leading to two amino acid substitutions. The Hra gene has been used to transform a variety of crops to herbicide resistance as described below.

Arabidopsis Mutants. Haughn and Somerville (26) obtained herbicide resistant *Arabidopsis* lines by screening a mutagenized seed population. *Arabidopsis* seeds were exposed to the mutagen EMS and then grown to maturity. Seeds collected from the resulting plants were plated on media containing 200 nM chlorsulfuron (a commercial sulfonylurea herbicide). Of 300,000 seeds screened, four seedlings germinated. One line (GH50) was further characterized. This line was homozygous with respect to the resistant phenotype, defining a single dominant nuclear mutation designated *Csr-1*. It was able to tolerate approximately 300-fold higher levels of herbicide than could wild type *Arabidopsis*. ALS activity in extracts from mutant plants was less sensitive to inhibition by sulfonylureas than was enzyme from wild type plant extracts. The cloned wild type *Arabidopsis* gene described above was used to isolate the ALS gene from a library made from the herbicide resistant line (Haughn, G.;

Smith, J.K.; Mazur, B.J.; Somerville, C. *Mol. Gen. Genet.*, in press). The mutant gene was sequenced and compared to the wild type gene. A single mutation, which changed proline 197 to a serine, was identified. A mutation at this position (to serine) has also been described in yeast selected for herbicide resistance (27). The cloned mutant *Arabidopsis* ALS gene has been used in transformation experiments by Haughn *et al.* (Haughn, G.; Smith, J.K.; Mazur, B.J.; Somerville, C. *Mol. Gen. Genet.*, in press), and as described below.

Introduction of Mutant ALS Genes into Plants

The Arabidopsis Gene. Genes encoding both sulfonylurea sensitive and sulfonylurea resistant forms of ALS were cloned into Ti plasmid-derived plant transformation vectors which also contained a gene conferring resistance to the antibiotic kanamycin. After transfer into *Agrobacterium*, each construction was used to transform a sensitive tobacco line, using a modified leaf-disk transformation protocol. Following selection on kanamycin, transformed shoots were excised and placed on rooting media. Several small leaves from each shoot were also placed on callus induction media containing kanamycin, herbicide, or neither compound. Seventeen of 19 plants transformed with the mutant *Arabidopsis* ALS gene were able to form callus on media containing 10 ppb chlorsulfuron, while none of the plants transformed with the wild type gene were able to form callus on the same media. With few exceptions, callus carrying the mutant ALS gene was able to grow in the presence of both kanamycin and herbicide. This was as expected, since the two genes were linked in the transformation vector. The kanamycin resistant callus cultures from several transformants were subcultured in order to generate large quantities of healthy callus tissue for further testing. Callus lines derived from transformants which had received the mutant ALS gene grew much better in the presence of 10 ppb chlorsulfuron than did lines derived from transformants which received the wild type gene, as shown in Figure 2. Also shown in Figure 2 are the results of ALS assays performed on extracts of transformed plants. At 100 ppb chlorsulfuron the enzyme activity from plants containing the sensitive ALS gene was almost completely inhibited, while activity in plants containing the resistant gene was inhibited only 30-60%. Each of the resistant transformants was forced to self-pollinate, and progeny tests were performed. These tests indicated that each transformant had received a single herbicide resistant ALS gene which was subsequently inherited in a simple Mendelian fashion.

The Tobacco Gene. The Hra line of tobacco is the plant mutant which has shown the highest level of resistance to the sulfonylureas. For this reason the cloned Hra gene has been used in transformation experiments. In one set of experiments the Hra gene was used to transform commercial tobacco cultivars. Figure 3 shows enzyme assays run on extracts from some of the transformants and on two wild type plants. ALS activity in the transformants was clearly more resistant to herbicide than was the activity from control plants. Tolerance to herbicide was also assayed by measuring the ability to form callus in the presence of herbicide, by measuring the ability of progeny seeds to germinate in the presence of

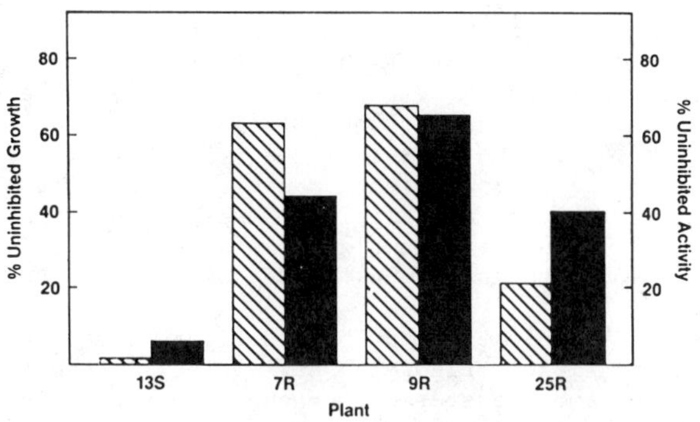

Figure 2. Assays of tobacco transformed with Arabidopsis ALS genes. Tobacco containing either the herbicide sensitive (S) or resistant (R) ALS gene was tested for the ability of transformed callus to grow in the presence of herbicide (slashed boxes) and the herbicide resistance of enzyme activity in plant extracts (solid boxes). Each measurement is expressed as a percentage of the value that was obtained in the absence of herbicide.

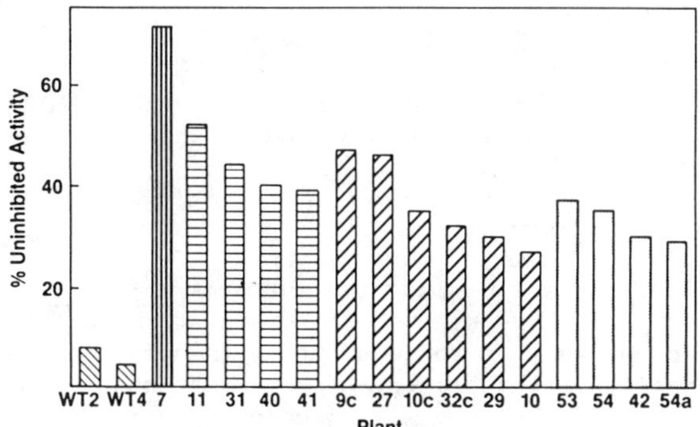

Figure 3. Assays of tobacco transformed with a mutant tobacco ALS gene. Enzyme activity in the presence of herbicide was measured in leaves of commercial tobacco cultivars transformed with the HRA gene (plants #7-54), and in untransformed plants (WT2, WT4). Activity is expressed as a percentage of the activity measured in the absence of herbicide.

increasing concentrations of herbicide, and by measuring phytotoxicity following foliar applications of various herbicides. Transformants were found to be more resistant than wild type plants by all of these criteria, although quantitative differences pointed to the necessity of assaying transformants by more than one method. Segregation analyses of progeny plants and blot hybridizations were used to determine the number of resistant ALS loci segregating in each transformed line. Most of the transformants had only one or two copies of the mutant gene. Line #7, which showed one of the highest levels of resistance of any of the transformants (see Figure 3) had four copies of the mutant allele. For breeding programs it is desirable to use a line which has a high level of resistance derived from a single mutant allele. Gene copy number as well as the position at which the mutant genes are integrated into the plant genome are expected to influence the degree of herbicide resistance of a given line.

Several of the plants described in Figure 3 are currently being evaluated in the field. Figure 4 shows an early result from this work. Wild type and transformed tobacco plants were tested for their ability to tolerate a 32 grams/hectare foliar application of a sulfonylurea herbicide. This dose was approximately four times the typical field application rate. Untransformed tobacco is extremely sensitive to this herbicide, as can be seen by the phytotoxic effects displayed by the the wild type plants in the figure. The transformants were unaffected by this level of herbicide application, demonstrating that the cloned Hra gene is able to confer useful levels of herbicide resistance in transgenic plants grown under field conditions.

The Hra gene has also been used to transform a number of heterologous species, and selectable levels of resistance have been obtained in each case. The Hra gene can probably be used to confer useful levels of herbicide resistance in most plant species.

In addition to being useful in the field, the herbicide resistance phenotype conferred by mutated ALS genes is a useful selectable marker in the laboratory.

Engineering of New Mutant ALS Genes

Altering Cloned Plant ALS Genes. The mutant *Arabidopsis* and tobacco genes described above have been shown to confer the herbicide resistant phenotype when introduced into both homologous and heterologous species. The cloned ALS genes can also be modified *in vitro* in order to modulate the level and/or specificity of herbicide resistance in transgenic plants. Several strategies can be used to achieve this goal. The alteration of regulatory elements such as promoters and enhancers can be used to increase/decrease mRNA levels, or to cause the message to be expressed in a tissue-specific, developmentally-specific, or inducible manner.

Mutations can also be incorporated into the ALS gene in order to alter the specific interactions between the enzyme and its inhibitors. Site directed mutagenesis has been used to incorporate mutations into cloned plant ALS genes based both on previously characterized mutants and on computer modeling information. These altered genes have been transformed into plants, and the transgenic plants have been tested for their sensitivity to a variety of herbicides. Transgenic plants can contain mutated copies of their native ALS genes, or mutant ALS genes from heterologous sources. The former may be preferable in order to expedite regulatory approval for commercial releases. In addition, native ALS genes may function better in some hosts then heterologous ones.

Expression of Plant ALS Genes in Microorganisms. Site directed mutagenesis has been used to incorporate a variety of mutations into several cloned ALS genes, as described above. However, the transformation, regeneration, and progeny testing are laborious processes. In order to take advantage of the power of microbial genetics to study plant ALS genes, a bacterial expression system was developed (Smith, J.K.; Schloss, J.V.; Mazur, B.J., submitted for publication). ALS genes (including the chloroplast transit peptide coding region) were cloned into expression vectors under the control of bacterial regulatory signals. The plant ALS genes were functionally expressed in *E. coli*, and the plant protein was able to complement a branched chain amino acid auxotrophy of the bacteria. This system has been used to determine the herbicide sensitivity of new mutant ALS proteins, as demonstrated in Figure 5. Bacteria expressing a wild type or mutant plant ALS gene were plated on minimal media, and filter paper disks impregnated with the active ingredients from two commercial herbicides were placed on the plate surface. A radial herbicide concentration gradient formed and, after allowing for bacterial growth, the zones of inhibition were compared. The mutant shown in Figure 5 was more resistant to the sulfonylurea Classic than was the wild type but it retained its sensitivity to the imidazolinone Scepter. This mutant thus exhibits selective herbicide resistance. Ultimately, interesting mutations identified in this system must be incorporated into plants for further testing. Preliminary work has shown that there is a good correlation between the phenotypes resulting from the mutant genes expressed in bacteria and the same mutant genes expressed in plants.

The *E. coli* expression system described here has also proven useful for purifying plant ALS enzymes. Because it is present in such small amounts, ALS has been difficult to purify from plant sources. Purifying the plant enzymes expressed in bacteria has provided material for use in enzymatic and structural studies as well as for the generation of immunological reagents.

Future Prospects

We have described the cloning, molecular characterization, expression, and reintroduction into plants of wild type and mutant ALS genes. Efforts to use information about the ALS target protein in order to manipulate herbicide sensitivity in crop species can proceed in two directions. In physical studies, the ALS protein itself can be characterized. Information about how the enzyme functions and how it specifically interacts with inhibitors can contribute to mutagenesis programs and to a more rational design of new herbicide candidates. Most currently used commercial herbicides were discovered in large synthesis/screening programs, but that "shotgun" approach is becoming increasingly unwieldy due to the enormous number of compounds that need to be tested before a single selective herbicide is found. In biological studies, new mutants

Figure 4. Field tests of transformed tobacco. Wild type tobacco (WT) and tobacco transformed with the HRA gene (SUR) were sprayed at 4X the typical field application rate with a commercial sulfonylurea herbicide preparation.

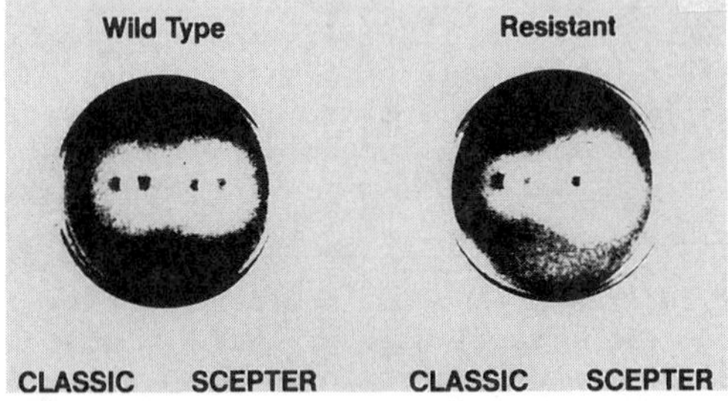

Figure 5. Disk assays of plant ALS genes expressed in bacteria. E. coli expressing a wild type or a herbicide resistant ALS gene were tested for their ability to grow in the presence of the active ingredients from two commercial herbicides.

can be characterized, mutations can be incorporated into ALS genes from a variety of species, and mutants can be tested for the ability to tolerate new or existing compounds. The biological and physical approaches are complementary, and together will further the development of new efficacious crop protection strategies.

Acknowledgments

We gratefully acknowledge our many colleagues at Du Pont who have contributed both directly and indirectly to the experiments described here. Special thanks are due to Tony Guida, Chok-Fun Chui, and Sharon Martin for molecular analyses of the plant ALS genes, and to Todd Houser and Chris Kostow for plant transformation experiments. We have enjoyed fruitful collaborations with George Haughn and Chris Somerville at Michigan State University on the isolation and characterization of the mutant *Arabidopsis* ALS gene, and with John Bedbrook, Kathy Lee, Jeff Townsend, and Pamela Dunsmuir at Advanced Genetic Sciences on the cloning and characterization of mutant tobacco ALS genes.

Literature Cited

1. LaRossa, R. A.; Schloss, J. V. *J. Biol. Chem.* 1984, *259*, 8753-8757.
2. Falco, S. C.; Dumas, K. D. *Genetics* 1985, *109*, 21-35.
3. Chaleff, R. S.; Mauvais, C. J. *Science* 1984, *224*, 1443-1445.
4. Ray, T. B. *Plant Physiol.* 1984, *75*, 827-831.
5. Shaner, D. L.; Anderson, P. C.; Stidham, M. A. *Plant Physiol.* 1984, *76*, 545-546.
6. LaRossa, R. A.; Falco, S. C.; Mazur, B. J.; Livak, K. J.; Schloss, J. V.; Smulski, D. R.; Van Dyk, T. K.; Yadav, N. S. In *Biotechnology in Agricultural Chemistry*; Le Baron, H. M.; Mumma, R. O.; Honeycutt, R. C.; Duesing, J. H., Eds.; ACS Symposium Series No. 334; American Chemical Society: Washington, DC, 1987; pp 190-203.
7. LaRossa, R. A.; Smulski, D. R. *J. Bacteriol.* 1984, *160*, 391-394.
8. Sweetser, P. B.; Schow, G. S.; Hutchison, J. M. *Pestic. Biochem. Physiol.* 1982, *17*, 18-23.
9. Friden, P.; Donegan, J.; Mullen, J.; Tsui, P.; Freundlich, M.; Eoyang, L; Weber, R.; Silverman, P. M. *Nucleic Acids Res.* 1985, *13*, 3979-3993.
10. Lawther, R. P.; Nichols, B.; Zurawski, G.; Hatfield, G. W. *Nucleic Acids Res.* 1979, *7*, 2289-2301.
11. Lawther, R. P.; Calhoun, D. H.; Adams, C. W.; Hauser, C. A.; Gray, J.; Hatfield, G. W. *Proc. Natl. Acad. Sci. USA* 1981, *78*, 922-925.
12. Newman, T.; Friden, P.; Sutton, A.; Freundlich, M. *Mol. Gen. Genet.* 1982, *186*, 378-384.
13. Squires, C. H.; DeFelice, M.; Wessler, S. R.; Calvo, J. M. *J. Bact.* 1981, *147*, 797-804.
14. Squires, C. H.; DeFelice, M.; Devereux, J.; Calvo, J. M. *Nucleic Acids Res.* 1983, *11*, 5299-5313.
15. Wek, R. C.; Hauser, C. A.; Hatfield, G. W. *Nucleic Acids Res.* 1985, *13*, 3995-4010.

16. Falco, S. C.; Dumas, K. D.; Livak, K. J. Nucleic Acids Res. 1985, 13, 4011-4027.
17. Mazur, B. J.; Chui, C.-F.; Falco, S. C.; Mauvais, C. J.; Chaleff, R. S. In The World Biotech Report 1985; Online International: New York, 1985; pp 97-108.
18. Chaleff, R. S.; Ray, T. B. Science 1984, 223, 1148-1151.
19. Jones, A. V.; Young, R. M.; Leto, K. J. Plant Physiol. 1985, 77, S293.
20. Miflin, B. J. Plant Physiol. 1974, 75, 827-831.
21. Magee, P. T.; Robichon-Szulmajster, H. de Eur. J. Biochem. 1968, 3, 502-506.
22. Ryan, E. D.; Kohlhaw, G. B. J. Bacteriol. 1974, 120, 631-637.
23. Karlin-Neumann, G. A.; Tobin, E. M. EMBO J. 1986, 5, 9-13.
24. Smith, H. In Nicotiana: Procedures for Experimental Use; Durbin, R. D., Ed.: US Dept. of Agriculture Technical Bulletin 1586; Academic Press: New York, 1979; p 3.
25. Chaleff, R. S.; Sebastian, S. A.; Creason, G. L.; Mazur, B. J.; Falco, S. C.; Ray, T. B.; Mauvais, C. J.; Yadav, N. S. In Molecular Strategies for Crop Protection; Alan R. Liss: New York, 1987; pp 415-425.
26. Haughn, G. W.; Somerville, C. R. Mol. Gen. Genet. 1986, 204, 430-434.
27. Yadav, N. S.; McDevitt, R. E.; Benard, S.; Falco, S. C. Proc Nat. Acad. Sci. USA 1986, 83, 4418-4422.

RECEIVED February 25, 1988

Chapter 3

5–Enolpyruvylshikimate 3–Phosphate Synthase

From Biochemistry to Genetic Engineering of Glyphosate Tolerance

G. Kishore, D. Shah, S. Padgette, G. della-Cioppa, C. Gasser, D. Re, C. Hironaka, M. Taylor, J. Wibbenmeyer, D. Eichholtz, M. Hayford, N. Hoffmann, X. Delannay, R. Horsch, H. Klee, S. Rogers, D. Rochester, L. Brundage, P. Sanders, and R. T. Fraley

Plant Molecular Biology Group, Monsanto Company, 700 Chesterfield Village Parkway, St. Louis, MO 63198

Glyphosate, the active ingredient of the nonselective, post-emergent, systemic herbicide Roundup®, has broad spectrum activity against a wide range of annual and perennial plants. Roundup® is currently used in a variety of agronomic and nonagronomic situations for vegetation control. However, lack of selectivity of this herbicide has prevented its use as an over-the-crop herbicide for efficient weed control. In this article, we describe two methods for engineering glyphosate tolerance into crop plants. In the first method, tolerance is achieved by overproduction of 5-enol-pyruvylshikimate 3-phosphate synthase (EPSPS), the enzyme target for glyphosate, involved in the biosynthesis of aromatic amino acids in both plants and microbes. The second method for engineering tolerance is based on expression of glyphosate tolerant mutant EPSPS gene. Transgenic plants producing the mutant enzyme are tolerant to Roundup®. With both methods, the tolerance of transgenic calli to glyphosate is significantly lower when the enzyme is localized in the cytosol instead of chloroplasts.

Glyphosate is the active ingredient of the nonselective, post-emergent, systemic, foliar applied herbicide, Roundup® (1). Roundup® kills a wide range of both annual and perennial plants. In addition to being nontoxic to mammals and fish, glyphosate is rapidly inactivated by interactions with soil and is also readily metabolized by soil microorganisms (2). Despite these outstanding environmental and weed control characteristics, glyphosate has only limited utility during the active growth season of crops and vegetables, because it kills the crops as well as the weeds. The herbicide is primarily used in vegetation control, low or no-till farming and in weed control during crop growth using special applicators for herbicide delivery. Recent developments in plant

0097–6156/88/0379–0037$06.00/0

genetic engineering have provided new techniques for introduction of herbicide tolerance genes into crop plants (3). In this article, we discuss the status of genetic engineering of Roundup® tolerance to plants. Our efforts towards engineering Roundup® tolerance have been facilitated by an understanding of the mechanism of herbicidal action of glyphosate.

Jaworski (4) reported that growth inhibition of both plant and microbes by glyphosate could be reversed by aromatic amino acids. Further work of Amrhein and his coworkers revealed that glyphosate inhibits the shikimate pathway enzyme, 5-enolpyruvylshikimate 3-phosphate (EPSP) synthase (5). This enzyme catalyzes the reaction shown in Figure 1. Glyphosate-treated plant and bacterial cultures accumulate shikimate and/or shikimate 3-phosphate (S3P), confirming that inhibition of EPSPS is at least a part of the *in vivo* mechanism of action of this herbicide (6, 7).

EPSPS has been isolated from both microorganisms and plants, and several of its properties have been studied. The bacterial and plant enzymes are monofunctional with molecular mass of 44-48 kD (8-15). The fungal enzyme is a part of the multifunctional arom complex which catalyzes four other reactions of the shikimate pathway (16). While the bacterial enzymes show differences with respect to glyphosate sensitivity, the plant enzymes exhibit a much more narrow range of sensitivity (17). This accounts for the susceptibility of most plant species to glyphosate.

Based on steady state kinetic studies, EPSPS initially forms a complex with S3P which interacts with the second substrate, phosphoenol pyruvate (PEP) to form the ternary enzyme*S3P*PEP complex (18), Anderson, K.; Sikorski, J.; Johnson, K. *Biochemistry*, in press). During turnover, inorganic phosphate (Pi) is released followed by EPSP. Glyphosate inhibits EPSPS competitively with respect to PEP and uncompetitively with respect to S3P (19). Although binding of PEP to EPSPS can be detected in the absence of S3P, binding of glyphosate occurs only in the presence of S3P. In the ternary complex of enzyme*S3P*glyphosate, the Kd for S3P is lower than in the binary enzyme*S3P complex. Thus the binding of S3P to the enzyme is enhanced in the presence of glyphosate.

While glyphosate inhibits EPSPS competitively with respect to PEP and thus may occupy the PEP binding site on the enzyme, it is interesting to note that glyphosate does not inhibit any other PEP-dependent enzymatic reaction (19, 20). Mechanistic studies indicate that during the EPSPS reaction, the C-2 carbon of PEP interacts with a nucleophile generating a tetrahedral carbon at C-3 of PEP (21-24). It has also been suggested that the C-2 of PEP may be polarized, generating a carbonium ion and that the cationic protonated nitrogen of glyphosate may mimic this carbonium ion. This aspect of the mechanism of interaction between glyphosate and EPSPS remains unknown. In addition to the formation of a tetrahedral carbon at C-3 of PEP during the EPSPS reaction, it is the C-O bond of PEP that is cleaved, and not the O-P bond. This is in contrast to most other PEP utilizing enzymes. These features may account for the specificity of glyphosate for inhibition of the EPSPS enzymatic reaction (25).

Based on the knowledge that glyphosate inhibits EPSPS, two mechanisms were evaluated for genetic engineering of glyphosate

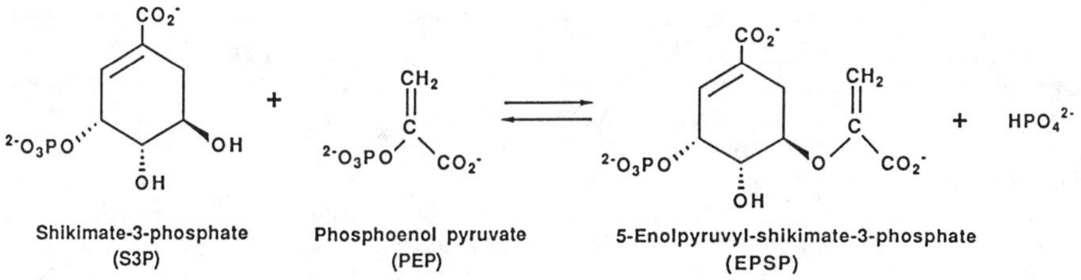

$$\text{S3P} + \text{PEP} \rightleftharpoons \text{EPSP} + HPO_4^{2-}$$

Shikimate-3-phosphate (S3P)

Phosphoenol pyruvate (PEP)

5-Enolpyruvyl-shikimate-3-phosphate (EPSP)

Figure 1 EPSP Synthase Catalyzed Reaction.

tolerance. The first mechanism was based on overproduction of wild type EPSPS; glyphosate tolerance may therefore be derived from the residual EPSPS activity of the uninhibited enzyme (26). In addition, the reduced cellular concentration of free glyphosate due to complex formation with EPSPS may also contribute towards tolerance to the herbicide. The second mechanism was based on isolation of glyphosate tolerant mutant EPSPS enzymes and expression of these mutant genes in plants (27, 28). Contrary to the first mechanism, the contribution of glyphosate binding to mutant EPSPS is minimal and tolerance is derived only if EPSPS is the sole target for glyphosate. As discussed in this article, chloroplast-targeted mutant EPSP synthases confer higher level of glyphosate tolerance than the wild type enzyme to transgenic tobacco plants.

Overproduction of EPSPS has been observed in several plant cell cultures tolerant to glyphosate (12, 13, 29). In the case of glyphosate-tolerant Corydalis cultures, Smart *et al.* demonstrated by 2 D-gel electrophoresis, the overproduction of other proteins besides EPSPS. Since the levels of activity of several shikimate pathway enzymes were unaltered in the tolerant cell line compared to the parent cell line, it was concluded that these amplified proteins may not be involved in aromatic amino acid biosynthesis. It is possible that the other proteins may not have a role in the tolerance mechanism. Alterations in protein profiles between glyphosate-sensitive and tolerant petunia cell lines have also been observed. With the glyphosate tolerant carrot cell line, in addition to overproduction of EPSPS, the levels of aromatic amino acids were found to be enhanced (29). Based on the results with plant cell cultures, it was therefore not clear if overproduction of EPSPS was sufficient to obtain glyphosate tolerance in plants.

Availability of the petunia cell line overproducing EPSPS significantly aided in the purification of the enzyme to homogeneity and elucidation of its amino terminal amino acid sequence (12, 26). The sequence corresponding to amino acids 8 to 13 was utilized for synthesis of three families of heptadecameric oligonucleotide probes (each 32 fold degenerate) and were designated EPSP1, EPSP2 and EPSP3 respectively. By Northern blot analysis, EPSP1 was found to have the correct sequence of oligonucleotides to encode the EPSP synthase mRNA. Using EPSP1, a λgt10 cDNA library of the glyphosate-tolerant petunia cell line was screened and a full length cDNA clone of petunia EPSPS was isolated (26). Additional studies indicated that the increased activity of EPSPS in the glyphosate tolerant petunia cell line was due to amplification of the EPSPS gene resulting in an increased synthesis of EPSP synthase mRNA and hence protein. A similar increase in the gene copy number has been demonstrated for glyphosate tolerant carrot somatic hybrid cells (30) whereas in Corydalis, no gene amplification has been detected (Amrhein, N. Personal Communication).

Nucleotide sequence determination of the petunia EPSPS cDNA clone revealed that it encoded a precursor protein (preEPSPS) with an additional 72 amino acids at the amino terminal end of the EPSPS amino acid sequence determined for the purified protein. Since it had been suggested that enzymes of the shikimate pathway may be localized in the chloroplasts (11, 31-35), it was of interest to determine if the overproduced EPSPS in the glyphosate tolerant petunia cell line was localized in the cytosol or plastids. Analysis of the plastidic and cytosolic fractions of the glyphosate tolerant petunia cell line revealed that the overproduced EPSPS was plastid-localized (36). These studies suggested that the amino terminal extension of preEPSPS may be involved in the translocation of the cytosolically synthesized protein to the plastids, analogous to the import of other proteins by chloroplasts (37-39).

In order to verify the role of the amino terminal extension of preEPSPS in its import by isolated chloroplasts, the cDNA of petunia EPSPS was cloned into a T7/SP6 transcription system for the *in vitro* synthesis of EPSPS mRNA (40). Translation of the mRNA in the presence of 35S-labeled methionine resulted in the synthesis of radiolabeled preEPSPS. The molecular weight of this protein was ~8kD higher than the molecular weight of EPSPS purified from petunia cell lines (55 vs 48 kD). Incubation of the preEPSPS with chloroplasts isolated from lettuce leaves, in the presence of ATP and light, resulted in a rapid uptake of the protein from the medium to the chloroplasts (40). The imported protein had a lower molecular mass (48 kD) indicating its rapid processing to the mature form of EPSPS, similar to that isolated from the glyphosate tolerant petunia cell line. Similar import and processing could also be demonstrated with chloroplasts prepared from the leaves of other plants. This indicates the conservation of both import and processing mechanisms in chloroplasts of different plants.

Examination of the EPSPS activity of the preEPSP synthase revealed that the preenzyme was catalytically active (40). No significant differences between the preenzyme and the mature enzyme could be detected either with respect to activity or glyphosate sensitivity. These studies suggest that the catalytic and chloroplast transit peptide domains of preEPSP synthase are distinct and independently folded. More recent studies indicate that the import of preEPSP synthase by isolated chloroplasts is inhibited by S3P plus glyphosate (41). The extent of maximal inhibition is only about 70-80%. The residual import rate could be reflective of the rate of import of the preenzyme*S3P*glyphosate complex. Since complex formation between these ligands and the preenzyme is conformationally restrictive, it is evident that conformational changes in the preprotein occur during its passage through the chloroplast membranes. Whether inhibition of import of the preenzyme is a physiological mode-of-action of glyphosate under *in vivo* conditions has not been determined.

More recently, the nucleotide sequences of tomato and Arabidopsis EPSPS genes have been determined (42, 43). It is interesting to note that the mature plant EPSPS shows a high degree of conservation at the amino acid level (>85%). Comparison of the plant EPSPS sequences with bacterial and fungal EPSPS sequences reveals that there is a 38% identity between plant and fungal enzymes while between the bacterial and plant enzymes, there is a 54% identity. The transit peptides of the plant enzymes are not conserved to a similar extent; between the Arabidopsis and petunia transit peptides, ~23% identity is observed. The same pattern of conservation has been observed with other chloroplast targeted plant proteins (44).

Following its isolation, the petunia EPSP synthase cDNA clone was engineered into the Ti plasmid vector system suitable for plant transformation. In this construct, the CaMV 35S promoter constituted the 5' end of the gene while the 3' flanking sequence including the polyadenylation signal was derived from the nopaline synthase gene (26). Transformation of petunia leaf discs using this construct resulted in the production of calli which tolerated glyphosate treatment at 0.5mM and 1.0mM. Under identical conditions, calli transformed with a control vector lacking the petunia EPSPS cDNA did not survive glyphosate treatment. The extent of overproduction of EPSPS in calli receiving the EPSP synthase gene was 40-80 fold. These experiments demonstrated that overproduction of EPSPS constituted a viable mechanism for the engineering of glyphosate tolerance to plant cells.

In order to determine if this mechanism could also confer tolerance to Roundup® at the whole plant level, petunia plants were regenerated from these transformed calli. The transformed plants were sprayed with 0.8lbs/acre of Roundup®. Within two weeks following the spray, the control plants were killed while the transgenic plants overproducing EPSPS were only slightly affected. These plants however displayed slight chlorosis in the growing tips and the newly emerged leaves. This is not surprising since glyphosate accumulates primarily in the metabolic sink regions of both shoots and roots (45). The extent of overproduction of EPSP synthase may not be therefore sufficient to confer complete tolerance in these tissues of the sprayed plant.

If inhibition of EPSPS is the primary effect of glyphosate, it should be possible to engineer glyphosate tolerance by the use of mutant enzymes. The effectiveness of this approach is not only dependent upon the level of tolerance of the mutant enzyme but also the V/K ratio of the mutant enzyme compared to the wild type. Glyphosate tolerant mutants which display very high level tolerance also show significant increases in the Km for PEP (25, 46-48). Two of the most studied glyphosate tolerant mutant enzymes are those from *Klebsiella pneumoniae* and *Escherichia coli*. A third glyphosate tolerant mutant from *Salmonella typhimurium* has also been described; this enzyme does not appear to be as tolerant to glyphosate as the other two mutants (49). The *K. pneumoniae* enzyme exhibits an I-50 of ~50 mM for glyphosate. Its appKm for S3P is 193 mM and appKm for PEP is 140 mM. This is in contrast to the values of 45 mM and 9 mM for the Km's of S3P and PEP respectively for the wild type enzyme. The mutant enzyme was more acidic than the wild type with pI value of 4.1 compared to 4.6. Based on specific activity determinations, it appears that the mutant enzyme has about one third of the activity of wild type. Interestingly, by SDS gel electrophoresis, a slight increase in the molecular mass of the mutant enzyme was also detected. The amino acid change responsible for the glyphosate tolerance with this mutant is not known.

The mutant enzyme from *E. coli* exhibits similar changes in kinetic constants as the *K. pneumoniae* enzyme. Its Km for S3P is altered ~4 fold (80 mM vs. 19 mM) while that for PEP is altered ~20 fold (220 mM vs. 10 mM). The Ki for glyphosate of the mutant enzyme was 4 mM compared to 0.5 mM for the wild type. The Vmax of the mutant enzyme was ~60% of wild type EPSPS. Unlike the *K. pneumoniae* mutant, the *E. coli* mutant did not exhibit any changes in either the molecular weight or isoelectric point. Based on the *E. coli* mutant EPSPS, a number of plant glyphosate tolerant EPSPS enzymes have been constructed by site directed mutagenesis of plant EPSPS genes. One of the kinetically-characterized petunia mutant EPSPS showed no change in Km for S3P, an 80 fold increase in its Km for PEP and ~65% of the Vmax of the wild type. In the reverse reaction, this mutant displayed no change in Km for EPSP but a 10-fold increase in its Km for Pi. The mutant had a Ki (glyphosate) of 3 mM (25). Based on these results, it can be concluded that in this mutant the binding of the phosphate moiety of PEP is altered. This region of the enzyme must also be critical for interaction of glyphosate with EPSPS since the binding of glyphosate is affected to a greater extent than that of PEP. It is not clear as to which moiety of glyphosate is involved in interaction with this region of the active site although it is tempting to speculate that it is the phosphonate moiety. In addition, it is not understood if the differences in binding of PEP versus glyphosate are due to selective steric problems associated with binding of glyphosate or loss of a few recognition sites due to conformational changes. Since the kinetic constants for the mutant enzymes with respect to PEP are affected, it is to be expected that these enzymes will have to be overproduced depending on the intracellular concentration of PEP and the Km(PEP) for the mutant enzyme.

The *E. coli* mutant EPSPS gene was engineered into a plant expression vector (pMON8078) and used for transformation of tobacco leaf discs. In this construct, the mutant EPSPS gene was driven by the CaMV 35S promoter and the 3'-polyadenylation signal was derived from the nopaline synthase gene.

Calli producing the mutant *E. coli* EPSPS did not show any significant glyphosate tolerance compared to untransformed controls (G. Kishore et al. Unpublished data). Analysis of the calli however revealed the presence of the mutant enzyme (Table I) indicating that the lack of tolerance was not due to inadequate expression of the bacterial mutant gene. Tobacco plants expressing the bacterial mutant EPSP synthase were regenerated from transformed calli and sprayed with 0.4 lbs/ac of Roundup®. While these plants displayed tolerance to the herbicide, the level of tolerance was not significant. The herbicide caused significant apical damage, growth of lateral shoots and inhibition of growth rate. These experiments established that the cytosolic production of the *E. coli* mutant EPSPS did not confer adequate glyphosate tolerance to tobacco calli.

Since aromatic amino acid biosynthesis occurs in the chloroplast of plants, and the petunia EPSPS was demonstrated to contain the information for its translocation to the chloroplast, it was of interest to determine if delivery of the bacterial mutant enzyme to chloroplast of transgenic plants affected the level of tolerance to Roundup®. A hybrid gene was synthesized containing

Table I
EPSPS specific activites of calli transformed with EPSPS genes with and without chloroplast transit sequences

		ESPS Specific activity, nmol/min•mg	
Construct		0 mM glyphosate in assay mixture	0.5 mM glyphosate in assay mixture
pMON542	$glp^{r}tp^{+}$	60	33
pMON8078	$glp^{r}tp^{-}$	71	68
pMON546	$glp^{s}tp^{+}$	198	0
pMON8083	$glp^{s}tp^{-}$	44	0
pMON505	Control	8	0

the transit peptide of petunia EPSP synthase plus the first 27 amino acids of the mature petunia enzyme fused to the remainder of *E. coli* mutant EPSP synthase. This gene was expressed *in vitro* under the control of T7 promoter. The *in vitro* synthesized polypeptide had EPSP synthase activity which was glyphosate tolerant and was rapidly imported by isolated chloroplasts (50). The imported chimeric protein was processed by proteolysis to a mature, catalytically active enzyme.

Tobacco leaf discs transformed with the hybrid petunia/E. coli mutant EPSP synthase gene produced calli which were tolerant to glyphosate. In this construct (pMON542) the hybrid gene was driven by the CaMV 35S promoter. The level of expression of the mutant EPSPS was comparable to that obtained with calli transformed with the pMON8078 vector (Table I). However unlike pMON8078 calli which did not survive 0.5 mM glyphosate, the pMON542 calli could survive glyphosate concentrations up to 10 mM. Plants expressing the hybrid gene were generated from transformed calli and sprayed with 0.4 lbs/ac of Roundup®. These plants did not show the apical damage observed with plants expressing the bacterial gene by itself and were not significantly affected by the herbicide. Subcellular fractionation studies revealed that the hybrid protein was localized in the chloroplasts of transgenic plants while the bacterial mutant protein was cytosol-localized (50). Interestingly, the level of expression of mutant EPSP synthase was similar in both cytosol-targetted and chloroplast-targetted transgenic plants. Thus, the chloroplast localized mutant EPSPS provides significantly improved glyphosate tolerance over that of the cytosol-localized mutanty.

It is clear from the above discussion ensure that glyphosate tolerance may be conferred to plants both by overproduction of wild type EPSPS as well as mutant EPSP synthases. It has been suggested that glyphosate may have multiple sites of action in plant cells (51-56). If this is true, mutant EPSPS enzymes would not confer glyphosate tolerance to plants, which is evidently not the case. It appears, therefore, that reports concerning the effect of glyphosate on other aspects of plant metabolism are due to secondary effects of the herbicide arising as a consequence of the inhibition of aromatic amino acid biosynthesis.

More recently, the petunia EPSPS gene has been demonstrated to confer Roundup® tolerance in transgenic tomato plants. The transgenic tomato plants were tested under field conditions for their productivity and vigor. No significant differences were observed in crop productivity between transgenic and control tomato plants. This demonstrates that introduction and expression of genes into plants does not affect crop yield.

The commercial implications of engineering herbicide resistance to crop plants are significant. While extending crop tolerance to the herbicide is one aspect of this development, other significant factors to note are: i. reduced dependency on herbicides, since the farmer can use the herbicide only when needed, ii. use of environmentally safe herbicides, and iii. development of new, safe herbicides (57). The tools of biochemistry and molecular biology have progressed significantly within the last decade to permit a more rigorous structure-function study of the herbicide-target protein interaction. This is expected to have a significant impact on development of new herbicides.

Acknowledgments

We thank Ms. Jamie Wehrheim for typing this manuscript. We are grateful to Dr. E. Jaworski for his support and encouragement throughout the course of these investigations.

Literature Cited

1. Baird, D.D.; Upchurch, R.P.; Homesley, W.B.; Franz, J.E. Proc. North Cent. Weed Control Conf. 1971, 26, 64-68.
2. Franz, J.E. In The Herbicide Glyphosate; Grossman, E. and Atkinson, D., Eds., Butterworths, London, 1985, pp. 3-17.

3. Fraley, R.T.; Rogers, S.G.; Horsch, R.B. CRC Crit. Revs. in Plant Sciences 1986, 4, 1-46.
4. Jaworski, E.G. J. Agric. Food Chem. 1972, 20, 1195-1198.
5. Steinrucken, H.C.; Amrhein, N. Biochem. Biophys. Res. Commun. 1980, 94, 1207-1212.
6. Hollander, H.; Amrhein, N. Plant Physiol. 1980, 66, 823-829.
7. Amrhein, N.; Deus, B.; Gehrke, P.; Steinrucken, H.C. Plant Physiol. 1980, 66, 830-834.
8. Lewendon, A.; Coggins, J.R. Biochem. J. 1983, 213, 187-191.
9. Duncan, K.; Lewendon, A.; Coggins, J.R. FEBS Lett. 1984, 165, 121-127.
10. Steinrucken, H.C.; Amrhein, N. Eur. J. Biochem. 1984, 143, 341-349.
11. Mousdale, D.M.; Coggins, J.R. Planta 1984, 160, 78-83.
12. Steinrucken, H.C.; Schulz, A.; Amrhein, N.; Porter, C.A.; Fraley, R.T. Arch. Biochem. Biophys. 1986, 244, 169-173.
13. Smart, C. C.; Johanning, D.; Muller, G.; Amrhein, N. J. Biol. Chem. 1985, 260, 16338-16346.
14. Ream, J.; Steinrucken, H.C.; Sikorski, J.A.; Porter, C.A. Plant Physiol. Suppl. 1986, 80, 47.
15. Padgette, S.R.; Huynh, Q.K.; Borgmeyer, J.E.; Shah, D.M.; Brand, L.A.; Re, D.B.; Bishop, B.F.; Rogers, S.G.; Fraley, R.T.; Kishore, G.M. Arch. Biochem. Biophys. 1987, 258, 564-573.
16. Duncan, K.; Edwards, M.R.; Coggins, J.R. Biochem. J. 1987, 246, 375-386.
17. Schulz, A.; Kruper, A.; Amrhein, N. FEMS Microbiol. Lett. 1985, 28, 297-301.
18. Boocock, M.R.; Coggins, J.R. FEBS Lett. 1983, 154, 127-133.
19. Steinrucken, H.C.; Amrhein, N. Eur. J. Biochem. 1984, 143, 351-357.
20. Anton, D.L.; Hedstrom, L.; Fish, S.M.; Abeles, R.H. Biochemistry 1983, 22, 5903-5908.
21. Bondinell, W.E.; Vnek, J.; Knowles, P.F.; Sprecher, M.; Sprinson, D.B. J. Biol. Chem. 1971, 246, 6191-6196.
22. Grimshaw, C.E.; Sogo, S.G.; Copley, S.D.; Knowles, J.R. J. Am. Chem Soc. 1984, 106, 2699-2700.
23. Grimshaw, C.E.; Sogo, S.G.; Knowles, J.R. J. Biol. Chem. 1982, 257, 596-598.
24. Floss, H.G. Rec. Adv. Phytochem. 1986, 20, 13-56.
25. Kishore, G.M.; Shah, D.M. Ann. Rev. Biochem. 1988, 57, 627-663.
26. Shah, D.M.; Horsch, R.B.; Klee, H.J.; Kishore, G.M.; Winter, J.A.; Tumer, N.E.; Hironaka, C.M.; Sanders, P.R.; Gasser, C.S.; Aykent, S.; Siegel, N.R.; Rogers, S.G.; Fraley, R.T. Science 1986, 233, 478-481.
27. Comai, L.; Facciotti, D.; Hiatt, W.R.; Thompson, G.; Rose, R.E.; Stalker, D.M. Nature 1985, 317, 741-744.
28. Kishore, G.M.; Padgette, S.R.; della-Cioppa, G.; Re, D.; Brundage, L.; Shah, D.; Gasser, C.S.; Sanders, P.R.; Klee, H.;

Horsch, R.; Hoffman, N.; Fraley, R.T. Fed. Proc. 1987, 46, 2055.
29. Nafziger, E.D.; Widholm, J.M.; Steinrucken, H.C.; and Killmer, J.L. Plant Physiol. 1984, 76, 571-574.
30. Hauptmann, R.M.; della-Cioppa, G.; Smith, A.G.; Kishore, G.M.; Widholm, J.M. Mol. Gen. Genet. In Press.
31. Feierabend, J.; Braseel, D. Z. Pflanzenphysiol. 1982, 82, 334-346.
32. Weeden, N.F.; Gottlieb, L.D. J. Hered. 1980, 71, 392-396.
33. Bickel, H.; Palme, L.; Schultz, G. Phytochem. 1978, 17, 119-124.
34. d'Amato, T.A.; Ganson, R.J.; Gaines, C.G.; Jensen, R.A. Planta 1984, 162, 104-108.
35. Ganson, R.J.; d'Amato, T.A.; Jensen, R.A. Plant Physiol. 1986, 82, 203-210.
36. della-Cioppa, G.; Hauptman, R.M.; Fraley, R.T.; Kishore, G.M. Curr. Top. Plant Biochem. and Physiol. 1986, 5, 194.
37. Chua, N-H.; Schmidt, G.W. Proc. Natl. Acad. Sci USA 1978, 75, 6110-6114.
38. Ellis, J.R. Ann. Rev. Plant Physiol. 1981, 32, 111-137.
39. Van Den Broeck, G.; Timko, M.P.; Kausch, A.P.; Cashmore, A. R.; Van Montagu, M.; Herrera-Estrella, L. Nature 1985, 313, 358-363.
40. della-Cioppa, G.; Bauer, S.C.; Klein, B.K.; Shah, D.M.; Fraley, R.T.; Kishore, G.M. Proc. Natl. Acad. Sci. USA 1986, 83, 6873-6877.
41. della-Cioppa, G.; Kishore, G.M. EMBO J. 1988, In press.
42. Klee, H.J.; Muskopf, Y.M.; Gasser, C.S. Mol. Gen. Genet. 1987, 210, 437-442.
43. Gasser, C.S.; Winter, J.A.; Hironaka, C.M.; Shah, D.M. J. Biol. Chem. 1988, In press.
44. Schmidt, G.W.; Mishkind, M.L. Ann. Rev. Biochem. 1986, 55, 879-912.
45. Mollenhauer, C.; Smart, C.C.; Amrhein, N. Pest. Biochem. Physiol. 1987, 29, 55-65.
46. Schulz, A.; Sost, D.; Amrhein, N. Arch. Microbiol. 1984, 137, 121-123.
47. Sost, D.; Schulz, A.; Amrhein, N. FEBS Lett. 1984, 173, 238-241.
48. Kishore, G.M.; Brundage, L.; Kolk, K.; Padgette, S.R.; Rochester, D.; Huynh, Q.K.; della-Cioppa, G. Fed. Proc. 1986, 45, 1506.
49. Stalker, D.M.; Hiatt, W.R.; Comai, L. J. Biol. Chem. 1985, 260, 4724-4728.
50. della-Cioppa, G.; Bauer, S.C.; Taylor, M.L.; Rochester, D.E.; Klein, B.K.; Shah, D.M.; Fraley, R.T.; Kishore, G. M. Bio/Technology 1987, 5, 579-584.
51. Roisch, U.; Lingens, F. Hoppe-Seylers Z. Physiol. Chem. 1980, 361, 1049-1058.
52. Rubin, J.L.; Gaines, C.G.; Jensen, R.A. Plant Physiol. 1982, 72, 833-839.
53. Bode, R.; Ramos, C.M.; Birnbaum, D. FEMS Microbiol. Lett. 1984, 23, 7-10.

54. Ganson, R.J.; Jensen, R.A. Arch. Biochem. Biophys. 1988, 260, 85-93.
55. Hoagland, R.E.; Duke, S.O.; Elmore, C.D. Physiol. Plant. 1979, 46, 357.
56. Lee, T.T. Weed Res. 1980, 20, 365.
57. Fraley, R.T.; Kishore, G.M.; Gasser, C.S.; Padgette, S.R; Horsch, R.B.; Rogers, S.; della-Cioppa, G.; Shah, D.M. Brit. Crop Prot. Conf.-Weeds 1987, 2, 463-471.

RECEIVED June 3, 1988

CONTROL OF PLANT DISEASES

Chapter 4

Trichothecenes and Their Role in the Expression of Plant Disease

Horace G. Cutler

Richard B. Russell Center, Agricultural Research Service, U.S. Department of Agriculture, Athens, GA 30613

In 25 years, the concept of mechanisms whereby fungal plant pathogens induce changes in host plants has changed. Originally, it was accepted that fungal spores landed on a suitable plant surface, germinated, produced an infection peg which physically penetrated the tissue, perhaps producing the enzymes cutinase, cellulase, and pectinase to disrupt cells. Subsequent ramification of inter and intra cellular mycelium induced morphological change and death. Recent evidence supports the hypothesis that fungi may be directly, or indirectly pathogenic to plants by the production of phytotoxic metabolites. Significant among these are the simple and macrocyclic trichothecenes which induce a number of responses in plants that range from blotching, chlorosis, distortion, necrosis, and apparent chimera production to "viral" effects. Trichothecenes bind to 60s ribosomes and may be either initiator or transpepidation inhibitors in eukaryotes. The high specific activity of the trichothecenes and their activity in plants may account for their role in the etiology of plant diseases and other disorders. The odd case of *Baccharis megapotamica*, which is resistant to certain trichothecenes, versus *B. halimifolia* which is not resistant, will be discussed at the gross and macromolecular level.

Until recently, developments in plant pathology and chemistry have been out of synchrony. Field observations of plant pathogens and their etiology have been meticulously recorded as a distinct, thorough discipline. In contrast, the isolation and identification of natural products from microorganisms that possess biological activity has been a separate domain for certain chemists with respect to plant pathogens and non-pathogens. Within the past decade especially, the concept that pathogens and other soil organisms (saprophytes) produce specific toxins that are discreet molecules capable of inducing pathogenic responses in target plants has become evident. An amalgamation of the physico-chemical aspects of pathogenesis better explains the total events that take place when organisms attack and establish a foothold in host plants. In this discussion I intend to start with the physical aspects of pathogenesis and finish at the molecular and macromolecular level.

Direct Physical Invasion. A general scheme for infection by a pathogen may consist of a spore landing on a plant surface and becoming lodged because of anatomical features. Thereupon, the spore germinates, sends out a germ tube, or in certain cases an appressorium, followed by a penetration peg that enters the cuticle on the tissue surface and gains entry to the underlying cells. Haustoria may branch from the main hyphae into the cell. The mycelium then ramifies throughout the tissues and, eventually, physical disruption of the cell contents brings about cessation of cellular functions, inducing death (1). In conjunction with these physical activities it has been shown that cellulases, lignases, and pectinases aid the invading organism in its establishment. Further developments in the physico-enzyme theory have been discussed in some detail by Kolattukudy et al. (2) who grew *Fusarium solani* f. sp. *pisi*, a pathogen of pea (*Pisum sativum* L.), using apple cutin as the only carbon source to culture the fungus, in elegant experiments. The fungus thrived on the substrate, indicating the extra cellular presence of cutinase, and the enzyme was successfully isolated. The procedure has since been used for a number of pathogens and the chemical characteristics of the cutinases from the assorted organisms appears similar (2). The properties of the enzymes, the substrates upon which they exert their influence, and the roles of cutinase(s) in pathogenesis have been detailed (2). In the latter case, some incisive experiments were executed with pea stems upon which were placed *F. solani pisi* spores and their germination and penetration, or lack thereof, observed by scanning electron microscopy. At the exact locus at which penetration occurred, the target area was treated with ferritin-conjugated antibodies to the enzyme and it was unequivocally demonstrated that cutinase was produced by the spore precisely at the spore target area. Also, *Fusarium* spores on pea stems in the presence of specific antibodies or organic phosphates were incapable of inducing stem lesions. Further evidence shows that when spores land on a cutin surface and germinate, a small amount of cutinase is exuded that breaks the cutin polymers into monomers. This, in turn, elicits spores to produce more cutinase with the end result that penetration is enhanced.

A parallel situation may exist with pectinase activity on

carbohydrates (2). That is, having penetrated the cutin barrier the fungus comes into contact with carbohydrate polymers which are then digested into oligomers and monomers. Pectinase production may be stimulated and the type of cycle noted with cutinase may take place. *In vitro* studies show that if *Fusarium solani* spores are placed in a pectin-containing medium, pectinase production is stimulated and peaks in an 8 hour period (2). And again, support for the production of pectinase and this see-saw mechanism is substantiated by work with *F. solani* spores, pea stems and the use of antibodies. While the evidence is not as complete for pectinase cycling as it is for cutinase, the case is leading to the conclusion that both events have a similar pattern. In conjunction with cutinase and pectinase, the production of cellulase by certain pathogenic fungi is well established and a listing of those organisms that are particularly adept at synthesizing cellulase may be found in the American Type Culture Collection Catalog (3).

An highly imaginative study has recently examined leaf surface structures and the relationship to the development of pathogen establishment (4). In this case, the organism examined was the bean rust fungus, *Uromyces appendiculatus*, whose spores normally germinate on the leaf surface sending out hyphae which are directionally oriented to stomates gaining entrance, by means of specialized structures, and thereby establishing itself. The physical messenger for the fungus is a simple ridge on the leaf surface that has an ideal height of 0.5µmeters and in the bean plant host, *Phaseolus vulgaris* L., the stomatal lip which is adjacent to this guard cell has this approximate measurement. The researchers, using electron beam lithography, were able to manufacture silicon wafers that had ridges ranging from 0.25 to 1.0 µmeters in height and upon these, spores of *U. appendiculatus* were germinated and their orientation observed. Results showed that orientation of the germ tubes was effected in a manner similar to that seen in the natural state with ridges that were 0.5 µmeters high and spaced no further than 0.6-6.7 microns apart and, furthermore, the angle between the ridge and the base substrate was shown to be necessarily acute. Thus, there is a perception by the invading organism of its physical surroundings but whether this is genetically controlled within the invading organism remains to be elucidated. Be that as it may, there is no doubt that physical events play a distinct role in the establishment of a pathogen and that key enzymes are an integral part of the disease cycle. But those events do not explain diverse effects that are noted as a result of the successful invasion of an organism, and these effects which include necrosis, chlorosis, malformations, stunting and other abnormalities of growth and development are better explained by the production of toxins with high specific activity, by the pathogen, and their introduction into the host. While there are many diverse secondary fungal metabolites that could be discussed, the trichothecenes are among the most interesting and varied in activity.

Indirect Pathogenesis (Allelochemical). The possibility that saprophytes produce toxic metabolites into the rhizosphere first came to my attention over fifteen years ago. I had been invited to look at a field of tobacco in northeast Georgia by a group of Extension Service scientists. Other problems in the field were quickly solved but one remaining problem vexed me and, at the time, defied solution. Certain irregular areas of the field had tobacco plants that were consistently stunted while appearing, in all other aspects, to be normal. There were neither abnormalities, nor chlorosis, nor necrosis, and the areas might have been described as those that had not received sufficient fertilizer especially if it had been hand broadcast in a circular but erratic motion. However, the fertilizer had been applied mechanically. The only clue was that in all the stunted areas there were pieces of chopped corn (*Zea mays* L.) that had been left from the previous crop and incorporated into the soil. Significantly, those stems sections supported a monoculture of *Trichoderma*, and it was thriving in areas in which stunting was taking place. However, my classical training in pathology eschewed the possibility of a connection between the organism on the cornstalks and the reduction in growth, (which would subsequently lead to decreased tobacco yields), until some years later.

Our research on biologically active natural products from fungi had led to the observations that many plant pathogens and non-pathogens produce metabolites that are highly toxic to plant growth and some of these were fungi that grew on assorted substrates (saprophytes). In common with all fungal natural products chemists, it was not long before the trichothecenes came to our attention and were isolated from some of our fungal accessions. Additionally, several colleagues had isolated *Trichoderma* species, among these being *T. viride* the producer of trichodermin (5,6) and studies showed that trichodermin was a potent inhibitor of plant growth. Etiolated wheat coleoptiles were inhibited 91, 80 and 15% ($P < 0.01$) at 10^{-3}, 10^{-4}, and 10^{-5}M, respectively, (Table 1) relative to controls and intact greenhouse grown beans, corn, and tobacco plants were greatly affected by sprays containing trichodermin at various concentrations (7). Ten-day-old bean plants responded quickly to treatment with the metabolite and within 24 hours at 10^{-2}M exhibited wilting of the primary leaves while there were no visible effects at 10^{-3} and 10^{-4}M. But within 3 days, the 10^{-2}M treated leaves were desiccated: those of the 10^{-3}M treatments exhibited diffuse necrotic areas that remained visible for one week and strongly resembled a viral symptom. Tobacco responded somewhat differently to bean and when six-week-old plants were treated there were no visible effects with 10^{-2} to 10^{-4}M solutions until 5 days following treatment: then the responses were dramatic. The 10^{-2}M treated plants suddenly became flaccid and symptoms of water stress were evident compared to control plants. A week after treatment those plants were dying and the 10^{-3}M treatments had patches of water stressed leaves, some slightly necrotic areas and young leaf growth was inhibited.

The 10^{-4}M treated plants appeared normal, though at two weeks following treatment the 10^{-2}M treated plants were dead, and the 10^{-3}and10^{-4}M treatments were stunted. By 28 days the 10^{-3}M treated plants were inhibited 83% and the 10^{-4}M were inhibited 72% relative to controls, but in all other respects the plants looked normal. That is, the plants in these experiments looked remarkably like those seen in the tobacco field in northeast Georgia.

Further experiments included feeding six-week-old tobacco plants with trichodermin through the roots in aqueous solution (7). Responses were even more dramatic than those obtained with sprays and at 48 hours following treatment the 10^{-2}M treated plants had necrotic spots on the leaves and margins were wilted. At 10^{-3} and 10^{-4}M these expressions were more pronounced. All treatments died by 86 hours and by 15 days the controls were still healthy. Thus, it was demonstrated that ground water containing trichodermin was far more lethal to plants than spray solution and dilute solutions of trichodermin in the rhizosphere would be likely to inhibit plant growth at dilutions of less than 10^{-4}M even though the end point in these experiments was not calculated. Of course, the vital links in the events are missing. The pieces of contaminated corn stalk were never brought to the laboratory from the tobacco field, the organism cultured, identified, extracted for phytotoxic metabolites and those metabolites reapplied to tobacco to induce the responses observed. It is doubtful in the present climate that the farm is operational. What is important is that as a result of the initial on site observation trichodermin, a metabolite of *T. viride*, which is not a plant pathogen but which is a common soil saprophyte, was shown to produce pathogenic symptoms in plants when applied exogenously. A further piece of evidence that led to the study of other trichothecenes was the relative inactivty of trichodermin on corn when it was placed in the leaf sheaths in intimate contact. Chlorosis was induced in 24 hours at 10^{-2} and 10^{-3}M and, at the higher concentration, there was some necrosis. Within a week all plants appeared normal and there no stunting so it appeared that responses to the metabolite were genus dependent. That is, as we shall see later, some plant species were apparently able to relatively detoxify the molecule.

Direct Chemical Action of Trichothecenes. Over a quarter of a century ago, Brian et al. (8) described the action of diacetoxyscirpenol on plants when 2.73 X 10^{-5}M solutions of the metabolite inhibited stem growth and scorched leaves in two pea varities, lettuce, winter tares and other economic crops when applied exogenously. But carrot, beet root, mustard and wheat were not affected at the same application rate signalling genus and species specificity for the metabolite. However, at rates that ranged from 1.37 x 10^{-6}

to 2.73 x 10^{-8}M elongation of cress roots was stimulated and, in that case, some promotorial mechanism was turned on. In contrast, scirpentriol, which differs in the R_2, R_3 positions from diacetoxyscirpenol by the placement of hydroxyl groups instead of acetyl functions but which is in all other aspects identical (Fig. 1), was toxic to cress roots and did not promote cress root elongation at low concentration. In earlier studies, Bawden and Freeman (9) demonstrated the selective action of trichothecin in experiments with bean (Phaseolus vulgaris L.) and tobacco (Nicotiana glutinosa L.) when 1.33x10^{-4}M of the metabolite damaged bean leaves but four times that concentration induced no apparent effects in tobacco. Trichothecin has different functional groups to both scripentriol or diacetoxyscirpenol and more closely resembles trichodermin with the major functional group difference being a carbonyl in position R_4 in trichothecin compared with a proton in trichodermin. T-2 toxin, which has received, undeservedly, considerable attention recently because of its implication as a biological warfare agent, has also been reported to stunt and reduce fresh weight of pea seedlings when roots were placed in solutions of the metabolite that ranged from 2.15x10^{-5} to 2.15 x 10^{-6}M for 20 minutes, then washed and planted (10). Again, T-2 toxin differs structurally from diacetoxyscirpenol at the R4 position where, in T-2 toxin, α-isoval, βH replaces the proton (Fig. 1). In all these cases the responses observed in plants are, upon treatment with the trichothecenes, very similar to those seen in plants. Among those genera that have been claimed to produce trichothecenes are Acremonium (11), Cylindrocarpon (11), Dendrostilbella (12), Fusarium (11), Myrothecium (11), Trichoderma (11), Trichothecium (11), and of these Cylindrocarpon, Fusarium, Myrothecium, and Trichothecium are plant pathogens.

The strucure-activity response is even more subtly expressed in the etiolated wheat coleoptile (Triticum aestivum L. Wakeland) that has been described in detail elsewhere (13). Subtle modifications in structure cause marked changes in the inhibitory responses induced (Table I). With the exception of 15-acetoxy-T- 2-tetraol, neosolaniol and isoneosolaniol which are inactive at 10^{-5}M (Fig. 2), all of the trichothecenes so far tested are active at 10^{-5}M. A further one-third are active at 10^{-6}M and the most potent, which are active at 10^{-7}M, are verrucarins A and J, and trichoverrin B. Neosolaniol, which inhibits 100% at 10^{-3} and 61% at 10^{-4}M has an acetate at C15 while isonesolaniol, which has an hydroxyl group at C15 inhibits coleoptiles 80 and 51% at 10^{-3} and 10^{-4}, respectively.

A detailed study of the effects of exogenously applied macrocylic trichothecenes on intact bean, corn, and tobacco plants recalls that they may induce a variety of responses that are tissue dependent and range from no effect to complete phytotoxicity (14). Initial experiments have conclusively shown that in the etiolated wheat coleoptile bioassay verrucarin A, verrucarin J, and trichoverrin B are exception-

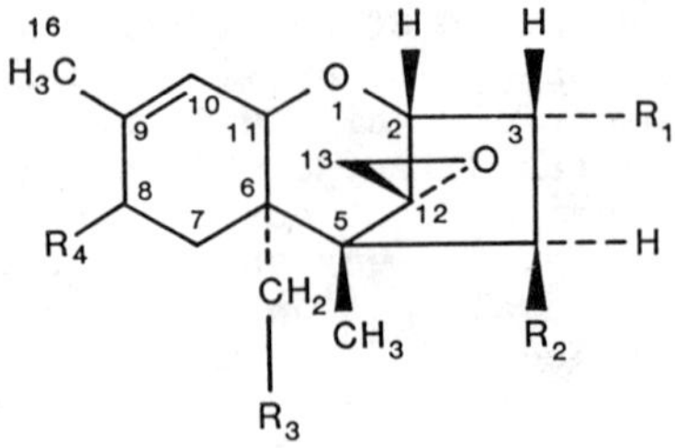

	R_1	R_2	R_3	R_4
Trichodermin	H	ßOAc	H	H_2
Diacetoxyscirpenol	αOH	ßOAc	OAc	H_2
Scirpentriol	αOH	ßOH	OH	H_2
Trichothecin	H	ßObutyl	H	=O
T-2 Toxin	αOH	ßOAc	OAc	α-iso-val; ßH

Figure 1. Some simple trichothecenes.

TABLE I. INHIBITION OF WHEAT COLEOPTILES (Triticum aestivm L.,cv. Wakeland), BY TRICHOTHECENES

COMPOUND	INHIBITION (%) BASED ON MOLAR CONCENTRATION				
	10^{-3}	10^{-4}	10^{-5}	10^{-6}	10^{-7}
TRICHOTHECENES					
TRICHODERMIN	91	80	15	0	0
NEOSOLANIOL MONOACETATE	87	69	45	16	-
3'-HYDROXY-HT-2 TOXIN	100	81	19	0	-
15-ACETOXY-T-2 TETRAOL	60	0	0	0	-
3'-HYDROXY-T-2 TRIOL	96	54	37	0	-
HT-2 TOXIN	100	62	14	0	-
T-2 TOXIN	100	100	50	39	-
RORIDIN A	100	81	40	0	-
ISORORIDIN E	100	100	60	0	-
BACCHARINOL B4	91	82	60	19	0
VERRUCARIN A	100	97	80	41	37
VERRUCARIN J	100	100	82	61	43
TRICHOVERRIN B	100	81	62	40	16
NEOSOLANIOL	100	61	0	0	-
ISONEOSOLANIOL	80	51	0	0	-

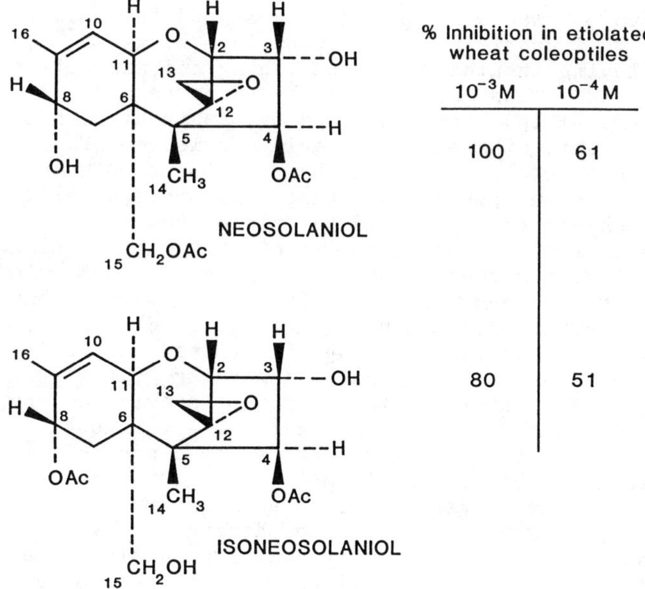

Figure 2. Structure activity relationship of neosolaniol and isoneosolaniol.

ally potent inhibitors of cell growth and, primarily, elongation. A plant growth inhibitor that controls many plant cell functions, abscisic acid (15), has been used as an internal standard and is 100 times less active than verrucarin A (16). The most prolonged response in higher plants was noted in corn (*Zea mays* L.). Single applications of trichoverrin B, an open chain congener, into the leaf sheath at 10^{-2}M induced an 85% inhibition 5 weeks after application and, as with trichodermin in tobacco, a distinct lag phase (72 hours) was observed. A synopsis of the study (14) indicates that of the macrocyclic trichothecenes studied, roridin A was the most toxic to intact greenhouse-grown plants and again effects were not seen until 3-5 days after treatments.

One noted effect that may be of practical value for genetic engineers, was the morphological changes induced in leaves and leaf tissue. These were strongly expressed in tobacco plants treated with isororidin E and baccharinol B_4 at 10^{-2}M, 48 hours following treatment with blotching and distortion of the second leaves. These appeared to be chimeras and the tissues curiously resembled those obtained by treatment with the mutagenic agent methanesulfonic acid ethylester (EMS), a compound that acts by adding methyl functions to guanine in DNA bringing about functional changes in treated cells and producing chimeras. It should be pointed out that treatment with this mutagen does not consist of spraying the plants but rather by treating seeds with a 0.3% aqueous solution of EMS and allowing the seeds to imbibe for 16-24 hours. This is followed by a rinse (3 hours) with frequent changes of water and culminates with the planting of the seed in a suitable medium (17). The mortality rate in morning glory (*Ipomea nil*) is high (c.70%), but the mutation rate, as evidenced by chimeras, is large in the M1 generation. It is obvious that equivalent sets of experiments need to be conducted with the trichothecenes substituting them for EMS and allowing, as one does for EMS, the M1 generation to self and produce the M2 progeny, or to take samples of the chimeras, obtain protoplasts for fusion and/or propagation in tissue culture. Synthetic modifications of various trichothecene molecules for this purpose are manifold, the simplest being specific reduction of carbonyl groups, alteration of hydroxyl groups and the addition of other molecules that intrinsically have mutagenic properties. But it should be reiterated that different genera, species, and plant parts behave in different, though reproducible ways. These compounds are extremely active and it may be presumed that if the materials could be directly introduced to the active sites then the specific activity of the molecules would increase by several orders of magnitude. However, resistance or susceptibility does occur in plant species and this leads to the next topic.

Plant Pathogenesis, Resistance. Partial pathogenesis, or partial resistance of plants may be perceived as an odd turn of phrase by most pathologists but these are suitable working terms for observations made with both plant pathogens and those metabolites that produce "pathogenic" responses in plants. A series of studies have been carried out by Miller et al. (18,19,20,21) with Fusarium graminearum Schwabe, a pathogen that causes head blight of wheat. Physically, the organism produces shrivelled grains, thereby reducing yields, and additionally quantities of the potent mycotoxin deoxynivalenol (3α,7α,15-trihydroxy-12,13-epoxytrichothec-9-en-8-one) also known trivially as vomitoxin. Infection may occur at anytime during the life cycle of the plant but a very vulnerable period is at the time of anthesis and, it appears, pollen may be a stimulating factor for Fusarium (22). Deoxynivalenol is a potent inhibitor of plant growth, as is 3-acteyldeoxynivalenol, and this has been elegantly demonstrated in experiments using etiolated wheat coleoptiles where concentration of 10^{-6}M were phytotoxic. Other metabolites of F. graminearum, specifically butenolide, culmorin, dihydroxycalonectrin and sambucinol, which are major metabolites (23, 24) were also phytotoxic to wheat coleoptiles (25) but were not of the same order of magnitude of activity. Thus it was shown that the metabolites of F. graminearum were phytotoxic without the presence of the fungal mycelium or spores. Futhermore, when wheat cultivars that were known to be resistant in field trials to F. graminearum were used as bioassay materials in the wheat coleoptile bioassay the effects of deoxynivalenol and 3-acetyldeoxynivalenol were considerably less pronounced. The resistant cultivar Sumai #3 was able to withstand ten times the amount of deoxynivalenol and one hundred times the amount of 3-acetyldeoxynivalenol; the cultivar Frontana was able to tolerate one thousand times the concentration of 3-acetyldeoxynivalenol compared to susceptible cultivars. Hence, in vitro experiments with assorted resistant and susceptible cultivars of wheat demonstrate that the ultimate responses obtained in the field using these cultivars, are identical. That is, phytotoxicity. At least, this makes the trichothecenes highly suspect, as agents released by invading pathogens, in the etiology of pathogensis.

Resistance, or apparent resistance, may be classified under three major headings. First resistance based on the morphology of the host which is purely physical. For example, spike morphology (19), thickness of cuticle and relative abundance of stomata. Second, resistance to hyphal invasion which may depend not only on the physical characteristics of plants but also on the production of metabolites by the host in response to the presence of hyphae in the cells. These compounds, the phytoalexins, are fungistatic and are unique in their ability to control the growth and development of invading species. Third, is the ability of select species within a genus to degrade or tolerate trichothecenes.

In addition to possible phytoalexin production it appears that certain flavonoids and furanocoumarins inhibit the production of trichothecenes. Strong preliminary evidence indicates that when Fusarium sporotrichioides is incubated in the presence of flavones or furanocoumarins the production of

T-2 toxin is greatly inhibited. Again, there is molecular specificity because only one of fifteen flavonoids tested and eleven of twelve furanocoumarins inhibited toxin production (26). But when inhibition was induced, trichodiene (a trichothecene precursor) increased proportionately and the sequential biosynthetic steps in which oxygenation of the molecule normally occurs did not take place.

Gross degradation studies in wheat show that after initial infection with F. graminearum, levels of deoxynivalenol increased until 6 weeks (9.5ppm) and then dropped by 73% (27). This degradation may be attributed to plant enzymes in the host (27) which structurally modify the trichothecene molecule. Using suspension cultures of the resistant cultivar Frontana, Miller showed that ^{14}C deoxynivalenol was converted into three products (21). One was $^{14}CO_2$ and another was probably deoxynivalenol glycoside (Fig. 3). In the converse experiment, using suspension cultures of the susceptible cultivar Casavant, less than 5% of the exogenously applied deoxynivalenol was converted to one compound. Resistance, therefore, appears to be highly correlated to the ability of the host to degrade the toxin. Further degradation studies of the trichothecenes show that several possibilities may occur. The easiest way to detoxify trichothecenes is to open the 12,13 epoxide ring (11a). Another, but more complicated way chemically, is to make, in vivo, addition products. This is accomplished in Fusarium sulphureum which converts, presumably enzymatically, monacetoxyscirpenol to the α-glucopyanoside derivative (28). The glucose moiety is attached at the C_4 position and the relative size and conformation of the sugar portion is such that it may fold over the epoxide structure. Compared to other tricothecenes the glycoside was not toxic to brine shrimp at levels up to 200 times greater than normal and it did not induce dermatotoxic effects. It was toxic to rats at less than 90mg/kg and this toxicity may be attributed to hydrolysis of the glycoside to yield monoacetoxyscirpenol in the stomach (28). In F. graminearum, which produces deoxynivalenol, the glucose molecule is most probably attached to the C3 postion and again the glycoside may be relatively inactive in plant systems (29). Whether any other metabolites are formed from the trichothecenes can only be speculated upon at this time. For example, in vitro studies with anguidine incubated in the presence of uridine 5'-diphosphoglucuronic acid, β-naphthoflavone-induced hepatic microsomes obtained from Long Evans rats, magnesium chloride and dipotassium phosphate gave rise to 15-acetyl-3α- 1'β-D-glucopyranosiduronly)-scirpen-3, 4β,15-triol (30). To date, that metabolite has not been found in either microbial or plant systems.

The strongest evidence to support the association between the production of a trichothecene by a plant pathogen and the resulting death of the host is work carried out with Myrothecium roridum (31, 32, 33). This pathogen attacks muskmelon (Cucumis melo L.) and while it attacks mainly the fruit, with concomitant losses of up to 30%, all parts of the plant

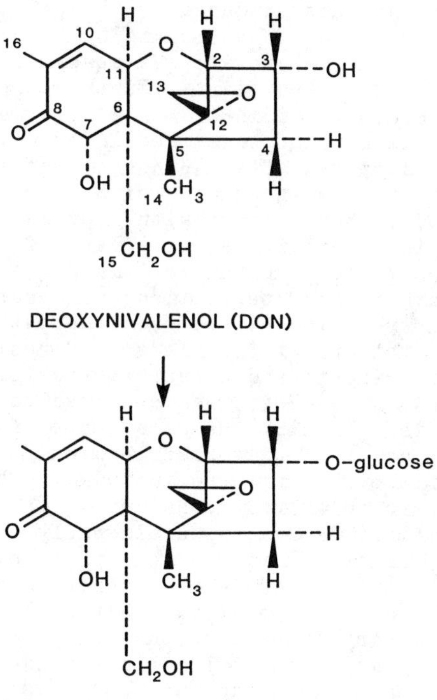

DEOXYNIVALENOL (DON)

Figure 3. Deoxynivalenol and the glucose derivative.

may become infected and necrosis ensues. In these studies, M. roridum was collected from an infected muskmelon field, cultured in liquid medium and, after fermentation, was suitably extracted. Only small amounts of the macrocyclic trichothecenes were produced, though it must be remembered that quantity bears no relationship to the specific activity of a molecule, and in addition to the trichoverroids, trichodermadienediols A (Fig. 4) and B (Fig. 5), roridin L-2 (Fig 6) and 16-hydroxyroridin L-2 (Fig. 7) were isolated. The relative amounts were low, in the order of 5-10mg/liter, and it must be assumed that under normal field conditions the amounts of trichothecenes produced by the pathogen would be of a low order of magnitude. When muskmelon seedling roots were fed with roridin A the metabolite was absorbed and translocated throughout the plant. Furthermore, roridin A was converted to 8β-hydroxyroridin A. In a later development, leaves of healthy muskmelon were inoculated with roridin A and necrotic lesions appeared after 48 hours, increasing with time. Also, leaves were inoculated with M. roridum and necrosis appeared 72 hours later with sporulation occurring 120 hours after inoculation. A third part of the experiment included both the M. roridum and roridin A, a clever idea, and in that case necrosis and sporulation was observed at 48 hours implying a strong relationship between the pathogen and its metabolites. At this point the element of resistance was placed in the experiments and a resistant cultivar (that is, less susceptible), Hales Best, was compared to a susceptible cultivar, Iriquois. Exactly the same types of inoculations were carried out as previously described and both roridin A and E and the trichoverroids were included. While the trichoverroids were relatively less active than the roridins both sets of metabolites caused considerably more necrosis in the susceptible cultivar Iriquois than the resistant Hales Best. Again, the trichothecene plus spore inoculation was most potent. The authors point out that while it is too early to invoke absolute involvement of the trichothecenes in pathogenicity they may play at least a secondary role by acting as a virulence factor during the invasion sequence (32).

Species Dependent Pathogenesis. So far, we have seen some cases of species dependent pathogenesis or relative susceptibility, and it may be wondered why the topic has suddenly been given a separate category. Suffice it to say that we now come to the curious case of the genus Baccharis, a Brazilian shrub that has some 400 members in South and Central America but only a few species in North America. Baccharis is a member of the Asteraceae of which the species megapotamica and coridifolia occur in Northern and Southern Brazil, respectively. A third species, halimifolia, is native to open woods, thickets, and marsh borders (note well), from the coast of Massachusetts to Florida, Texas and Mexico. All three species, with the exception of the flowers, appear morphologically dissimilar. Extracts of both B. megapotamica

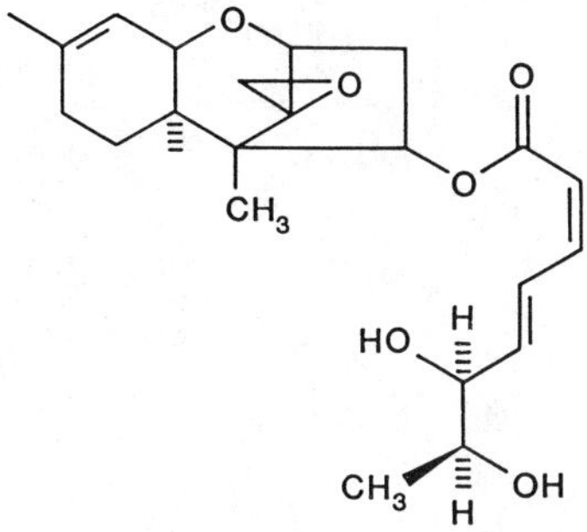

(Myrothecium roridum)

Figure 4. Trichodermadienediol A.

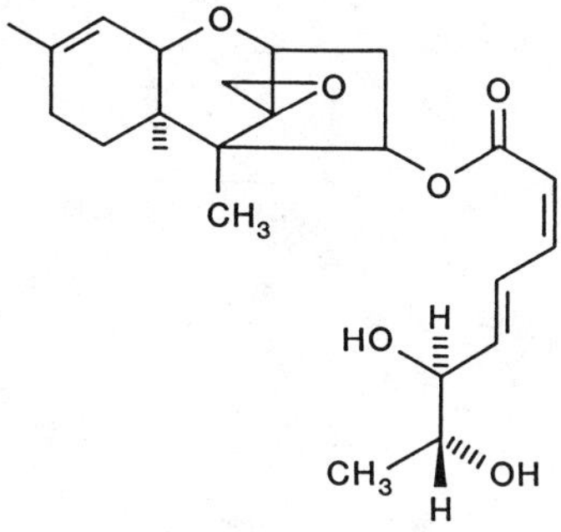

(Myrothecium roridum)

Figure 5. Trichodermadienediol B.

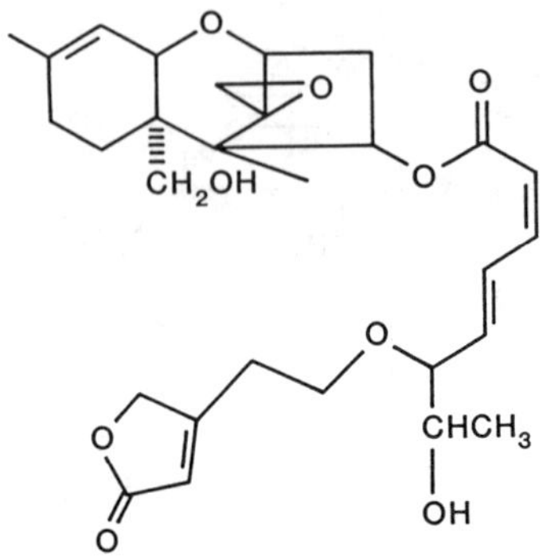

(*Myrothecium roridum*)

Figure 6. Roridin L-2.

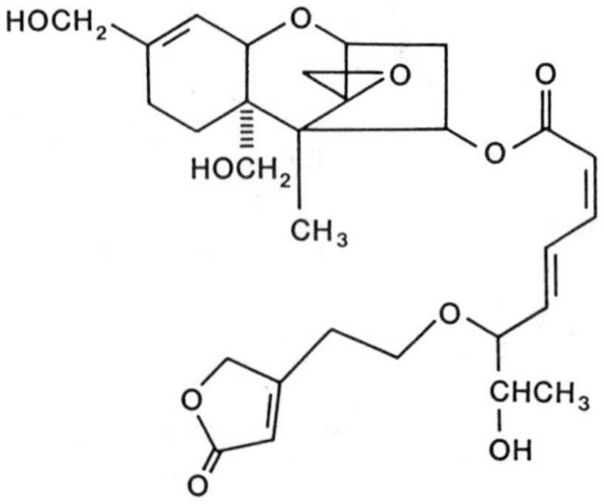

(*Myrothecium roridum*)

Figure 7. 16-Hydroxyroridin L-2.

(34) and B. coridifolia (35) were shown to contain appreciable quantities of macrocyclic trichothecenes. Because trichothecenes had always been found to be produced only by fungi, there was a presumption, in both cases, that the origin of the toxins in both species was a soil fungus, although visual inspection of the plant material gave no sign whatsoever of any fungal presence. This notion of fungal involvement was reinforced by the observation that B. megapotamica grown from seed produced no trichothecenes but, when fed roridin A, was capable of converting this normally highly phytotoxic compound (14) to baccharinoid B7, a toxin which occurs normally in the native plant (35). One striking aspect of this report is the apparent complete lack of sensitivity of B. megapotamica to roridins and baccharinoids; whereas, like all other plants tested (tobacco, beans, corn, artichokes, peppers and tomatoes), B. halimifolia is rapidly killed by these potent phytotoxins. A search for the possible trichothecene-producing fungi associated with the Brazilian species turned up a few macrocyclic trichothecene-producing isolates of Myrothecium fungi, both on the plant surfaces and in the rhizosphere of the Brazilian plants. However, the number of colonies of Myrothecium were small and infrequent, (36) and did not appear to be sufficient to account for the levels of trichothecenes in the plants.

A more careful study of B. coridifolia has now revealed a remarkable series of events. The trichothecenes (roridin A, D, and E and verucarins A and J are the major congeners in this plant) are to be found only in the female plants following pollination (37). This occurs not only in Brazil but has been duplicated in a Maryland greenhouse. Careful inspection of the plant tissue for endophytes was negative, which leads one to conclude the B. coridifolia and by analogy B. megapotamica as well, is synthesizing de novo macrocyclic trichothecenes. Not only are the toxins absent in both male and unfertilized female plants, but the toxins are to be found only in the inflorescences and specifically in the seed coats were they are concentrated to a level as high as 5% by dry weight. Furthermore, extracts of the seed of B. coridifolia contain the exact same set of macrocyclic trichothecenes most commonly produced by cultures of M. verrucaria and M. roridum (38). This suggests an interesting possibility that gene transfer has taken place between Myrothecium and Baccharis, a process undescribed in the literature but whose basis in theory has been suggested (39).

Here we see an individual case of species dependent chemical pathogenesis and this leads to some exciting research. For example, Jarvis has noticed that the materials concentrate in the gynoecium of the plants and, it appears, there may be a concentration of the trichothecenes in the seeds against a gradient. If this is so then it is possible to understand why the offspring may meet with little competition from other plant species in their natural habitat. But, more importantly, if other economic plants are resistant to the trichothecenes or may be so engineered, and if the same concentration effect occurs in the seeds then it is possible that the trichothecenes may be used to produce germplasm that contain a biodegradable herbicide.

Obviously the next question involves the transfer of resistance. If crosses are made between the *Baccharis* species will resistance be imparted to the offspring? What sort of Mendelian ratios might be found? Would this better be accomplished by protoplast fusion and tissue culture? Can this resistance be transferred to other genera? Is there a practical application (herbicide incorporation into seed) for this discovery? Do the genes that enable *B. megapotamica* to handle the trichothecenes exist in *B. halimifolia* and, if they do, what turns them on or off? There is some exciting work yet to be done in this area of research. And we have now entered the next topic.

Macromolecular Sites of Activity for the Trichothecenes. All the trichothecenes tested act, in eukaryotic systems, at the site of the 60s ribosome subunit and specifically inhibit peptidyl transferase. However, the trichothecenes are divided into two specific classes relative to this activity and are either initation inhibitors or elongation and/or termination inhibitors. The initiation inhibitors are: 15-acetoxyscirpendiol; 4-acetylnivalenol; diacetoxyscirpenol; HT-2 toxin; nivalenol; T-2 toxin; scirpentriol; verrucarin A (40, 11b). The elongation and/or termination inhibitors are: crotocin; crotocal; trichodermin; trichodermol; trichothecin; trichothecolone; verrucarol (40, 11b).

For convenience, the steps in protein synthesis are divided into four distinct phases. Sequentially they are initiation (Fig. 8), codon recognition (Fig. 9), peptide bond formation (elongation) (Fig. 10) and termination (Fig. 11) with protein release. In the initiation stage, a single strand of mRNA upon which is contained triplets of genetic code binds to the 40s ribosome in the presence of certain co-factors (Mg^{2+}, eIF-3; Mg^{2+}, GTP, eIF-1, eIF-2). Then a helical single strand of t-RNA, to which is attached an (NH_2)-methionine molecule at the 3'end, and which bears a complementary triplet genetic code to its counterpart m-RNA resting on the 40s ribosome moves into position and the corresponding (but opposite) genetic sequences, or bases, chemically bond. The 60s ribosome, in the presence of Mg^{2+} takes a proximal position to the 40s ribosome and the peptidyl (P) site is occupied on the 60s ribosome by the methionyl-tRNA molecule. During the codon recognition stage a further amino acid, for example alanyl-tRNA, bearing the correct triplet code moves into the aminoacyl (A) position on the 60s ribosome and bonds with m-RNA and the 40s ribosome in the presence of Mg^{2+}, $EF1^{\alpha}$, EF2, GTP. At elongation, methionine is transferred, in this example, to alanine in the presence of Mg^{2+}, K^{+} and peptidyl transferase and this is followed by release of the spent t-RNA (which requires Mg^{2+}) from the P position.

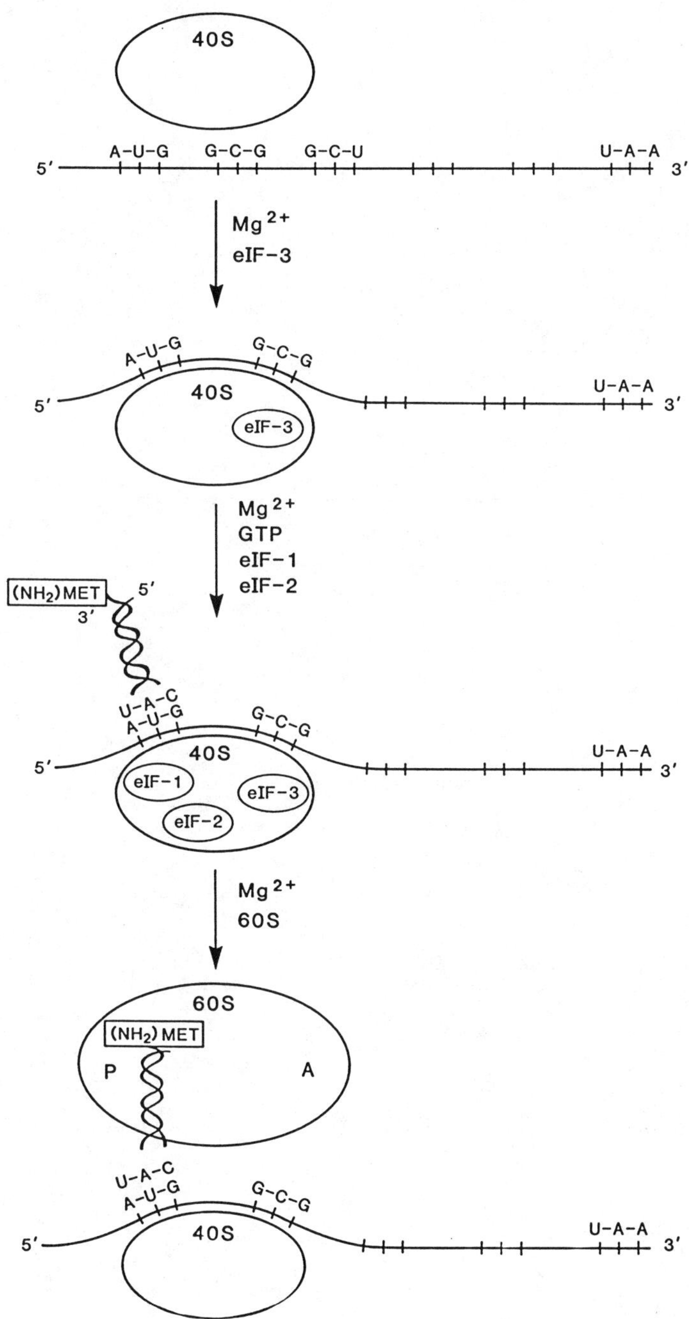

Figure 8. Protein synthesis: initiation.

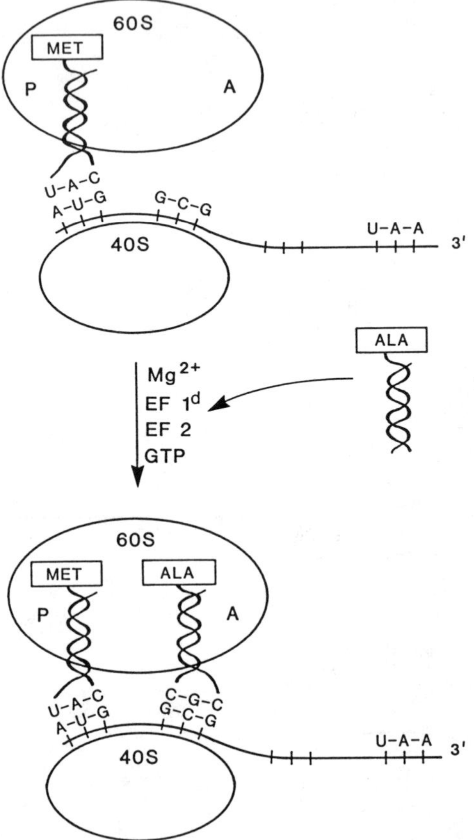

Figure 9. Protein synthesis: codon recognition.

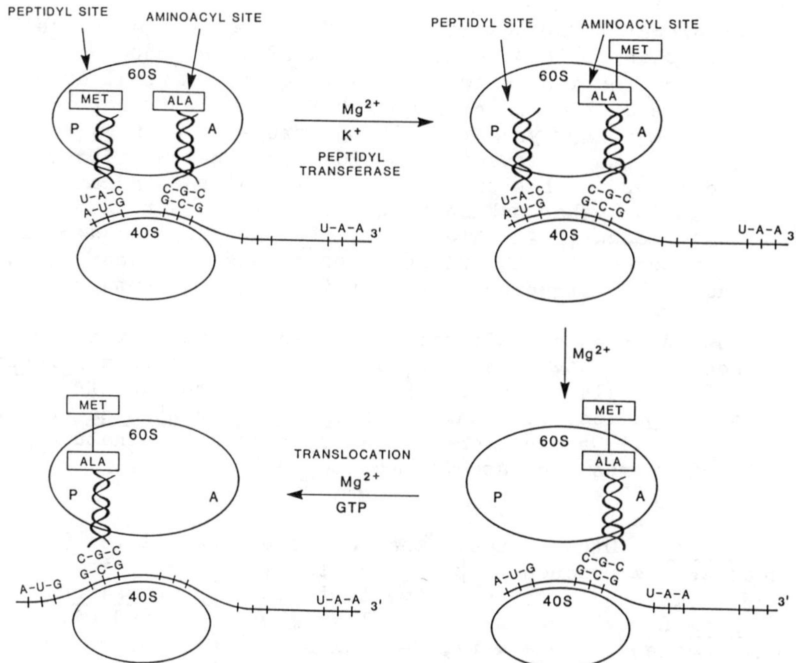

Figure 10. Protein synthesis: elongation.

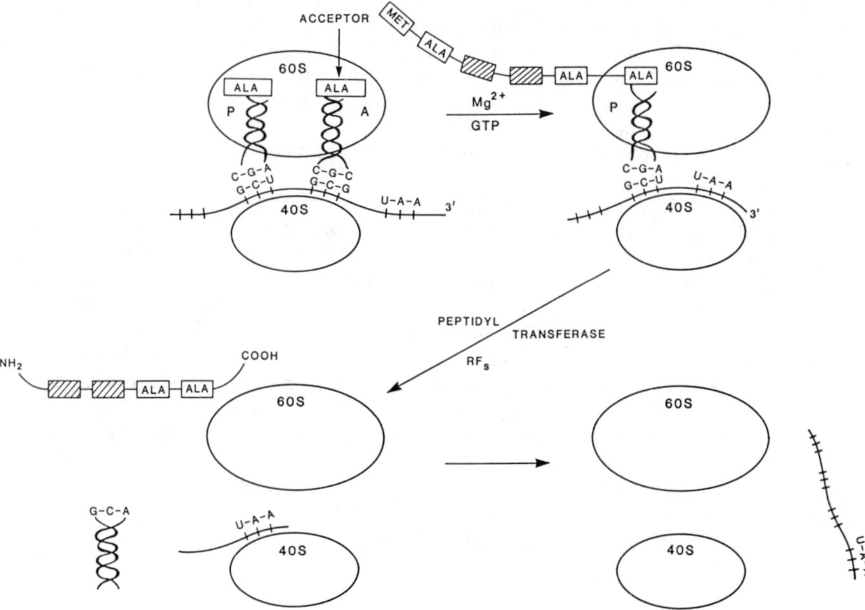

Figure 11. Protein synthesis: termination.

Then translocation of the dipeptide from the A site to the P site follows and this step is Mg^{2+} and GTP dependent.

During termination, after the additon of the final amino acid to the sequence, which requires Mg^{2+} and GTP, peptidyl transferase again plays a role in the presence of RF_s (release factors) to release the peptide from the 60s ribosome. The nascent peptide has an amine function at one end and a carboxyl at the other making it convient for other chemical reactions to occur. At this point, the mRNA becomes distal to the 40s ribosome, the 60s ribosome moves away, as does the now freed last t-RNA structure to enter the protein synthesis system.

A moot question at this point concerns peptidyl transferase. How different, or similar are the peptidyl transferases within a given plant species? How different are they from species to species, for example *Baccharis megapotamica* versus *B. halimifolia*? Or are there slight changes in the nature of the 60s ribosomes from species to species?

Other Events. Other effects that some, but not all, trichothecenes produce in plants cells is electrolytic leakage. In one set of experiments tomato (*Lycopersicon esculentum* L. cv. Supermarmande) leaves were treated with six trichothecenes. These were, T-2 toxin, diacetoxyscirpenol, deoxynivalenol, 3-acetyldeoxynivalenol, nivalenol and fusarenone. The T-2 toxin induced electrolyte leakage which increased with both concentration and exposure. Diacetoxyscirpenol also induced electrolyte leakage but the values were lower than those obtained with T-2 toxin. The other trichothecenes had no effects (41). Roridin E has also been implicated in electrolyte leakage and the response is concentration and duration dependent in muskmelon (33). Therefore, it appears that cell membrane permeabilty is changed in the presence of certain trichothecenes.

Conclusion: It is clear that much work remains to be done with pathogenic organisms (and non-pathogens) that produce trichothecenes at both the gross and macromolecular levels. Some of the findings may lead to revolutionary results, such as the production of natural herbicides and incorporation into seeds of economically important crops, and the complete understanding of changes at the DNA level so that resistant crop species may be easily produced by genetic manipulation. However, a word of caution is in order. The devastating effects of the trichothecenes in crops is readily apparent, yet the progression of our understanding of the interactions between the microorganisms, their metabolites, and the combined effect on gene expression has been painfully slow. How much slower will it be for those cases in which the effects are far more subtle?

Literature Cited

1. Wheeler, H. In "Plant Pathogenesis". Thomas, G.W.; Sabey, B.R.; Vaadia, Y.; Eds. Springer-Verlag, New York, 1975; pp. 5-16.
2. Kolattukudy, P.E.; Crawford, M.S.; Woloshuk, C.P.; Ettinger, W.F.; Soliday, C.L. In "Ecology and Metabolism of Plant Lipids." Fuller. G.; Nes, W.D.; Eds.; ACS Symposium Series 325, Washington, DC, 1987; Chapter 10.
3. In "American Type Culture Collection Catalogue of Fungi/Yeasts Seventeenth Edition, 1987"; Jong, S.C.; Gantt, M.J.; Eds.; American Type Culture Collection, Maryland. 1987; pp. 429.
4. Hoch, H.C.; Staples, R.C.; Whitehead, B.; Comeau, J.; Wolfe, E.D. Science 1987, 235, 1659.
5. Abrahamsson, S.; Nilsson, B. Acta Chem. Scand. 1966, 20, 1044.
6. Godtfredsen, W.O.; Vangedal, S. Acta Chem. Scand. 1965, 19, 1088.
7. Cutler, H.G.; LeFiles, J.H. Plant and Cell Physiol. 1978, 19, 177.
8. Brian, P.W.; Dawkins, A.W.; Grove, J.F.; Hemming, H.G.; Lowe, D.; Norris, G.L.F. J. Exp Bot. 1961, 12, 1.
9. Bawden, F.C.; Freeman, G.G. J. Gen. Microbiol. 1952, 7, 154.
10. Marasas, W.F.O. Moldy corn: Nutritive value, toxicity and mycoflora with special reference to Fusarium tricinctum (Corda) Snyder et Hansen. Ph.D. Thesis, Univ. of Wisconsin, 1969.
11. In "Protection Against Trichothecene Mycotoxins"; National Academy Press, Washington, DC. 1983. p.1.; 11a. p.73-75.; 11b. p.133 (Table 6-5).
12. Cutler, H.G.; Cole, P.D.; Arrendale, R.F.; Cox, R.H.; Roberts, R.G.; Hanlin, R.T. Agric. Biol. Chem. 1986, 50, 2667.
13. Cutler, H.G. In "The Science of Allelopathy."; Putnam, A.; Tang, C-S.; Eds.; J. Wiley and Sons, Inc., New York, 1986. Chapter 9.
14. Cutler, H.G.; Jarvis, B.B. Environmental and Experimental Botany. 1985, 25, 115.
15. Addicott, F.T.; In "Abscisic Acid". Praeger Pubs. (Holt, Reinhart and Winston), New York, 1983.
16. Cutler, H.G.; LeFiles, J.H.; Crumley, F.G.; Cox, R.H. J. Agric. Food Chem. 1978, 26, 632.
17. EMS Mutagensis of Seeds. Notes from C. Somerville. c.1986.
18. Miller, D.J.; Greenhalgh, R. Mycologia 1985, 77, 130.
19. Miller, J.D.; Young, J.C.; Sampson, D.R. Phytopath. Z. 1985, 113, 359.
20. Miller, J.D.; Young, J.C. Can. J. Plant Pathology. 1985, 7, 132.
21. Miller, J.D.; Arnison, P.G. Can. J. Plant Pathology 1986, 8, 147.

22. Cook, J.R. In "Fusarium: Diseases, Biology and Taxonomy"; Nelson, P.E.; Toussoun, T.A.; Cook, J.R.; Eds.; Pennsylvania State University Press, University Park, Pennsylvania, 1981; p.39-52.
23. Greenhalgh, R.; Meier, R.M.; Blackwell, B.A.; Miller, J.D.; Taylor, A.; Apsimon, J.W. J. Agric. Food Chem. 1984, 32, 1261.
24. Greenhalgh, K.; Meier, R.M.; Blackwell, B.A.; Miller, J.D.; Taylor, A.; ApSimon, J.W. J. Agric. Food Chem. 1986, 34, 115.
25. Wang, Y.Z.; Miller, J.D. J. Phytopath. In Press.
26. Desjardins, A.E.; Plattner, R.D.; Spencer, G.F. In Press.
27. Miller, J.D.; Young, J.C. Can. J. Plant Pathology. 1985, 7, 132.
28. Gorst-Allman, C.P.; Steyn, P.S.; Vleggar, R. J. Chem. Soc. Perkin Trans. 1985, 1, 1553.
29. Correspondence with J.D. Miller, 1987.
30. Roush, W.R.; Marletta, M.A.; Russo-Rodriquez, S.; Recchia, J. J. Am. Chem. Soc. 1985, 107, 3354.
31. Bean, G.A.; Fernando, T.; Jarvis, B.B.; Bruton, B. J. Natural Products. 1984, 47, 727.
32. Kuti, J.O.; Ng, T.J.; Bean, G.A. Phytopath. 1985, 75, 1378.
33. Kuti, J.O.; Ng, T.J.; Bean, G.A. Proc. 7th International Meeting Biodeterioration Soc. London; International Mycological Institute, London. In Press
34. Jarvis, B.B.; Çömezoğlu, N.S., Rao, M.M.; Pena, N.B. J. Org. Chem. 1987, 52, 45.
35. Habermehl, G.G.; Busam, L.; Heydel, P. Mebs, D.; Tokarnia, C.H.; Dobereiner, J.; Spraul, M. Toxicon. 1985 23, 731.
36. Jarvis, B.B.; Wells, K.M.; Lee, Y.-W.; Bean, G.A.; Kommedah, F.; Barros, C.S.; Barros, S.S. Phytopath. 1987, 77, 980
37. Jarvis, B.B.; Midiwo, J.O., Bean, G.A.; Aboul-Nasr, M.B.; Barros, C.S. Submitted for publication.
38. Tamm, C. Fortschr. Chem. Org. Naturst. 1974, 32, 63.
39. Lamboy, W.F. Evol. Theor. 1984, 7, 45.
40. Busby, Jr., W.F.; Wogan, G.N. In "Mycotoxins and N-Nitroso compounds: Environmental Risks."; R.C. Shank, Ed.; CRC Press Inc., Florida, 1981. Volume II, Table 2, p. 40.
41. Iacobellis, N.S.; Bottalico, A. Phytopath. medit. 1981, 20, 129.

RECEIVED February 18, 1988

Chapter 5

Potential Role of Phytoalexins in Aflatoxin Contamination of Peanuts

R. J. Cole, J. W. Dorner, P. D. Blankenship, and T. H. Sanders

National Peanut Research Laboratory, Agricultural Research Service, U.S. Department of Agriculture, Dawson, GA 31742

Peanut phytoalexins appear to be involved in resistance to drought-induced preharvest aflatoxin contamination of immature peanuts. Mature peanuts are considerably more resistant to environmentally-induced preharvest aflatoxin contamination of peanuts than are immature peanuts. The mechanism of this latter resistance is unknown. The identification of this resistance mechanism and other resistance may provide one approach to subsequent use of biotechnology to incorporate field resistance traits into commercially acceptable varieties. Biotechnology may also be a valuable approach to exploiting genetic resistance to preharvest aflatoxin found in wild species that have evolved in an arid environment.

The concept of plants being able to produce defensive substances, called phytoalexins, in response to infection was first proposed by Muller and Borger (1). Phytoalexins are presently defined as low molecular weight antimicrobial compounds that are synthesized by and accumulate in plants after their exposure to microorganisms.

The first report of phytoalexin production in peanuts appeared in 1972 when Vidhyasekaran *et al.* (2) reported that they detected phytoalexin production in response to invasion by storage fungi. It was observed that phytoalexin was produced to a greater extent in immature pods and to a lesser extent in mature pods. The chemical nature of these phytoalexins was not determined in this study. Keen (3) reported the production of two inducibly formed antifungal chemicals in peanuts (*Arachis hypogaea*) and inferred a phytoalexin role for these chemicals. In a follow-up study, Keen and Ingham (4) reported that these two phytoalexins from *A. hypogaea* were *cis*- and trans-isomers of 3,5,4'-trihydroxy-4-isopentenylstilbene (Figure 1A) or Arachidin II. Simultaneously, Ingham (5) isolated two related compounds from the fungal-infected hypocotyls of an African-grown cultivar of *A. hypogaea*. These were identified as *cis*- and *trans*-3,5,4'-trihydroxystilbene or resveratrol (Figure 1B). Previously, stilbene phytoalexins had been associated only with Gymnospermae.

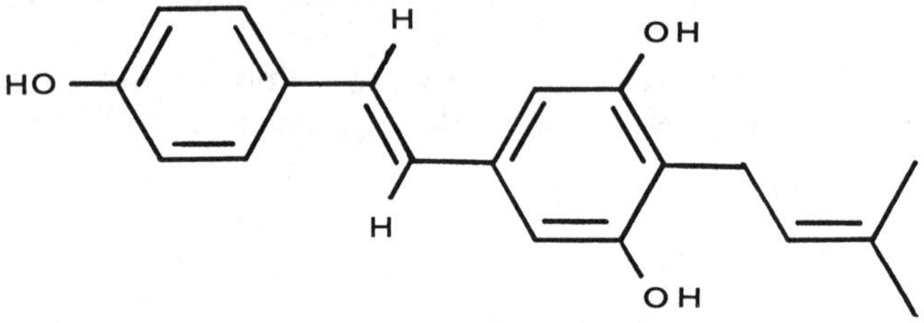

4-Isopentenylresveratrol

Figure 1A. Chemical structure of Arachidin II.

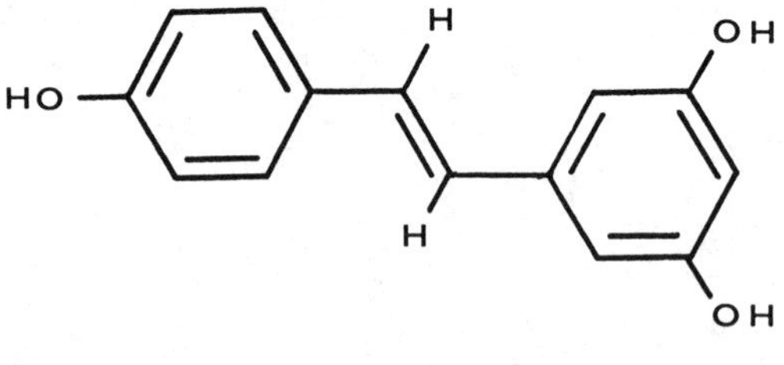

Figure 1B. Chemical structure of Resveratrol.

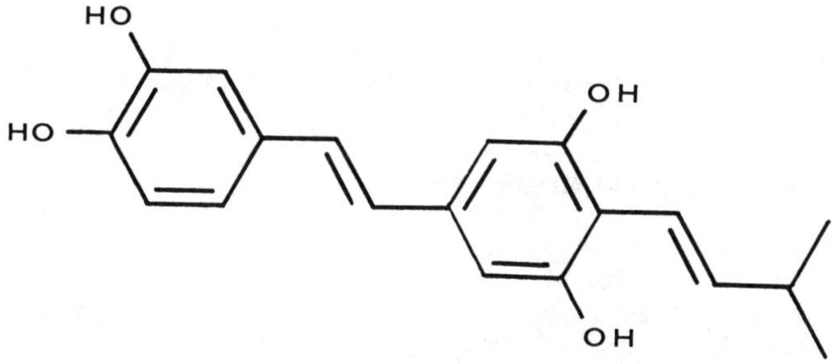

Figure 1C. Chemical structure of Arachidin I.

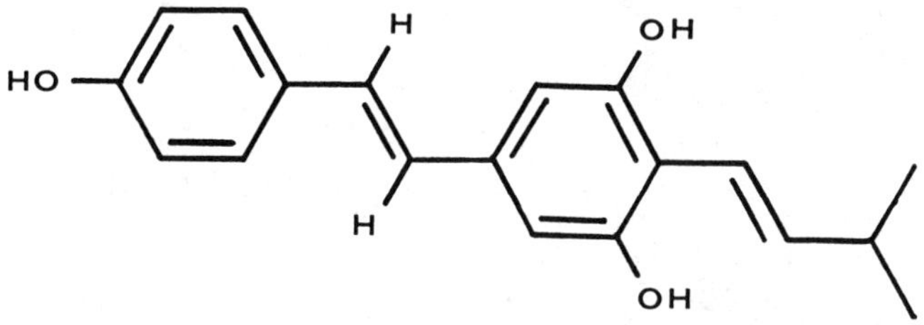

Figure 1D. Chemical structure of Arachidin III.

Aguamah et al. (6) isolated 3 antifungal compounds from A. hypogaea that were identified as 4-(3-methyl-but-1-enyl)-3,5,3',4'-tetrahydroxystilbene (Arachidin I, Figure 1C), 3,5,4'-trihydroxy-4-isopentenylstilbene (Arachidin II, Figure 1A), and 4-(3-methyl-but-1-enyl)-3,5,4'trihydroxystilene (Arachidin III, Figure 1D).

Narayanaswamy and Mahadevan (7) reportedly isolated cis and trans 3,5,4'-trihydroxy-4-isopentenylstilbene and two other unknown phytoalexins from the seeds of A. hypogaea (cultivar TMU7). These investigators later reported the isolation of phytoalexin from the leaves of A. hypogaea, although the chemical nature of the phytoalexins was not studied.

Recent studies by Wotton and Strange (8) provided circumstantial evidence for phytoalexin involvement in the resistance of peanuts to Aspergillus flavus. Their results indicated that resistance of peanut kernels to invasion by A. flavus was correlated with their capacity to synthesize phytoalexins as an early response to wounding. Also, conditions that promoted invasion of peanuts by A. flavus inhibited phytoalexin production. Thus, kernels of drought stressed plants, which are more susceptible to A. flavus than kernels of non-drought stressed plants, produced less phytoalexin in response to wounding by slicing than kernels from non-stressed plants.

This review summarizes a recent study that evaluated the role of stilbene-based phytoalexin resistance in peanuts using six environmental control plots (Dorner, J. W.; Cole, R. J.; Sanders, T. H.; Blankenship, P. D. Phytopathology, in press).

The use of biotechnology to incorporate phytoalexin-based genetic resistance from wild species that have evolved in an arid environment into commercially available varieties is a potentially viable approach to developing field resistance against A. flavus invasion and subsequent aflatoxin production in peanuts.

Materials and Methods

Culture and Treatment of Peanuts. Florunner variety peanuts were grown in the environmental control plot facility at the National Peanut Research Laboratory, Dawson, GA (9)(Dorner, J. W.; Cole, R. J.; Sanders, T. H.; Blankenship, P. D. Phytopathology, in press). Conventional cultural practices were used for all peanuts up to 96 days after planting (DAP) at which time drought treatments were imposed. One plot, which served as the control, received optimal moisture throughout the study. Two plots were equipped with thermostatically-controlled, lead-shielded heating cables to elevate the soil temperature to a mean of ca. 29°C (optimal for aflatoxin contamination)(9). The other three plots were equipped with heating cables (to elevate soil temperature) as well as epoxy-coated copper tubing through which cool water was circulated (to reduce soil temperature) as necessary to achieve a mean soil temperature of ca. 25°C (not as conducive for aflatoxin contamination)(10). Treatments began with control of soil temperature at 103 DAP for the 29°C treatment and 105 DAP for the 25°C treatment (Dorner, J. W.; Cole, R. J.; Sanders, T. H.; Blankenship, P. D. Phytopathology, in press).

Sampling of Peanuts. Samples from the different treatments were collected by hand-digging ca. 60 feet of row beginning at 120 DAP or 17 treatment days (TD), in the 29°C treatment and 121 DAP (18 TD) in the 25°C treatment. An initial sample from the irrigated treatment was taken 114 DAP and five other samples were collected throughout the treatment period with the final sample taken 184 DAP. Samples from the two drought treatments were taken at weekly intervals with the final sample from the 29°C treatment taken at 162 DAP (59 TD) and the final 25°C sample at 183 DAP (80 TD).

All undamaged pods were sand-blasted and each pod was placed into one of five maturity stages based on the method of Williams and Drexler (11) and described by Henning (12). The stages, in order of increasing maturity, were yellow 1, yellow 2, orange, brown, and black.

Immediately after maturity classification, water activity (a_w) and moisture were determined for samples at each maturity stage. Water activity (a_w) is defined as the fundamental property of aqueous solutions, and by definition a_w is also numerically equal to the corresponding relative humidity (R.H.) expressed as a fraction, i.e., R.H./100. Ten pods per maturity stage were handshelled, split, and placed in sealed sample dishes at 25°C for a_w measurements with a Beckman Model E2BFA hygroline sensor attached to a Beckman Model VFB 2 hygroline flat-bed recorder (Beckman Industrial Corp., Cedar Grove, NJ). For moisture determinations, three 50-pod samples from each maturity stage were handshelled, weighed, oven-dried for six hours at 130°C, and reweighed to determine moisture loss. Percent moisture was calculated as initial weight/final weight/initial weight.

Evaluation of Phytoalexin-Producing Potential. Six g of kernels (x3) from each maturity stage were sliced 1-2 mm thick, distributed in open 60 mm tissue culture dishes, and dusted with spores of a non-aflatoxin producing strain of *A. parasiticus* (CP 461; SRRC 2043) to elicit phytoalexin production. The open dishes were incubated in the dark at 25±1.0°C for four days in sealed dessicators over unsaturated NaCl solutions of a_w corresponding to that determined for each maturity stage. This was to maintain all peanuts at their preharvest a_w during the incubation.

Phytoalexins were extracted and analyzed via high performance liquid chromatography according to Dorner *et al*. (Dorner, J. W.; Cole, R. J.; Sanders, T. H.; Blankenship, P. D. *Phytopathology*, in press). Area counts of peaks corresponding to phytoalexin standards were summed to give total phytoalexin content for comparison of samples.

Aflatoxin Analyses. The remainder of the peanuts in each maturity stage were dried, shelled, and analyzed for aflatoxin by the method of Dorner and Cole (Dorner, J. W.; Cole, R. J. *J. Assoc. Off. Anal. Chem*., in press).

Results

Relationship Between Treatment and Kernel Moisture. Peanut kernels of all maturity stages from the irrigated control plot maintained an a_w of 1.0 throughout the experiment period. These peanuts consistently produced relatively high quantities of phytoalexins and no samples had significant aflatoxin contamination (>5 ppb) during the period of the experiment (Dorner, J. W.; Cole, R. J.; Sanders, T. H.; Blankenship, P. D. *Phytopathology*, in press).

As the study progressed, the a_w and moisture of drought-stressed peanuts decreased. The rate of moisture loss was faster in the 29°C treatment than in the 25°C treatment, particularly in the more immature stages (Dorner, J. W.; Cole, R. J.; Sanders, T. H.; Blankenship, P. D. Phytopathology, in press). However, the moisture loss was not uniform within a given maturity stage in a given treatment. For example, in the 141 DAP (38 TD) samples from the 29°C treatment, the moisture content of kernels in the yellow 2 maturity stage (next to most immature stage) ranged from 40.3% to 17.0% (a_w range of 1.00 to 0.92). The yellow 1 (most immature stage) and orange stage had similar ranges, but the moisture range in the mature stages (brown, black) was much more uniform. This phenomenon was observed in both the 29° and 25°C treatments, but as expected, the number of drier kernels within a maturity stage was greater in the 29°C treatment. In addition, in the yellow 2 and orange categories, pods containing very low moisture kernels had a characteristic "mustard-colored" appearance after sand-blasting, which proved to be a consistently sound indicator of low kernel moisture. Kernels in the yellow 1, yellow 2, and orange maturity stages were further segregated based on visual assessment of kernel moisture. All kernels from "mustard-colored" pods were separated into groups of high, medium, and low moisture before determinations of moisture, a_w, phytoalexin-producing potential, and aflatoxin contamination were made. These segregations were essential to testing the hypothesis that kernel a_w was the primary factor that directly or indirectly led to aflatoxin contamination. Since a_w and moisture of peanuts in the brown and black maturity stages were generally quite uniform, it was not necessary to further segregate kernels from those categories.

Comparisons of moisture content of particular maturity stages from the 29° and 25°C treatments were difficult, since there were up to four moisture contents for each maturity stage. For example, in the yellow 2 group from the 29°C treatment samples at 148 DAP (45 treatment days), the moisture content of the different segregations within the group were: "mustard-colored," 11.7%; "normal-colored" low moisture, 15.0%; medium moisture, 29.3%; high moisture, 37.3%. Each sampling from both the 29° and 25°C treatments yielded kernels of similar moistures within a maturity stage. However, the 29°C treatment consistently had more kernels at lower moistures than the 25°C treatment. Therefore, to arrive at a single moisture content for a segregated maturity stage, the amount of kernels at a given moisture was taken into account and a "weighted moisture" was used for each maturity stage (Dorner, J. W.; Cole, R. J.; Sanders, T. H.; Blankenship, P. D. Phytopathology, in press).

As expected, the more immature peanuts lost moisture sooner in the 29°C soil than in the 25°C soil. Generally, it took about a week longer for kernels in the cooler treatment to reach moistures attained in the warmer treatment. The data clearly showed a direct relationship between soil temperature and the rate of water loss from peanut kernels under late-season drought stress.

Relationship Between Kernel a_w and Phytoalexin Production. As kernel a_w decreased as a result of drought stress, the capacity of kernels to produce phytoalexins also decreased and eventually was lost. This trend was independent of maturity and soil temperature. When a_w

was high (0.97-1.00), kernels of all maturities and from both treatments produced abundant phytoalexins. No significant phytoalexin production was observed when a_w was below 0.95. This was consistent throughout all maturities and in both treatments. This illustrated the correlation between lower a_w and reduced phytoalexin production irrespective of maturity and soil temperature.

Relationship Between Phytoalexin Production and Aflatoxin Contamination. Many samples that had lost the capacity for phytoalexin production had no detectable aflatoxin, while in other similar samples aflatoxin levels ranged as high as 2400 ppb (Table I). Significantly however, no sample of peanuts that had moderate to high phytoalexin-producing capacity contained >5 ppb aflatoxin. Therefore, it can be stated that as long as peanut kernels had the capacity for phytoalexin production, aflatoxin was not formed (Table I). However, the loss of phytoalexin-producing capacity resulted in aflatoxin formation but in only a portion of these susceptible kernels (Dorner, J. W.; Cole, R. J.; Sanders, T. H.; Blankenship, P. D. Phytopathology, in press).

Relationship Between Kernel a_w and Aflatoxin Contamination. In order to evaluate the relationship between a_w and aflatoxin contamination, it was useful to determine the percentage of samples contaminated with aflatoxin both above and below a_w of 0.95 (the a_w below which the capacity to produce phytoalexins was lost). In the 29°C treatment, when a_w was >0.95, only 5% of all samples contained greater than 5 ppb aflatoxin. However, 50% of all samples with a_w <0.95 had greater than 5 ppb aflatoxin. In the 25°C treatment 7% of samples greater than 0.95 a_w had more than 5 ppb aflatoxin compared to 26% of samples when a_w was <0.95. These data are for all maturities combined.

When samples from all maturities and both treatments are combined, 94% had <5 ppb toxin when a_w was >0.95. This is compared to 65% having <5 ppb when a_w was >0.95. When 20 ppb is used as the cutoff, the percentages were 98% for a_w >0.95 and 73% for a_w <0.95. The difference in contamination between samples above and below a_w of 0.95 was greater in the more immature peanuts. For example, when treatments are combined, only 13% of the yellow 1 samples had >5 ppb aflatoxin when a_w was >0.95, but 67% of yellow 1's were >5 ppb when a_w was >0.95, but when a_w was <0.95, only 17% of the samples from each maturity were >5 ppb. Proceeding from yellow 1 (most immature) to black (most mature) when a_w was <0.95 the percentage of samples >5 ppb decreased accordingly - 67%, 50%, 24%, 19%, 17%.

A higher percentage of samples from the 29°C treatment had contamination than from the 25°C treatment. The percentages of samples from each maturity (yellow 1-black) containing >5 ppb aflatoxin when a_w was <0.95 were 80%, 50%, 50%, 50%, and 20%, respectively, in the 29°C treatment. Corresponding values from the 25°C were 57%, 50%, 8%, 0%, and 14%, respectively.

Therefore, it can be seen that when a_w falls below 0.95 where phytoalexin-producing capacity is lost, immature peanuts are more likely to become contaminated than mature peanuts, particularly in high temperature soil (Table I). However, in mature peanuts, low a_w and the loss of the capacity to produce phytoalexins does not result in the same degree of contamination (both in numbers and levels) as

Table I. Phytoalexin production and aflatoxin contamination in five maturities of peanuts from two drought treatments

Trt days	Total phytoalexins (Combined peak area counts)[a] Yellow 1	Yellow 2	Orange	Brown	Black	Total aflatoxins (ppb) Yellow 1	Yellow 2	Orange	Brown	Black
					29° Treatment					
17	17,000	10,700	8,600	4,800	9,200	0	0	0	0	0
24	0	13,700	11,800	2,900	3,200	0	0	0	0	0
31	0	0	0	0	0	0	0	0	0	0
38	0	0	0	0	0	861	15	0	0	6
45	0	0	0	0	0	1,190	624	0	12	0
51	0	0	0	0	0	2,419	118	329	0	0
59	0	0	0	0	0	785	1,915	33	291	0
					25° Treatment					
18	11,000	4,600	5,000	7,000	1,000	0	0	0	0	0
25	11,500	11,300	4,400	5,300	6,100	0	0	0	0	0
32	35,000	500	0	4,700	6,100	0	0	0	0	0
39	200	200	0	0	0	12	0	0	0	0
46	0	0	0	0	0	784	62	0	0	0
52	0	0	0	0	0	0	447	0	0	0
60	0	0	0	0	0	0	484	0	.0	0
66	0	0	0	0	0	5	8	13	0	0
73	0	0	0	0	0	1,189	234	173	0	0
80	0	0	0	0	0	0	291	0	0	82

[a]Data reported are the mean of three determinations.

(Dorner, J. W.; Cole, R. J.; Sanders, T. H.; Blankenship, P. D. Phytopathology, in press) it does in the immature peanuts (Table I). This indicates that after protection by phytoalexin production is lost (<0.95 a_w), other as yet unknown factors are providing a high degree of protection from preharvest aflatoxin contamination to mature peanuts.

Conclusions

Phytoalexins appear to be at least partly involved in resistance to drought-induced preharvest aflatoxin contamination of peanuts. Phytoalexin-based resistance appears to be more important in immature peanuts than in mature peanuts. Therefore, mature peanuts are more resistant to preharvest aflatoxin contamination than are immature peanuts. The exact mechanism of this resistance is not known.

A viable approach to obtaining field resistance to aflatoxin contamination in peanuts under late season drought stress may be found in wild peanut species that have evolved in an ecosystem that included seasonal periods of drought stress. They may have evolved a phytoalexin-producing enzyme system that remains operative at an a_w lower than commercial varieties (i.e., <0.95 a_w) in order to cope with seasonal drought periods. The application of biotechnological techniques could provide a mechanism to incorporate this genetic potential into commercially acceptable varieties. Other genetically controlled resistance mechanisms could also be exploited using biotechnology.

Literature Cited

1. Muller, K. O.; Borger, H. Arb. Biol. Aust. Reichsanst. (Berl.) 1940, 23, 189.
2. Vidhyasekaran, P.; Lalithakumari, D.; Govindaswamy, C. V. Indian Phytopathology 1972, 25, 240.
3. Keen, N. T. Phytopathology 1975, 65, 91.
4. Keen, N. T.; Ingham, J. L. Phytochemistry 1976, 15, 1794.
5. Ingham, J. L. Phytochemistry 1976, 15, 1791.
6. Aguamah, G. E.; Langcake, P.; Leworthy, D. P.; Page, J. A.; Pryce, R. J.; Strange, R. N. Phytochemistry 1981, 20, 1381.
7. Narayanaswamy, P.; Mahadevan, A. Acta Phytopath. Acad. Scientiarum Hungaricae 1983, 18, 33.
8. Wotton, H. R.; Strange, R. N. J. Gen. Microbiol. 1985, 131, 478.
9. Blankenship, P. D.; Cole, R. J.; Sanders, T. H.; Hill, R. A. Oleagineux 1983, 38, 615.
10. Cole, R. J.; Blankenship, P. D.; Hill, R. A.; Sanders, T. H. In Toxigenic Fungi - Their Toxins and Health Hazard; Kurata, H.; Ueno, Y., Eds.; Elsevier: New York, 1984; p 44.
11. Williams, E. J.; Drexler, J. S. Peanut Sci. 1981, 8, 131.
12. Henning, R. The Peanut Farmer 1983, 19, 11.

RECEIVED April 1, 1988

Chapter 6

Preinfection Changes in Germlings of a Rust Fungus Induced by Host Contact

Richard C. Staples[1] and Harvey C. Hoch[2]

[1]Boyce Thompson Institute, Cornell University, Ithaca, NY 14853
[2]New York State Agricultural Experiment Station, Cornell University, Geneva, NY 14456

Germlings of the bean rust fungus, Uromyces appendiculatus, invade their host plant through stomates using a set of specialized cells, the infection structures, that are stimulated to develop in response to a thigmotropic stimulus provided by the lip of the stomatal guard cell. The essential thigmotropic signal is a ridge or groove in the substrate having an abrupt change in elevation of 0.5 μm. Development of the infection structures is accomplished in three important phases, i.e. a very rapid cessation of germ tube elongation, the start of DNA replication, and development of the infection structures. Development of the infection structures includes two rounds of nuclear division, the synthesis of about 15 differentiation-related proteins, and the expression of at least five differentiation-specific genes.

Rust fungi are basidiomycetes which in nature are obligate parasites of higher plants, i.e. require living hosts to complete their life cycle. They are serious pests of crop plants. While several of the rusts have been cultured (1), survival of axenic rusts is poor outside the laboratory. Their very complex life cycles, -- long-cycle rusts have five types of spores with the sexual and asexual stages on different hosts (2) -- pose fascinating questions of host selectivity, spore initiation, environmental regulation of the life cycle, and host cues for fungal penetration. Any of these, were more known about them, could serve as a focus of control strategies.

Uredospores are asexual, dikaryotic spores, many species of which colonize their hosts by seeking and penetrating only via the stomates. This review describes the developmental responses induced in uredospore germlings by host contact which guides appressorium placement over the stomatal opening. Many researchers have contributed to these ideas beginning with the inventive work of Sidney Dickinson (3). Examples of other

0097–6156/88/0379–0082$06.00/0

important contributions to an understanding of differentiation include research by Heath and Heath (4,5), Maheshwari et al. (6), Rowell and Olien (7), and Wynn (8). The most recent treatise on the rust fungi is one on cereal rusts by Bushnell and Roelfs (9,10). Extensive reviews on the differentiation process include those by Emmett and Parbery, (11), Littlefield and Heath (2), Wynn and Staples (12), and Hoch and Staples (13).

Cytology of Infection Structure Development

Morphology. Uredospores of many of the rusts colonize their hosts by seeking and penetrating the stomates. The ridges and indentations of the leaf are signals which control both the direction of growth of the germ tube (orientation), and the site of appressorium development for penetration (differentiation). These fungal responses to contact stimuli are growth changes.

Orientation. The orientation response by *Uromyces appendiculatus* germlings on bean leaves in an area where the surface is regularly striated is shown in Fig. 1A. Uniformly spaced scratches on plastic surfaces also induce germ tube orientation (Fig. 1B). Orientation is the first visual effect of sensing by the uredospore germling, and orientation has been postulated to improve the chance that a germ tube will contact a stomate on the host leaf by reducing wandering (14).

Differentiation. When the germ tube contacts a stomate, the germling undergoes a process of differentiation in which a series of infection structures are formed beginning with the *appressorium* which sits over the stomatal opening (Fig. 2). Development of the appressorium is followed by the *peg*, which penetrates between the guard cells, the *vesicle*, which fills the substomatal cavity, and the *infection hyphae*, which ramify throughout the leaf tissue. *Haustorial mother cells* develop when the infection hyphae encounter the mesophyll of the host. It is the *haustoria* that develop within the mesophyll cells, and which obtain nourishment for the fungus.

Haustorial mother cells are thought to arise only when the infection hyphae make contact with host cells; however, haustorial mother cells do develop occasionally in the absence of the host (15). All of the infection structures prior to the haustorial mother cells develop as a result of the original contact by the germling with the stomate.

Chemicals can also induce infection structures, and each fungus seems to have its own requirement. For example, *Puccinia graminis tritici* responds to acrolein (16), while *U. appendiculatus* responds to K^+ (17). The thigmotropic response by *U. appendiculatus* to the stomate, however, is entirely physical as demonstrated by Wynn (8).

Dimensions of the Signal

Differentiation can be readily induced by a single scratch (18). This observation led to the preparation of templates

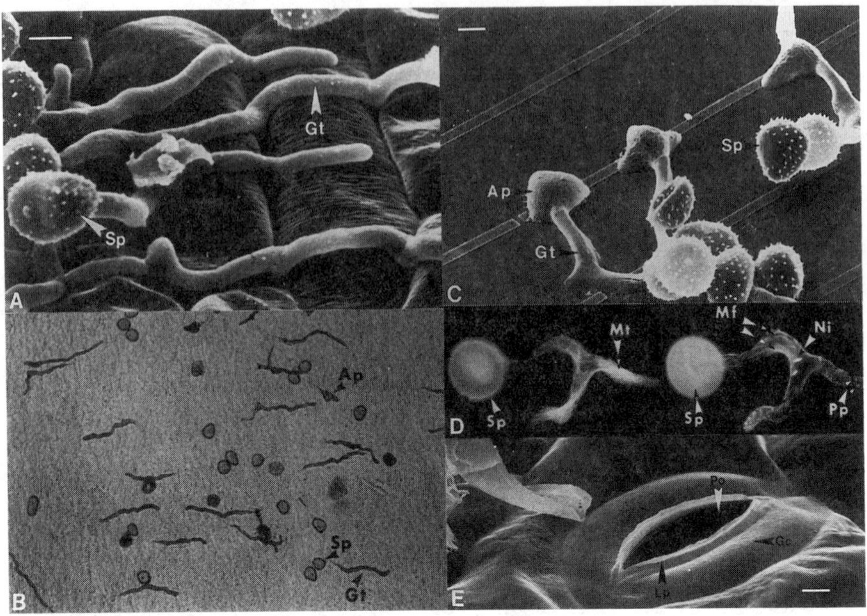

Figure 1. Photomicrographs showing the development of U. appendiculatus uredospore germlings and the stomatal pore of the bean leaf. (A) Growth of U. appendiculatus germlings on the surface of the host plant (Phaseolus vulgaris) showing the orientation response by the germ tubes to the topography of the leaf. Bar scale, 10 μm. After Hoch et al. (19). (B) Growth of U. appendiculatus germlings on the inductive surface of high-density polyethylene. The germ tubes are oriented perpendicular to the inherent features formed on the surface. Appressoria have begun to form on some of the germ tubes (4 h after start of germination). X190. After Staples et al. (34). (C) Polystyrene replica of precision ion-etched silicon wafer templates, containing ridges 0.5 μm high by 4.0 μm wide, was highly inductive for infection structure formation in U. appendiculatus germlings. Bar scale, 10 μm. After Hoch et al. (19). (D) Germling of U. appendiculatus stained for microtubules (left), and F-actin (right). Some of the microtubule and F-actin microfilament profiles (Mf) occupy adjacent sites. An arrow points to one of the two F-actin-containing nuclear inclusions (Ni) in the nuclei (one in each nucleus). Actin-containing peripheral plaques (Pp) are also shown. X1250. After Staples et al. (34). (E) Scanning electron micrograph of bean leaf stomatal guard cells (Gc) having prominent erect lips (Lp) that serve as the signal for appressorium formation in U. appendiculatus. The fungus eventually enters the leaf through the stomatal pore (Po) where infection of the host cell occurs. Bar scale, 10 μm. Ap, appressorium; Gt, germ tube, Sp, uredospore. After Hoch et al. (19).

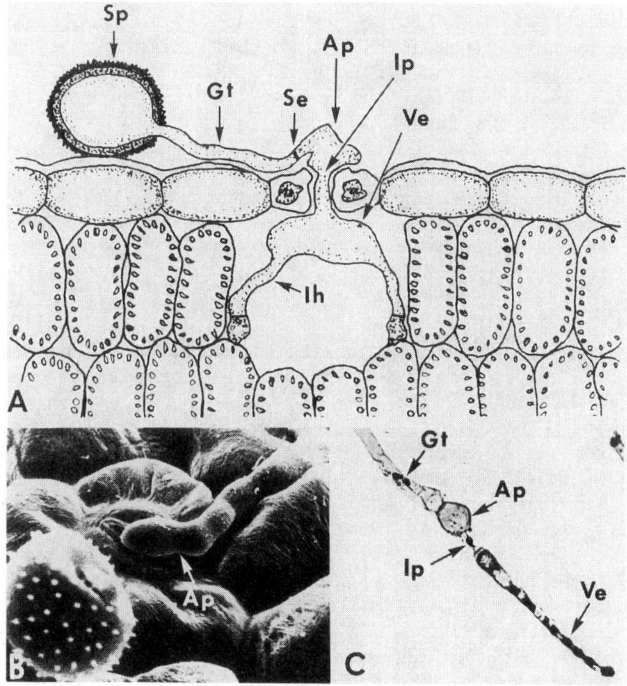

Figure 2. Infection structures developed by uredospore germlings of U. appendiculatus. (A) Diagram of the infection structures developed after germ tube contact with a stomatal guard cell. (B) SEM micrograph showing appressorium developed over a bean leaf stomate. (C) Typical infection structures of U. appendiculatus. X400. Ap, appressorium; Gt, germ tube; Ih, infection hyphae; Ip, peg; Se, septum; Sp, spore; Ve, vesicle.

microfabricated onto silicon wafers using electron-beam lithography (19). The differentiation signal was found to be a simple ridge having an optimum height of 0.5 ± 0.1 µm (Fig. 1C). Negative ridges (grooves) are equally effective.

Orientation and differentiation are separate responses. While differentiation occurs if the ridge is solitary and well separated from other ridges, it was found that germ tubes of the fungus are highly oriented when multiple ridges are present which are spaced more closely than 15 µm. Differentiation is inhibited and orientation is enhanced as the spacing is diminished.

The height of the differentiation signal is small compared with the diameter of the germ tube which averages 5 to 8 µm (Fig. 3). It is obvious that the fungus is able to distinguish minute differences in leaf surface topography (perhaps as small as ± 0.1 µm) in order to target the stomate.

Wynn (8) has suggested that on the bean plant, the lip of the stomatal guard cell was the essential differentiation signal to which the U. appendiculatus germling responded. Photographs by SEM of specimens of the bean stomate which were quick frozen then freeze-dried, revealed prominent guard cell lips oriented nearly perpendicularly to the cell surface (Fig. 1E). The lip frequently appeared to be somewhat ragged along the outer edge, and non-uniform in height. Mean height of the lip was 0.487 ± 0.07 µm, and we assume with Wynn (8) that the essential differentiation signal on the bean leaf is the lip of the stomatal guard cell.

Ridges 0.5 µm in height, so suitable for induction of differentiation by the U. appendiculatus germling, are not inductive for germlings of P. graminis tritici, the wheat stem rust fungus (Hoch, unpublished information), and neither were Wynn's plastic replicas of the bean leaf stomate (12). While appressoria are induced on germlings of P. graminis tritici by contact with the stomate (20), the stomates are quite different in shape. Using methyl methracylate spheres mixed into a polystyrene membrane, Dickinson (3) estimated that the signal for P. recondita (leaf rust of wheat) consists of multiple small ridges about 0.0012 µm in height and spaced about 0.12 µm apart. These estimations have not yet been verified using etched wafers, i.e. aggregates of the granules (longer dimensions) may have been inductive; however, Dickinson did demonstrate that membranes containing smaller granules were not effective. As these dimensions are quite different from those for U. appendiculatus, and inasmuch as our wafers would not induce P. graminis tritici germlings to differentiate, the developmental signals appear to differ among the rusts. The possibility exists that small changes in the architecture of the leaf would be a useful strategy for reducing the virulence of the pathogen.

The Thigmotropic Response

Time-scale of Response. A list of some of the events which occur during the development of infection structures, and the estimated times when the response occurs after contact with the inductive ridge, is shown in Table I. The following discussion is oriented around the time-scale which was assembled from a variety of sources for this purpose.

Table I. Time Scale of Differentiation For Bean Rust Uredospore Germlings[1]

Minutes	Activity Started
0	GERM TUBE APEX AT INDUCTIVE RIDGE
<1.0	STOP FORWARD GROWTH + APPRESSORIUM GROWTH BEGINS
15	FIRST ROUND OF DNA REPLICATION BEGINS + SYNTHESIS OF EARLY DS-PROTEINS + EXPRESSION OF EARLY DS-GENES
36	NUCLEAR MIGRATION INTO APPRESSORIUM
44	MITOSIS I BEGINS NUCLEI CONDENSE +
56	TELOPHASE ENDS (= 4 NUCLEI)
83	APPRESSORIUM SEPTUM COMPLETED
83 - 150	MORPHOGENETIC PAUSE
>200	CONSTRUCTION OF VESICLE EXPRESSION OF LATE DS-GENES + SYNTHESIS OF LATE DS-PROTEINS + MIGRATION OF NUCLEI + MITOSIS II (= 8 NUCLEI)

[1]These times are approximate only, and were assembled from experiments carried out for a variety of different purposes. The data on mitosis was obtained from Tucker et al. (18); DNA replication I (15 min) from Staples et al. (21). The morphogenetic pause and mitosis II were estimated from the behavior of a population of spores (DS, differentiation-specific).

Early Response Period. The germ tube elongates over the surface at 1 to 2 µm/min. As the width of the stomatal opening on a bean leaf usually is not greater than 3 µm, the germ tube must respond in less than a minute in order not to overshoot the opening. That the germling usually does not overshoot can be seen in Fig. 1C, where all of the appressoria are resting on top of the signal ridge which is 4 µm wide. Thus, the initial response to the signal is rapid, and as shown in the figure, where an overshoot has occurred, the fungus has a corrective mechanism for retracting over the target guided by the signal. Neither the mechanism by which the signal is perceived by the germling, nor the manner by which it responds, is known yet; however, it seems unlikely that gene expression is involved.

DNA Replication and Mitosis. One of the earliest responses to contact is the start of DNA replication and mitosis (21). The nuclei in the germ tube are haploid, and replication begins about 15 min after induction of differentiation (Table I). Nuclear division requires about 15 min, and is completed 30 min before the appressorial septum is completed (Table I; 18).

Morphogenetic Pause. At the end of telophase, the germling constructs a septum across the germ tube guided by a septal ring of F-actin (18). When this activity ceases, the germling appears to remain quiescent for about an hour or so. Activity begins again when the vesicle elongates out from the appressorium.

Protein Synthesis and Gene Expression. Development of the infection structures is accompanied by the synthesis of at least 15 differentiation-related (dr) proteins, i.e. proteins not present in the germling until differentiation is induced Fig. 4; (22). A downshift in the synthesis of some proteins also occurs during infection structure development.

To address the question of the number of genes expressed during development of the infection structures, we have screened a bean rust genomic library using a probe enriched for the dr-sequences prepared by a modified cascade hybridization procedure (23). We have obtained twenty differentiation-specific clones. From restriction mapping and Southern hybridization analyses, these clones can be grouped into six classes. **Class I** consists of fourteen homologous clones, **class II** consists of two homologous clones, and **classes III to VI** consist of a single clone each.

Northern analyses, used to confirm that these clones were specific to the differentiated stage of the fungus, showed that class I clones hybridized with transcripts that were equally abundant in differentiated and nondifferentiated germlings. The remaining classes, however, hybridized with transcripts that were more abundant in differentiated germlings. Thus, it appears that the contact stimulus activates a small gene set of at least five genes that are dedicated to the differentiation process. We have also obtained clones of genes whose expression is downshifted, but we have not characterized them yet.

Using hybrid arrest-translation procedures, we have identified a 23 kDa polypeptide which is coded for by clone 24 of class II. One of the dr-proteins has an approximate molecular weight of 23 kDa; however, its identity with this peptide must await the preparation of a specific antibody.

Although the studies on homology are still rather limited, the differentiation-specific clones appear to be homologous only with DNA from other rusts and not with DNA from more distantly related fungi. So far, we have probed a BamH1 digest of total DNA from two rusts (*Uromyces vignae*, and *U. appendiculatus*), two plant pathogenic ascomyces which develop appressoria on their hosts (*Cochliobolous heterostrophus* and *Colletotrichum lindemuthianum*) and the saprophyte, *Aspergillus nidulans*, which does not differentiate. Three clones representing classes II, IV and VI, hybridized to DNA from the two related rusts but not to the DNA from the other three more distantly-related fungi.

All of the clones isolated so far hybridize with transcripts which appear in germlings at the time when the vesicles form, although the class VI clone also hybridizes with a transcript which appears during development of the appressorium. Thus, the ds-genes appear to have a role in construction of the infection structures. A more precise timing of expression will have to await additional studies.

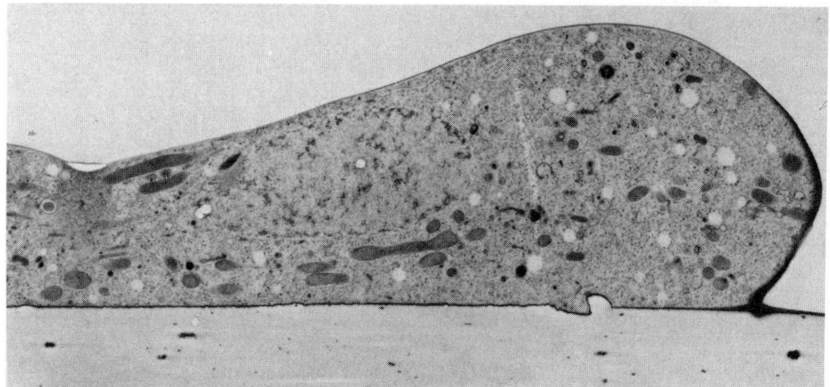

Figure 3. A developing appressorium sectioned transversely to the direction of a scratch on a Mylar substrate which was previously coated with a thin layer of palladium/gold. X7,300. After Bourett et al. (35).

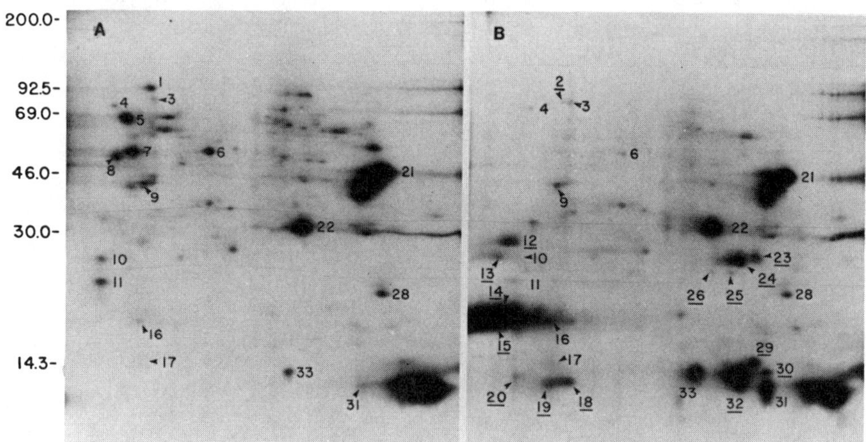

Figure 4. Autoradiograms of proteins separated by electrophoresis on two dimensional slab gels of 15% polyacrylamide. An ampholine-generated pH gradient was employed in the first dimension, while a 15% SDS gel was employed in the second dimension. The position of the molecular weight markers (kDa) is shown along the left-hand margin. Uredospore germlings on collodion membranes were misted with [^{35}S]-methionine for 1.5 h before harvest and extraction of the proteins. (A) Nondifferentiated uredospores germinated 6 h, (B) differentiated uredospores germinated 6 h with appressoria and vesicles. Paraffin oil was incorporated into the collodion membrane to induce differentiation. Underscored numbers indicate proteins synthesized specifically during differentiation. Not all of the proteins have been numbered.

Elements of the Sensing Mechanism

Extracellular Matrix. Unless germlings of the rust fungi are attached to host surfaces, infection structures either do not develop or are abnormally oriented (12). Protein in the extracellular matrix of the germ tube appears to be an important component of the sensing mechanism because proteases, e.g. Pronase E, but not other hydrolases, loosen the germ tube from the surface and prevent thigmodifferentiation (24,25). Pronase E did not affect germination or germling growth, but adhesion to a polystyrene surface was significantly reduced. The simplest explanation for the effects of the proteases is that extracellular proteins, possibly glycoproteins, are required for binding the germ tube to the inductive surface. The extracellular matrix appears to be amorphous by TEM, and nothing is known yet about its possible association with the plasmalemma or intracellular components.

Cytoskeleton. *U. appendiculatus* germlings have an extensive network of cytoplasmic microtubules oriented parallel with the long axis of the hypha that are especially abundant near the hypha-substrate interface Fig. 1D, left; (26). Microtubule disrupting agents such as cold, demecolcine, griseofulvin, and nocodazole, effectively depolymerize the microtubules in the germ tube but not the actin microfilaments (27). Enhanced microtubule arrays are observed in the presence of either Taxol or D_2O, agents known to favor microtubule stabilization (28).

F-actin is observed in three important configurations: as filaments, associated with filasomes, and positioned within the nuclear matrix as an inclusion (Fig. 1D, right). Treatment of the germlings with cytochalasin E leads to the disappearance of cytoplasmic filaments but has no perceivable effects on the filasomes (27).

What possible role may the cytoskeleton have in transmission of the differentiation signal? The evidence is only circumstantial so far. The location of the microtubules near the cell-substrate interface means that the microtubules are located where they ought to be to have a role in reception. Furthermore, nocodazole- or griseofulvin-treated germlings, in which the microtubules are depolymerized, fail to develop appressoria when the germ tube contacts a surface scratch (Hoch, H.C., unpublished). This suggests that the microtubules must be intact in order for the germling to respond to an inductive signal.

The microfilament cytoskeleton must also be intact for differentiation to occur, but it probably does not have a direct role in signal transduction as postulated for the microtubules. For example, Tucker et al. (18) have demonstrated that treatments of germlings with brief pulses of cytochalsin E which depolymerize the microfilaments, and consequently distort the shape of the germ tube, do not inhibit the start of mitosis in the appressorium. Even the time-course was not altered. Removal of the drug restores cell shape. However, development of the appressorial septum no longer occurs, so that completion of appressorium construction probably does depend on the integrity of the microfilaments.

Transmembrane Signaling. High exogenous levels (10 mM) of both cAMP and cGMP induce DNA replication and nuclear division in *U. appendiculatus* germlings efficiently, but the infection structures are inhibited from forming (29). However, more modest levels (10^{-7} M), as well as exogenously applied stimulators of adenylate cyclase and inhibitors of cAMP-dependent phosphatase, induce the complete sequence of DNA replication, nuclear division and development of the infection structures. The data suggest that *U. appendiculatus* germlings utilize a cAMP- and cGMP-dependent cascade in metabolism, including a cyclic nucleotide-dependent protein kinase which phosphorylates a 54 kDa protein (Epstein, unpublished). We have been unable to distinguish between the nucleotides as the preferred messenger.

Phosphatidic acid (PA), rapidly produced during receptor-stimulated breakdown of phosphoinositides, mobilizes Ca^{+2}, raises the pH of the cytoplasm, and stimulates DNA synthesis in activated mammalian cells (30). Recently, we have found that the the dipalmatoyl derivative of PA, as well as several diacylglycerols, e.g. 1,2-dipalmitoyl-sn-glycerol, induce both mitosis and infection structure development in *U. appendiculatus* uredospores (Staples, unpublished). Furthermore, external Ca^{+2} also stimulates DNA replication and infection structure development, although the stimulation is more effective if low levels of K^{+} (1 mM) are present (17). In addition, calmodulin, the calcium receptor protein, has now been isolated from uredospores (31), so that a calcium-related second messenger system almost certainly is present. As Favre and Turian (32) have recently isolated protein kinase C from mycelia of *Neurospora crassa*, the pathway may be common in fungi as well as animal cells. At this time, then, it seems likely that transmembrane signaling from a surface stimulus could occur in *U. appendiculatus* in conjunction with a second messenger system, by one or both of the cyclic nucleotide and the phosphoinositidyl pathways.

G-proteins are a family of guanine nucleotide-binding proteins that functionally couple a wide array of membrane receptors to biochemical effector systems that regulate second messengers (33). Since both physical stresses such as membrane depolarization and stretching will initiate G-protein activity (33), a system of G-protein-related second messengers would provide a mechanism for the very rapid response that uredospore germlings make to thigmotropic signals. We note that GTP is as effective as K^{+} as an inducer of differentiation *U. appendiculatus* germlings (Staples, unpublished).

A Perspective

Much research is now aimed toward an understanding of host responses to infection. These responses typically include various resistance mechanisms, toxin effects, induced susceptibility, and race specificity. The present research, and the many contributions by others which preceded it, suggests that host recognition, and adaptation to host morphology by the pathogen, is exquisite.

The rust fungi are serious threats to crops, and have caused widespread losses. Use of resistant crop varieties is at present the most reliable control measure for rusts where these can be developed; however, new varieties must be continually developed in order to provide protection against virulent new races of the rust pathogen as they arise. The preinfection changes in the rust fungi that were reviewed here might be exploited to provide more stable plant protection. For example, there are two preinfection sensing responses that might be manipulated, i.e. orientation and differentiation. Further study of how the synthesis of the differentiation-response proteins is controlled, and of how the expression of their proteins is regulated, could tell us much more about how a parasite senses and responds to its host.

Acknowledgments

The research by the authors reviewed here was supported in part by grants DCB-8401409 and PCM-8315713 from the US National Science Foundation.

Literature Cited

1. Maclean, D.J. The Rust Fungi; Scott, K.J.; Chakravorty, A.K., Eds.; Academic: London and New York, 1982; pp 37-120.
2. Littlefield, L.J.; Heath, M.C. Ultrastructure of Rust Fungi; Academic: New York, 1979.
3. Dickinson, S. Phytopathol. Z. 1970, 69, 115-24.
4. Heath, I.B.; Heath, M.C. Can. J. Bot. 1979, 57, 1830-37.
5. Heath, M.C.; Heath, I.B. Can. J. Bot. 1978, 56, 648-61.
6. Maheshwari, R.; Hildebrandt, A.C.; Allen, P.J. Can. J. Bot. 1967, 45, 447-50.
7. Rowell, J.B.; Olien, C.R. Phytopathology 1957, 47, 650-55.
8. Wynn, W.K. Phytopathology 1976, 66, 136-46.
9. Bushnell, W.R.; Roelfs, A.J., Eds.; The Cereal Rusts; Academic: New York, 1984; Vol. I.
10. Bushnell, W.R.; Roelfs, A.P., Eds.; The Cereal Rusts; Academic: New York, 1985; Vol. II.
11. Emmett, R.W.; Parbery, D.G. Annu. Rev. Phytopathol. 1975, 13, 147-67.
12. Wynn, W.K.; Staples, R.C. Plant Disease Control: Resistance and Susceptibility; Staples, R.C.; Toenniessen, G.H., Eds.; Wiley Interscience: New York, 1981; pp 45-69.
13. Hoch, H.C.; Staples, R.C. Ann. Rev. Phytopath. 1987, 25, 231-47.
14. Lewis, B.G.; Day, J.R. Trans. Br. Mycol. Soc. 1972, 58, 139-45.
15. Mendgen, K. Naturwissenschaften 1982, 69, 502-03.
16. Macko, V.; Renwick, J.A.A.; Rissler, J.F. Science 1978, 199, 442-43.

17. Staples, R.C.; Grambow, H.-J.; Hoch, H.C. Exp. Mycol. 1983, 7, 40-46.
18. Tucker, B.E.; Hoch, H.C.; Staples, R.C. Protoplasma 1986, 135, 88-101.
19. Hoch, H.C.; Staples, R.C.; Whitehead, B.; Comeau, J.; Wolf, E.D. Science 1987, 235, 1659-62.
20. Staples, R.C.; Grambow, H.-J.; Hoch, H.C.; Wynn, W.K. Phytopathology 1983, 73, 1436-39.
21. Staples, R.C.; Gross, D.; Tiburzy, R.; Hoch, H.C.; Webb, W.W. Exp. Mycol. 1984, 8, 245-55.
22. Staples, R.C.; Yoder, O.C.; Hoch, H.C.; Epstein, L.; Bhairi, S. Biology and Molecular Biology of Plant-Pathogen Interactions; Bailey, J.A., Ed.; Springer-Verlag: London, 1986; pp 331-41.
23. Timberlake, W.E. Biology and Molecular Biology of Plant-Pathogen Interactions; Bailey, J., Ed.; Springer-Verlag: Berlin, 1986; pp 343-63.
24. Epstein, L.; Laccetti, L.; Staples, R.C.; Hoch, H.C.; Hoose, W.A. Phytopathology 1985, 75, 1073-76.
25. Epstein, L.; Laccetti, L.B.; Staples, R.C.; Hoch, H.C. Physiol. Mol. Plant Pathol. 1987, 30, 373-88.
26. Hoch, H.C.; Staples, R.C. Mycologia 1983, 75, 795-824.
27. Hoch, H.C.; Staples, R.C. Eur. J. Cell Biol. 1983, 32, 52-58.
28. Hoch, H.C.; Staples, R.C. Protoplasma 1985, 124, 112-22.
29. Hoch, H.C.; Staples, R.C. Exp. Mycol. 1984, 8, 37-46.
30. Moolenaar, W.H.; Kruijer, W.; Tilly, B.C.; Verlaan, I.; Bierman, A.J.; deLaat, S.W. Nature, 1986, 323, 171-73.
31. Laccetti, L.; Staples, R.C.; Hoch, H.C. Exp. Mycol. 1987, 11, 231-235.
32. Favre, B.; Turian, G. Plant Sci. 1987, 49, 15-21.
33. Berridge, M.J. Sci. Amer. 1985, 253, 142-52.
34. Staples, R.C.; Hoch, H.C.; Epstein, L.; Laccetti, L.; Hassouna, S. Can. J. Plant Pathol. 1985, 7, 314-22.
35. Bourett, T.; Hoch, H.C.; Staples, R.C. Mycologia 1987, 79, 540-49.

RECEIVED February 4, 1988

Chapter 7

Structural, Biogenetic, and Activity Studies on Metabolites of Fungal Pathogens

Albert Stoessl[1], Z. Abramowski[2], H. H. Lester[2], G. L. Rock[1], J. B. Stothers[3], G. H. N. Towers[2], and R. C. Zimmer[4]

[1]Agriculture Canada, Research Centre, London, Ontario N6A 5B7, Canada
[2]Department of Botany, University of British Columbia, Vancouver, British Columbia V6T 2B1, Canada
[3]Department of Chemistry, University of Western Ontario, London, Ontario N6A 5B7, Canada
[4]Agriculture Canada, Research Station, P.O. Box 3001, Morden, Manitoba R0G 1T0, Canada

Cercospora traversiana, a hitherto chemically unexamined fungal pathogen of fenugreek, produces substantial amounts of a diterpene aldehyde, traversianal, which was shown, mainly by ^{1}H- and ^{13}C-nmr, to have the carbon skeleton of a fusicoccane but the functionalities of the sesterterpenoid ophiobolanes. The assigned structure was confirmed by the ^{13}C-couplings of the compound isolated from cultures supplemented with $[1,2\text{-}^{13}C_2]$acetate. Further labelling studies, with $[2\text{-}^{2}H_3, 2\text{-}^{13}C_1]$acetate and $[2\text{-}^{2}H_3, 1\text{-}^{13}C_1]$acetate have indicated that the biosynthetic route to traversianal is unlike that to the fusicoccanes, but either involves or closely resembles that to the ophiobolanes. Although many fusicoccanes and ophiobolanes are important phytotoxins, traversianal showed only limited activity in this regard. Instead, it was found to have pronounced mycotoxic and molluscicidal properties. Comparison of its activity with that of dothistromin has provided more evidence that part, but only part of the toxicity of the latter is due to its photoactivity.

It is now widely accepted that phytotoxic secondary metabolites of fungi or bacteria may be causal factors in plant disease. This has been established with a high degree of certainty for an increasing array of host-specific or host-selective phytotoxins (1-5) but for the much more numerous non-specific phytotoxins (2-7), there are only relatively few instances for which correlations with disease rest on persuasive evidence. Two examples of this kind are the photodynamically active perylenequinone cercosporin (8, 9) and the linear difuroanthraquinone dothistromin (10). These two compounds, while structurally quite dissimilar, are further linked by the fact that both are metabolites of *Cercospora* spp. It is intriguing that, with the single known exception of *C. personata* (11, 12), pigment-producing *Cercospora* spp. generate either cercosporin and its isomers, or dothistromin together with small amounts of related com-

0097-6156/88/0379-0094$06.00/0

pounds, but not both types (12, 13). When one of us discovered a *C. traversiana* infection in a field of fenugreek (*Trigonella foenum-graecum* L.) (14), it was therefore of interest to determine to which of the two groups, if either, this species belonged, particularly since no chemical work at all appeared to have been recorded for this fungus.

As it turned out, the fungus did not produce any detectable pigment on the two or three culture media that were examined. Instead, it furnished substantial amounts of an easily isolatable, crystalline, colourless compound with a structure that called for biosynthetic and biological investigation.

Isolation and Structure of Traversianal

The compound, which we have named traversianal, crystallized directly on concentration of ethyl acetate extracts of potato-dextrose agar cultures of the fungus. Further amounts could be obtained by chromatography of the mother liquors over silica, leading to overall yields of about 100 mg/L after 21 days' growth. After recrystallization from benzene, traversianal melted at 234-5°C and had an optical rotation of $[\alpha]_D = -248°$. Microanalytical data and a high resolution mass spectrum established its empirical constitution as $C_{20}H_{28}O_3$. It could therefore be surmised that traversianal was a diterpene and this was substantiated by the ^{1}Hmr spectrum which contained resonances from one secondary and one tertiary methyl, and one isopropenyl group. An additional, biogenetic methyl group was present as an aldehyde group attached to a double bond and responsible for ir absorption at 1678 cm^{-1} as well as the uv chromophore of the compound (λ_{max} 234 nm, ϵ 13,000). A complete analysis of the spectrum was accomplished principally with the aid of 2D (COSY) spectra and a series of decoupling experiments, with the results summarized in Table I. The sum of this information led directly to part structures A, B, and C. The ^{13}Cmr spectra supported these inferences. The assignments of the pertinent resonances (see part structures) were substantiated by HETCOR experiments. In addition, the ^{13}Cmr spectra disclosed the presence of a quaternary carbinyl (79.1 ppm) and of a cyclopentanoid keto group (220.6 ppm) which also accounted for ir absorption at 1740 cm^{-1}. All 20 carbons and all functional groups of the molecule were therefore identified and from its molecular composition, it followed further that traversianal contained two rings in addition to that of the cyclopentanone residue. The chemical shifts of the protons, H-4α and H-4β, indicated that this methylene group (part structure B) was directly attached to the cyclopentanoid carbonyl. With the further assumption that no carbon-carbon rearrangements had intervened in the biosynthesis, the identified fragments could now be most plausibly assembled as in structure **1**.

To confirm this conclusion, traversianal was isolated from cultures of the fungus on liquid potato-dextrose medium supplemented with [1, 2-$^{13}C_2$]acetate, 10 mL 0.041 M solution being added to each 100 mL of cultures on days 10 and 13. The ^{13}Cmr spectrum of the compound isolated on day 16 showed that all carbons were highly enriched (4.4% specific incorporation), in a pattern that was in precise accord with expectation. The relevant parameters are indicated

Table I. 300 MHz ^{1}Hmr data for **1** in C_6D_6

	H-n	m[a]	J (Hz)	demonstrated couplings to:
8.74	H-18	s	-	-
6.22	H- 8	t	8.75	H-9α ; H-9β
4.86	H-16a	dt	2.5, 1.40	17-Me
4.80	H-16b	dt	2.5, 0.70	
3.20	H-9β	dt	11.9, 8.4	H-8, H-9α , H-10
2.91	OH	s	-	-
2.83	H-14	dt	11.5, 9.1	H-10, H-13α , H-13β , H-16b
2.66	H-4α	dd	19.6, 1.4	H-3, H-4β
2.40	H-2	ddd	12.5, 8.4, 3.9	H-1α , H-1β , H-3
2.33	H-4β	dd	19.6, 8.8	H-3, H-4α
1.96	H-9α	dd	11.9, 9.1	H-8, H-9β
1.80	H-3	[dt][b]	[7.7, 2.2][b]	H-2, H-4α , H-4β
1.75	H-13β	m	-	H-12α , H-12β , H-14
1.65	H-13α	[ddd][b]	[13.3, 13.3, 9.1][b]	
1.64	17-Me	dd	1.4, 0.7	H-16a, H-16b
1.35	H-1β	dd	14.4, 3.9	H-1α
1.32	H-10	m	-	-
1.27	H-12	m	-	-
1.25	H-12	m	-	-
1.13	20-Me	s	-	-
1.00	H-1α	dd	14.0, 12.2	H-2, H-1
0.68	19-Me	d	7.5	H-3

[a] apparent multiplicity; [b] assignment tentative

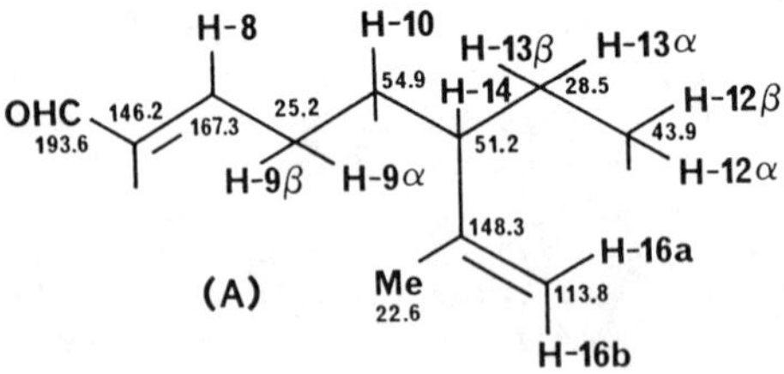

(A)

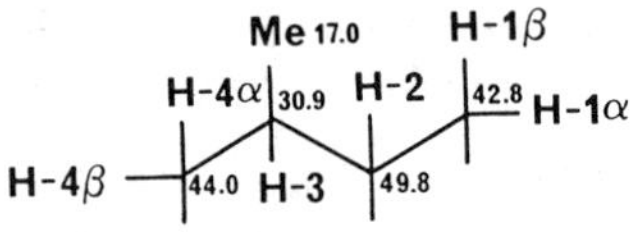

(B)

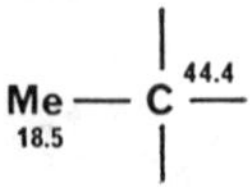

(C)

in **2**, where heavy bonds represent intact acetate residues, as identified by the spacings J of corresponding pairs of satellites, while the heavy dots represent enriched singlet carbons derived from C-2 of acetate by decarboxylation at the mevalonate stage of the biosynthesis. In addition to this evidence, the high enrichment level allowed a ^{13}C COSY experiment which revealed adventitious couplings of C-2 with C-3 and C-6, C-4 with C-3 and C-5, C-8 with C-7 and C-9, C-10 with C-11 and C-14, C-13 with C-12, and C-15 with C-14 and C-16.

Biosynthetic Studies

The planar structure **1** was therefore undoubtedly correct. Ignoring for the time being the stereochemistry, previously reported diterpenes which contain this tricyclic carbon framework comprise the fusicoccanes and cotylenins from Fusicoccum amygdali (15) and a Cladosporium (16) respectively, as represented by fusicoccin A (**3**), the fusicoplagins (17) and anadensin (18) from liverworts, as represented by fusicoplagin A (**4**), and several compounds such as roseanolone (**5**) and roseadione (**6**) from the higher plant Hypoestes rosea (19, 20). The same tricyclic carbon framework but with an extended side-chain is found also in the sesterterpenoid ophiobolins such as ophiobolin A (**7a**) from Cochliobolus (Helminthosporium) and Cephalosporium spp. (21, 22), ophiobolin G (**8**) from Aspergillus ustus (23, 24) and ceroplastol (**9**) and related compounds from the scale insect Ceroplastes albolineatus (25, 26).

The fusicoccins have been the subject of penetrating biosynthetic studies (27-29) which established the route shown in Scheme 1, of which the most remarkable and pertinent features are the loss of H-10 and the hydride shifts from C-14 to C-15, from C-2 to C-6, and from C-6 to C-7. In contrast, these protons do not migrate in the biosynthesis of those ophiobolins that have been investigated in this respect, in similarly elegant studies (30, 31). These also revealed a remarkable hydride shift but from C-8 to C-15 as in Scheme 2. We therefore set out to determine which, if either, of these two pathways was operative in the biosynthesis of **1**. Two feeding experiments were undertaken for this purpose. In the first, the culture medium was supplemented with $[2-^{2}H_{3}, ^{13}C_{1}]$acetate, using the same experimental protocol as earlier. Examination of the ^{13}Cmr spectrum (Figure 1) of the enriched traversianal that was isolated revealed clearly α-shifted resonances for C-18, C-8, C-16, C-14, C-2, C-4, C-12 and the three methyl groups but the C-10 resonance was obscured by that of the solvent (CD_2Cl_2). The data nevertheless were clearly incompatible with a fusicoccin-type biosynthetic sequence where hydride loss or migration would preclude the observation of α-shifts and deuterium couplings for C-2, C-10, and C-14. On the other hand, as far as it was observable, the labelling pattern was as expected for an ophiobolin-type pathway. In order to clarify the question of deuterium substitution at C-10 and also to obtain further evidence, a second experiment was undertaken in which the aldehyde was isolated from liquid cultures supplemented with $[2-^{2}H_{3}, 1-^{13}C_{1}]$acetate. The sample exhibited unambiguous β-shifts and couplings for C-15, C-7, C-11, C-1, C-3, C-13, and C-9 (Figure

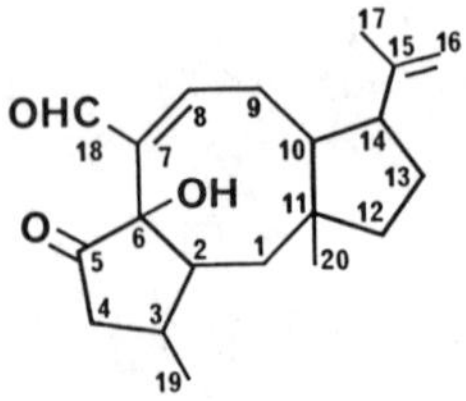

(1)

(2)

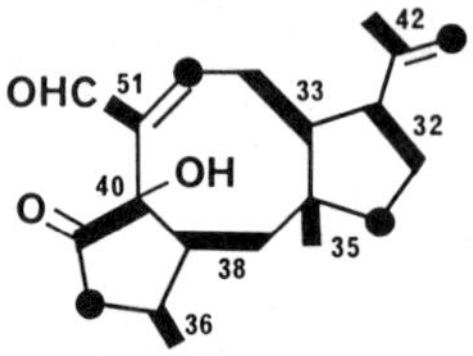

(3)

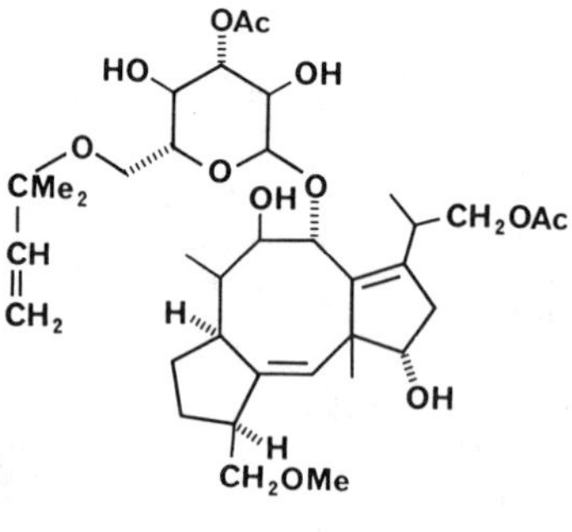

(4)

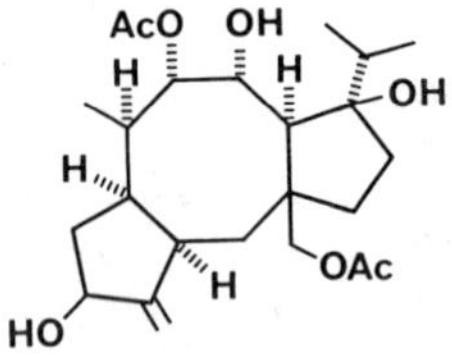

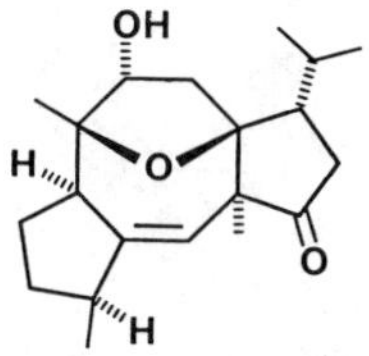

(5)

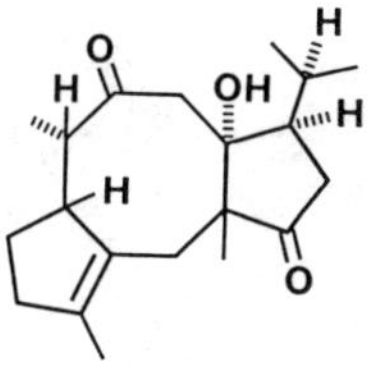

(6)

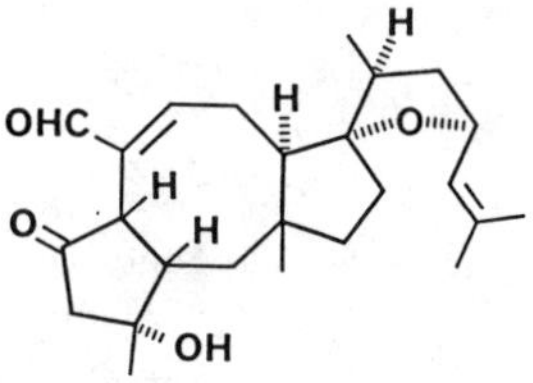

(7a)

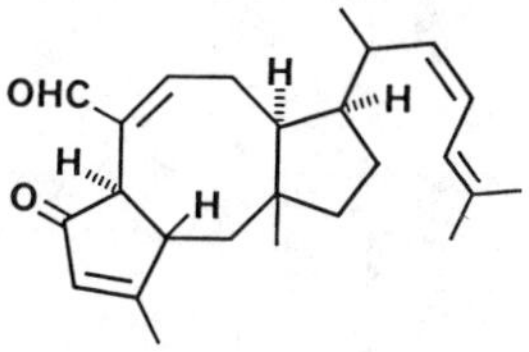

(8)

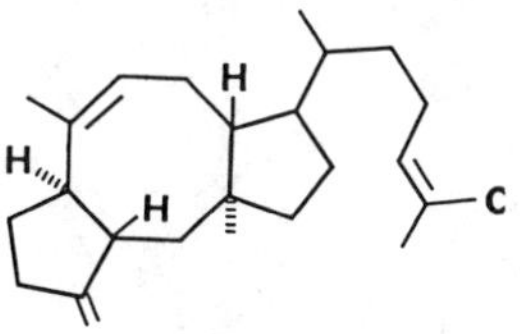

(9)

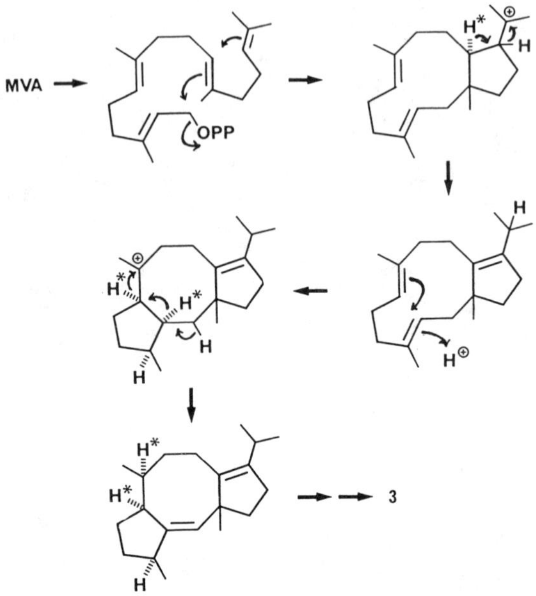

Scheme 1. Biosynthetic route to fusicoccin A.

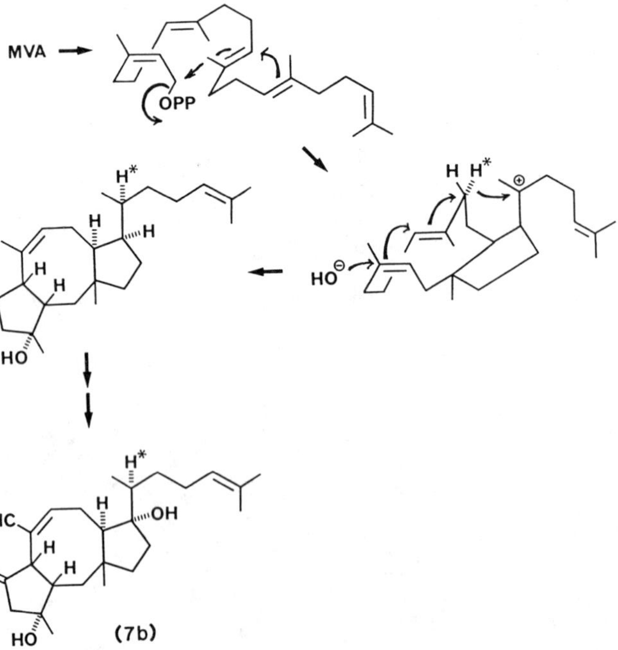

Scheme 2. Biosynthetic route to ophiobolin B.

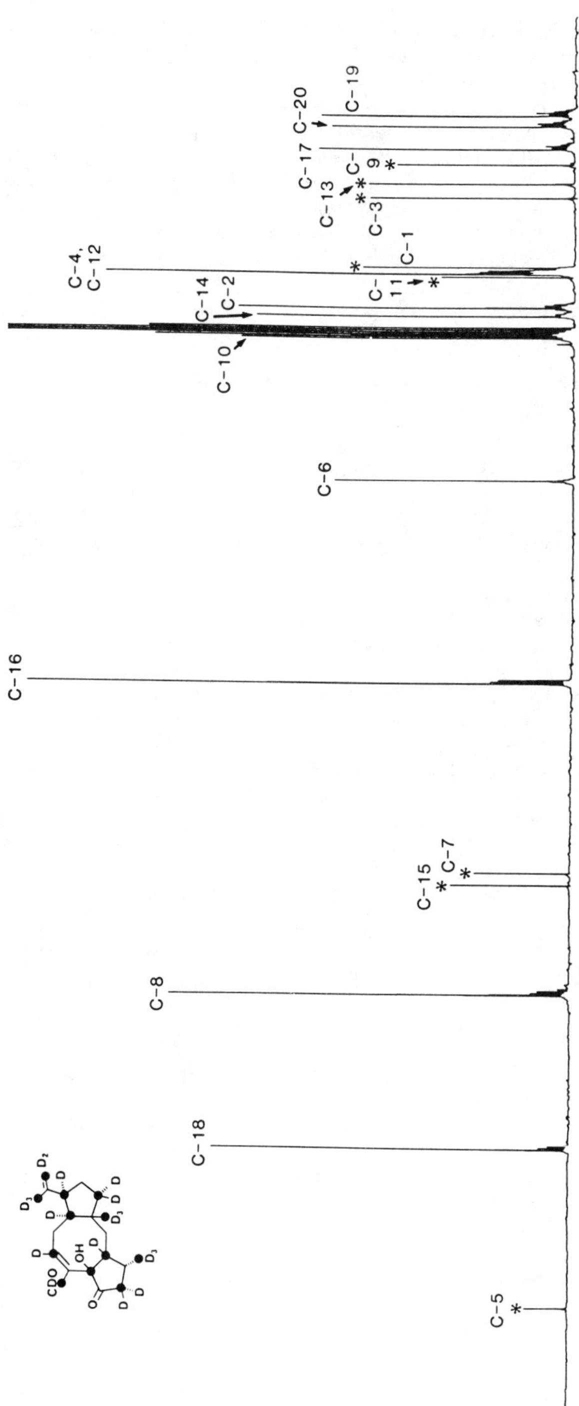

Figure 1. ^{13}Cmr spectrum of 1 enriched from [2-$^{2}H_3$, 2-$^{13}C_1$]-acetate (*denotes unenriched carbons).

2). The α-shift for C-1 is particularly pertinent because it establishes retention of deuterium at C-2, and that of C-9 demonstrates that C-10 was also labelled. Further information was obtained from the ^{2}Hmr spectrum of the sample in CH_2Cl_2 (Figure 3a), with chemical shifts and integrals corresponding to the presence of three ^{2}H at C-17, C-19, and C-20, two ^{2}H at C-4, C-12, and C-16, relative to one ^{2}H at C-2, C-8, C-10, C-14, and C-18. The ^{1}Hmr spectrum in CD_2Cl_2 (Figure 3b) is shown for comparison. All of these data were consistent with an ophiobolin-type pathway except that they could not reveal whether a C-8 to C-15 hydride shift had occurred at some intermediate stage. Further, it remained to be shown that the stereochemistry of 1 was in accord with the stereochemical consequences of the ophiobolin route (Scheme 2).

Stereochemistry and Biosynthetic Origin

The stereostructures of ophiobolin A (**7a**), ophiobolin G (**8**), and ceroplastol (**9**) have been ascertained by X-ray crystallography (21, 23, 25). It should be noted again that these three compounds were isolated from three different organisms and also differ from each other in stereochemical features. Traversianal (**1**) is analogous to ophiobolin B (**7b**) and hence also ophiobolin A (**7a**) (22), as well as to **8** and **9** , in that H-8 gives rise to a triplet due to nearly equal coupling (J 8.7 Hz) to both H-9α and H-9β. For **1** in CD_2Cl_2, the couplings of these protons to H-10 are 11.9 and 1.0 Hz respectively while that of H-10 to H-14 is 9 Hz. Unfortunately, corresponding data for **8** and **9** are not available from the literature. Nevertheless, from a consideration of Dreiding models, the data for **1** are consistent with α-configurations for both H-10 and H-14 as in **7** and **8.** A more difficult problem was the configuration of the C-3 methyl and the ring junctions of the cyclopentanone.Differential n.O.e. experiments were employed to settle these questions. They indicated that the methyl attached to C-11 was in spatial proximity to H-9β and H-2, hence both this methyl and H-2 are in β-configurations. The methyl attached to C-3 interacted with H-1_{ax} (δ 1.34, $J_{1,2}$ 12.3 Hz) and H-4_{equ} (δ 2.52, $J_{3,4}$ 2.9 Hz) but not with H-2, implying α-configurations for C-19, H-4_{equ}, and H-1_{ax}. This inference was in excellent accord with the observation of further n.O.e.'s between H-1_{ax} and H-10, and between H-10 and H-14, already assigned α-configurations. With the assumption of a cis-ring junction at C-6, the conformation depicted in **10** is consistent with all of these data and also the observed coupling constants for H-4β-H-3 (J, 7.9 Hz), H-3-H-2 (J, 7.9 Hz), and H-2-H-1 (J, 3.6 Hz). In contrast, as far as could be judged from Dreiding models, a trans-ring junction at C-6 (α- OH) does not allow any stable conformation that would be consistent with all of the observed couplings and n.O.e's.

The α-configuration assigned to the C-3 methyl of traversianal is opposite to that in the classical ophiobolins A-F and also in the recently reported 6-epiophobolin A (32), and suggests that the biosynthetic route to the compound is a different one. An obvious possibility is shown in Scheme 3 in which **1** is represented as a true diterpene. However, while this is its most probable origin,

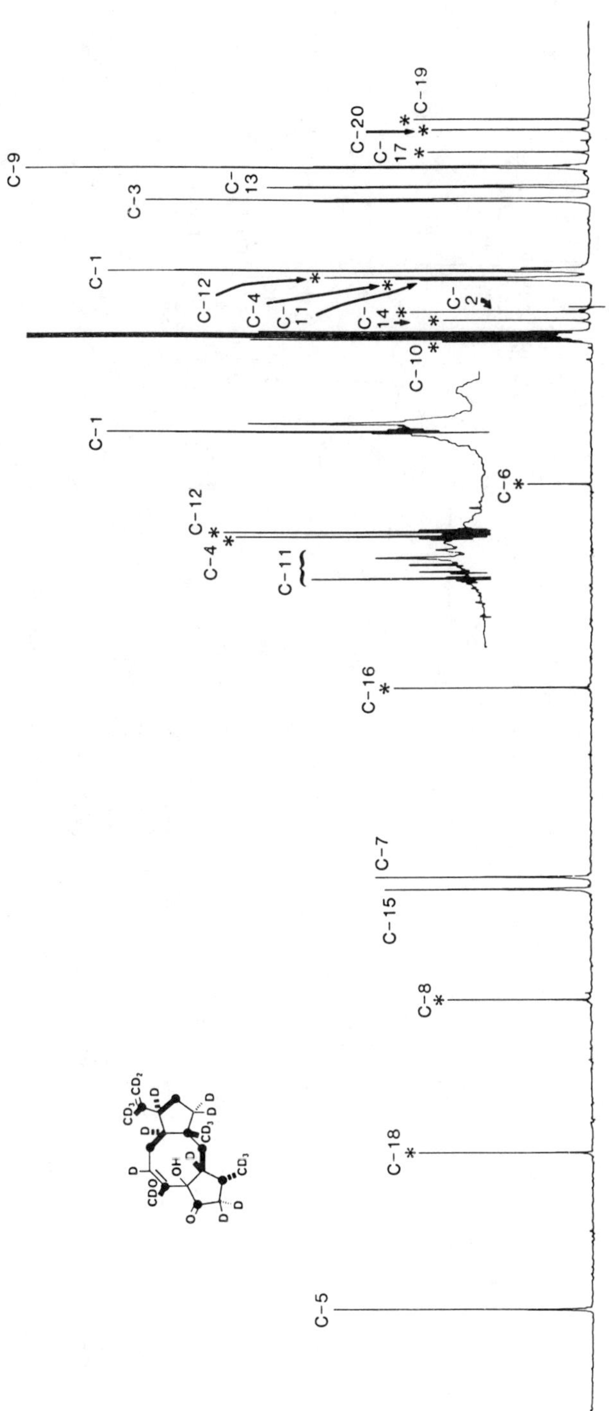

Figure 2. ^{13}Cmr spectrum of 1 enriched from [2-$^2H_3$, 1-$^{13}C_1$]-acetate (*denotes unenriched carbons).

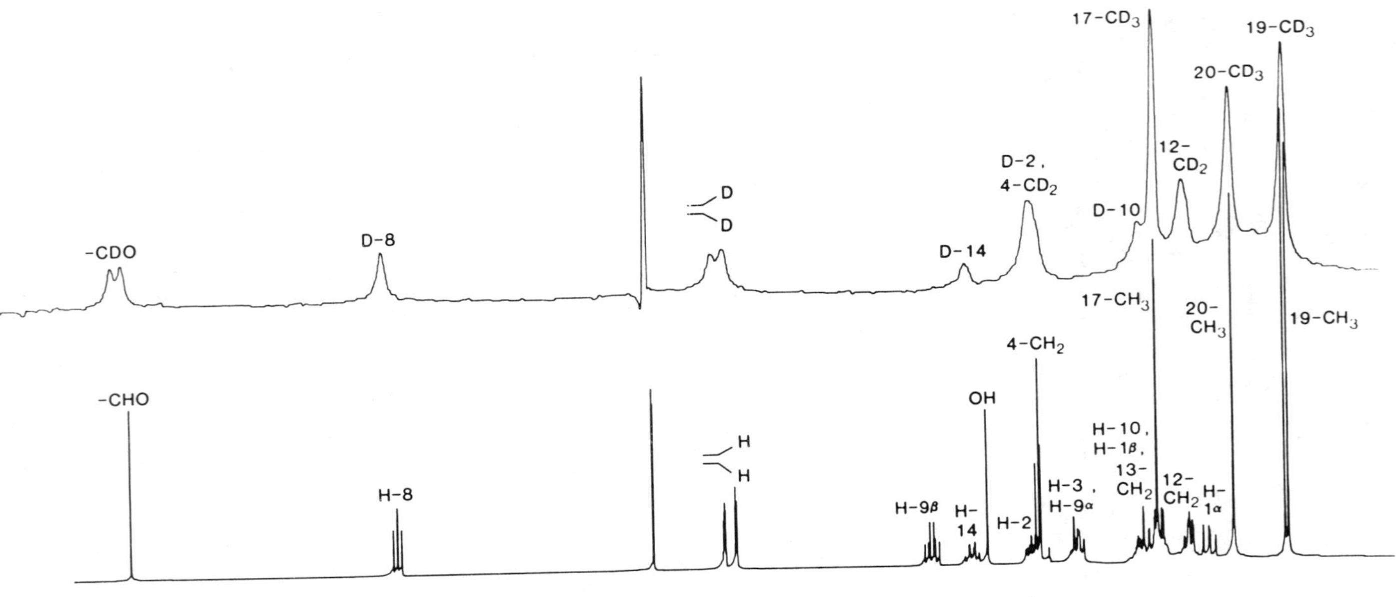

Figure 3a. ^{2}Hmr spectrum of ^{2}H-enriched **1** in CH_2Cl_2.

b. ^{1}Hmr spectrum of **1** in CD_2Cl_2.

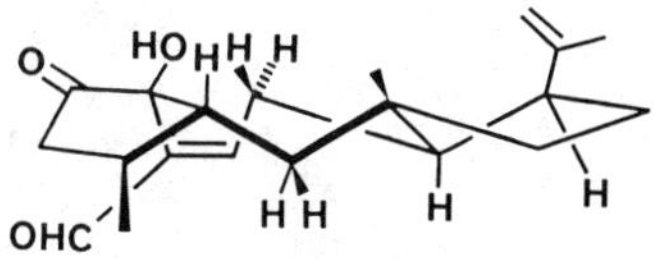

(10)

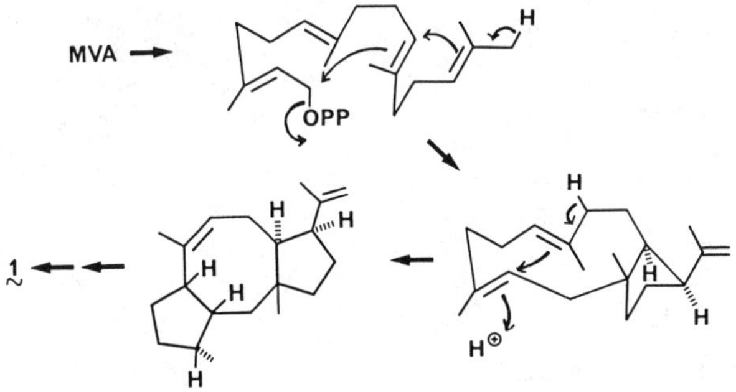

Scheme 3. Probable biosynthetic route to traversianal.

derivation by side-chain degradation of a sesterterpene cannot be entirely discounted.

Biological Activities of Traversianal and Dothistromin: A Comparison

Both the fusicoccins and the ophiobolins have attracted much attention as phytotoxins that may be involved in plant diseases (5, 6, 23, 24) and hence, because of its structural similarities to these compounds, it was clearly important to examine whether traversianal has similar biological properties. It was also of interest, from the outset, to compare the activity spectrum of traversianal with that of dothistromin. The latter compound was available from earlier work with *C. arachidicola* (33) and a comparison, it was thought, might lead to the discovery of host-specific effects. In general, by providing a better understanding of the processes of plant disease, studies of fungal phytotoxins are requisite as a basis for rational disease control measures (see also chapter by H. G. Cutler in this section). There is also a potential, already realized at least partially in several instances (e.g., bialaphos (7)) that naturally occurring phytotoxins can be utilized for weed control, either by direct application as herbicides, or by furnishing structure-activity clues for the synthesis of such compounds. The following is an account of our still ongoing studies on traversianal and dothistromin.

Neither compound caused wilt, chlorosis, or necrosis in standard tests for these effects with wheat, tomato, pea, peanut,or fenugreek. However, we noticed in these tests that dothistromin accumulated as a red zone at the cut ends of seedling stems that were immersed in its solution. The apparent absence of a toxic effect might therefore arise only from the failure of the compound to penetrate to critical sites. This could conceivably be the case also for traversianal and further work is needed to clarify the position.

Definite indications for phytotoxicity were obtained when beetroot discs were incubated (34) with solutions of either toxin. In both cases, there was considerable efflux of pigment (betacyanin) that was not observed for appropriate controls.

In another test for phytotoxicity, seed of wheat, radish, and tomato were incubated in the dark on filter paper moistened with 10^{-4} M solutions of traversianal. There was no inhibition of germination but radicle elongation was slightly stimulated in each instance. In subsequent tests, with pea and fenugreek, the seeds were allowed to germinate in the dark but exposed to light afterwards. Both traversianal and dothistromin were tested in this fashion. Results for the former were essentially the same as in the first series of tests, and dothistromin also induced slight root elongation for germinated pea seed exposed to light. Startlingly, however, for fenugreek, dothistromin caused almost complete inhibition of root elongation but only in the presence of light.

The phenomenon was strong evidence that dothistromin was photoactive, as observed previously by others (10, 35) and also in accord with results obtained by us from antibiosis tests. In these, dothistromin gave rise to clear inhibition zones, but only in the presence of light, when tested with a number of fungi, yeasts, and bacteria. It was inactive in most instances, and much less active in all others in darkness. Traversianal had no significant activity in any of these tests.

The third piece of evidence for the photoactivity of dothistromin came from haemolysis tests on human red blood corpuscles (36). These were lysed very readily (1 h) by 2.5 x 10^{-4} M dothistromin but only if both light and oxygen were present. Only very little lysis occurred (ca 10%) either in the dark or under nitrogen. Traversianal also was active in the haemolysis test, inducing 50-93% lysis at 5 x 10^{-5} M, but the effect developed only relatively slowly (16-24 h). Because of the absence of any chromophore above 300 nm, the activity of traversianal clearly cannot be light-dependent. Both compounds were further tested for toxicity to animal systems by the brine shrimp (*Artemia salina*) bioassay (37). Traversianal proved to be highly active, causing 70 - 90% mortality at 1.25, and 90 - 100% at 2.5 x 10^{-6} M. The activity of dothistromin was less pronounced but, significantly, appeared to be independent of the presence or absence of light.

Traversianal, though not dothistromin, was also tested for toxicity to snail (*Biomphalaria* sp.) and was found to be unusually toxic, causing 100% mortality at 3.2 x 10^{-5} M (10 p.p.m.) and 25% at 3.2 x 10^{-6} M. This activity level is equal to or greater than that observed for helenalin, the most active of several compounds tested previously in the UBC laboratory (38), as part of a program on potentially medicinally useful natural products. Molluscicides, particularly naturally occurring ones (39), are of much current interest as agents for the control of snails as vectors of schistosomiasis (40).

In summary, there are indications that traversianal is phytotoxic, and there is strong presumptive evidence that it might be a potent mycotoxin. It is clearly imperative that it should be further tested in the latter respect in animal systems, such as chick and mice, that were not available to us. It also deserves further study as a potent molluscicide.

Another matter of importance is the fact that dothistromin is photoactive in the presence of oxygen. This suggests firstly, that dothistromin might substitute for cercosporin as a virulence factor in the pathogenicity of some *Cercospora* spp. Secondly, to the extent to which its activity is light and oxygen dependent, its role must be that of a general and not a host-specific phytotoxin. It is therefore important to recall that the brine shrimp bioassay gave clear, and the beetroot experiments tentative, indications that dothistromin also exerts toxic effects that are not light-dependent. This too, will require much further study.

Biotechnological Aspects

As can be seen from the preceding section there is some potential that traversianal might be useful as a molluscicide in the control of schistosomiasis, or that it might give structural leads for the development of such agents. Other potential applications would be less direct. Thus, the structural and biogenetic differences between the various phytotoxic compounds should facilitate the identification of the responsible genes and eventually, the transfer of these genes from one fungal species to another. If nothing else, this would assist materially in determining whether the compounds are causally implicated in the pathogenicity of their progenitors. The differences in the biosynthetic pathways, e.g. between those to traversianal and dothistromin, or to traversianal and fusicoccin A, might then make it possible to develop agents that block specific steps on such routes, thereby affording practical, discriminating and hence safe control measures. Similarly, transfer of the genes for the synthesis of a non-specific phytotoxin, e.g. dothistromin, to a fungus that is host-specific to a noxious weed, would afford a biologically safe and effective herbicide.

Acknowledgments

It is a pleasure to thank V.M. Richardson and S. Wilson for measuring many of the nmr-, and D. Hairsine for the mass spectra. Financial support from the Natural Sciences and Engineering Research Council of Canada is gratefully acknowledged.

Literature Cited

1. Kono, Y., Knoche, H.W., Daly, J.M. In Toxins in Plant Disease; Durbin, R.D., Ed.; Academic: New York, 1981; Chapter 6.
2. Macko, V. In Toxins and Plant Pathogenesis; Daly, J.M.; Deverall, B.J., Eds.; Academic: Sydney, 1983; Chapter 2.
3. Scheffer, R.P.; Livingston, R.S. Science 1984, 223, 17-21.
4. Strobel, G.A. Annu. Rev. Biochem. 1982, 51, 309-33.
5. Ballio, A. In Toxins in Plant Disease; Durbin, R.D., Ed.; Academic: New York, 1981; Chapter 10.
6. Stoessl, A. In Toxins in Plant Disease; Durbin, R.D., Ed.; Academic: New York, 1981, Chapter 5.
7. Stoessl, A. In Handbook of Microbiology, 2nd Ed., Vol. 8. Laskin, A.I.; Lechevalier, H.A., Eds.; CRC Press: Boca Raton, 1987, Chapter 5.
8. Macri, F.; Vianello, A. Plant Cell Environ. 1979, 2, 267-71.
9. Daub, M.E.; Briggs, S.P. Plant Physiol. 1983, 71, 763-66.
10. Shain, L.; Franich, R.A. Physiol. Plant Pathol. 1981, 19, 49-55.
11. Ramanujam, M.P.; Swamy, R.M. Phytopathol. Mediterr. 1984, 23, 83-84.
12. Assante, G.; Locci, R.; Camarda, L.; Merlini, L.; Nasini, G. Phytochemistry 1977, 16, 243-47.
13. Lynch, F.J.; Geoghegan, M.J. Trans. Br. Mycol. Soc. 1977, 69, 496-97.
14. Zimmer, R.C. Can. Plant Dis. Survey 1984, 64, 33-35.
15. Ballio, A.; Brufani, M.; Casinovi, C.G.; Cerrini, S.; Fedeli, W.; Pellicciari, R.; Santurbano, B.; Vaciago, A. Experientia 1968, 24, 631-35.
16. Sassa, T.; Takahama, A.; Shindo, T. Agric. Biol. Chem. 1975, 39, 1729-34.
17. Hashimoto, T.; Tori, M.; Taira, Z.; Asakawa, Y. Tetrahedron Lett. 1985, 26, 6473-76.

18. Huneck, S.; Baxter, G.; Cameron, A.F.; Connolly, J.D.; Rycroft, D.S. Tetrahedron Lett. 1983, 24, 3787-88.
19. Okogun, J.I.; Adesomoju, A.A.; Adesida, G.A.; Lindner, H.J.; Habermehl, G. Z. Naturforsch. C: Biosci. 1982, 37c, 558-61.
20. Adesomoju, A.A.; Okogun, J.I.; Cava, M.P.; Carroll, P.J. Phytochemistry 1983, 22, 2535-36.
21. Nozoe, S.; Morisaki, M.; Tsuda, K.; Iitaka, Y.; Takahashi, N.; Tamura, S.; Ishibashi, K.; Shirasaka, M. J. Am. Chem. Soc. 1965, 87, 4968-70.
22. Canonica, L.; Fiecchi, A.; Galli Kienle, M.; Scala, A. Tetrahedron Lett. 1966, 1329-33.
23. Cutler, H.G.; Crumley, F.G.; Cox, R.H.; Springer, J.P.; Arrendale, R.F.; Cole, R.J.; Cole, P.D. J. Agric. Food Chem. 1984, 32, 778-82.
24. Cutler, H.G. In Bioregulators for Pest Control; Hedin, P.A., Ed.; ACS Symposium Series No. 276; American Chemical Society: Washington, D.C., 1985; pp. 455-68.
25. Iitaka, Y.; Watanabe, I.; Harrison, I.T.; Harrison, S. J. Am. Chem. Soc. 1968, 90, 1092-93.
26. Rios, T.; Gomez, F.G. Tetrahedron Lett. 1969, 2929-30.
27. Banerji, A.; Hunter, R.; Mellows, G.; Sim, K.-Y.; Barton, D.H. R. J. Chem. Soc. Chem. Commun. 1978, 843-45.
28. Barrow, K.D.; Jones, R.B.; Pemberton, P.W.; Phillips, L. J. Chem. Soc. Perkin Trans. 1, 1975, 1405-10.
29. Randazzo, G.; Evidente, A.; Capasso, R.; Colantuoni, F.; Tuttobello, L.; Ballio, A. Gazz. Chim. Ital. 1979, 109, 101-04.
30. Canonica, L. In Phytotoxins in Plant Diseases, Wood, R.K.S.; Ballio, A.; Graniti, A., Eds.; Academic: New York, 1972; Chapter 10.
31. Nozoe, S.; Morisaki, M. J. Chem. Soc. Chem. Commun. D. 1969, 1319-20.
32. Kim, J.-M.; Hyeon, S.-B.; Isogai, A.; Suzuki, A. Agric. Biol. Chem. 1984, 48, 803-05.
33. Stoessl, A. Can. J. Microbiol. 1985, 31, 129-33.
34. Chattopadhyay, A.K.; Samaddar, K.R. Physiol. Plant Pathol. 1976, 8, 131-39.
35. Youngman, R.J.; Elstner, E.F. In Oxygen Radicals in Chemistry and Biology, Bors, W.; Soran, M.; Tait, D., Eds.: Gruyter: Berlin, New York, 1984; pp. 501-08.
36. Mew, D.; Wat, Ch.-K.; Towers, G.H.N.; Levy, J.G. J. Immun. 1983, 130, 1473-77.
37. Meyer, B.N.; Ferrigni, N.R.; Putnam, J.E.; Jacobsen, L.B.; Nichols, D.E.; McLaughlin, J.L. Planta Med. 1982, 45, 31-34.
38. Marchant, Y.Y.; Balza, F.; Abeyesekera, B.F.; Towers, H.G.N. Biochem. Syst. Ecol. 1984, 12, 285-86.
39. Kloos, H.; McCullough, F.S. Planta Med. 1982, 46, 195-209.
40. The Control of Schistosomiasis, Report of a WHO Expert Committee, World Health Organization Technical Report Series 728, World Health Organization: Geneva, 1985.

RECEIVED February 14, 1988

Chapter 8

Insecticidal Metabolites from *Fusarium avenaceum*, a Fungus Associated with Foliage of *Abies balsamea* Infested by Spruce Budworm, *Choristoneura fumiferana*

George M. Strunz and Douglas B. Strongman

Canadian Forestry Service–Maritimes, P.O. Box 4000, Fredericton, New Brunswick E3B 5P7, Canada

Hyphae from cultures of *Fusarium avenaceum*, a fungus associated with spruce budworm-infested balsam fir foliage, were toxic to spruce budworm larvae when ingested. The toxic principle isolated from hyphal extracts was identified as enniatin complex, rich in enniatin A/A_1. A minor *F. avenaceum* product, ergosterol peroxide, caused damage to spruce budworm cell cultures, but did not display toxicity in feeding assays at the range of concentrations tested. Ecological implications of the presence of insecticidal fungi on the foliage of spruce budworm host-trees are addressed.

The spruce budworm, *Choristoneura fumiferana* (Clemens), is a phytophagous insect, whose principal hosts are balsam fir, *Abies balsamea* (L.) Mill., and three spruce, *Picea* species, which together dominate the coniferous forests of eastern North America. Populations of the insect periodically irrupt to epidemic proportions, resulting in widespread destruction of susceptible host trees, with major ecological and recently, economic consequences. The outbreaks are followed by dramatic declines of the insect populations, and there is evidence indicating that, at least during the last two centuries, these oscillations have occurred quite regularly with a frequency of 30-40 years (1, 2). A detailed understanding of the processes that cause the collapse of an epidemic spruce budworm population is a desideratum for both the ecologist and the forest manager, and intensive research on natural spruce budworm mortality factors is being conducted in several laboratories. The study described in this report originated in a suggestion that fungi colonizing the foliage of the spruce budworm's host trees might produce toxic metabolites, adversely affecting the insects and possibly contributing to the decline of an outbreak population (E. Eveleigh, T. Royama, Canadian Forestry Service - Maritimes; D.B. Strongman, N. Whitney, Univ. of New Brunswick, unpublished discussions, 1984). This hypothesis receives some support from a growing literature on plant-fungal interactions that result in protection of the plant against the depredation of phytophagous insects and other herbivores (3-6).

0097–6156/88/0379–0110$06.00/0
Published 1988 American Chemical Society

Essential ground-work for the investigation was laid by the 1985 study of Miller et al. (7), in which fungi were isolated both from foliage of budworm-infested balsam fir trees and from budworm-free trees. Although the presence of spruce budworm markedly affected the foliage mycoflora, more than doubling the number of fungal species isolated, the majority of isolates were identified as common colonizers of plant surfaces. However, approximately 6% of the isolates from the budworm-infested foliage were unusual needle phylloplane fungi that were not found on insect-free foliage. Significantly, some of the fungi in this group are toxigenic, entomopathogenic or both (7), and a detailed examination of some of the more active insecticidal isolates is presently in progress. Our observations to date concerning one of the group, Fusarium avenaceum (Corda ex Fr.) Sacc. (FA 120)[1] are described in this report.

Fusarium avenaceum has a global distribution as a pathogen of cereal crops and many other host plants, including conifers (8, 9). It is not however, considered to be an important disease organism in the spruce-fir forests of eastern North America (10) (Magasi, L.P., Canadian Forestry Service - Maritimes, personal communication.) The isolation of FA 120 from balsam fir foliage appears to be associated with the presence of spruce budworm larvae (7). The fungus is known to colonize insects, having been isolated from pupae of Oulema gallaeciana, a cereal leaf beetle (11). It has, furthermore been observed to grow saprophytically on spruce budworm larvae (Strongman, D.B., unpublished data, 1985).

After its selection in the initial screening process, FA 120 was grown in submerged culture on a medium containing glucose, yeast extract, and peptone. Incubation of the fungus under suitable conditions for 14 days afforded 8 g of hyphae (dry weight) per litre of culture medium. Hyphal matter was uniformly incorporated into a meridic insect diet (12) for bioassay, by grinding the two lyophilized materials together and then reconstituting the diet by addition of water (A.W. Thomas, Canadian Forestry Service - Maritimes, personal communication, 1985). Spruce budworm larvae, reared from the second instar for 14 days on diet containing 1% (wet weight) ground hyphae of FA 120 suffered more than 50% mortality, and surviving insects were enervated and retarded in their development relative to controls (of which only 7% died). An extract of the culture filtrate was also somewhat toxic to spruce budworm larvae, but this was not investigated further because the amount of material available was small and its activity modest.

Standard procedures were followed for isolation of the toxic principles from mycelium of FA 120. A methylene chloride extract of the freeze-dried hyphae was initially partitioned between hexane and aqueous methanol to separate lipids from more polar material. Bioassay-monitored chromatographic fractionation of the hexane-soluble material led to the isolation of a fraction (ca. 5% of the hyphal weight) which could account for much of the toxicity of the hyphae of FA 120 to spruce budworm larvae. The spectroscopic and chemical properties of this material were characteristic of the enniatins, a group of cyclic hexadepsipeptide ionophore antibiotics produced by several plant pathogenic Fusarium species, including F. lateritium

[1]A culture of FA 120 is deposited in the herbarium at the Biosystematics Research Institute, Ottawa, Ontario (DAOM #196490)

and F. sambucinum (13-18) (Strongman, D.W., Strunz, G.M., Giguère, P., Yu, C.M., and Calhoun, L., J. Chem. Ecol. In press, 1988). Whereas solvent partition left most of the enniatin fraction among the lipids, the balance of this material was found, with fewer co-metabolites, in the smaller aqueous methanol extract.

The enniatins are composed of three N-methyl-L-amino acid residues linked in an alternating fashion with three residues of D-2-hydroxyisovaleric acid to form an 18-membered cyclic structure **1**, in which the R groups may be Pr^i, Bu^s, or Bu^i. They are usually isolated as complexes which may contain enniatin A (1; $R^1=R^2=R^3=Bu^s$), A_1 (1; $R^1=R^2=Bu^s$, $R^3=Pr^i$), B_1 (1; $R^1=Bu^s$, $R^2=R^3=Pr^i$) and B (1; $R^1=R^2=R^3=Pr^i$) (19). The natural occurrence of an isomer of enniatin A, designated enniatin C (1; $R^1=R^2=R^3=Bu^i$), was inferred from the detection of N-methylleucine among the hydrolysis products of some

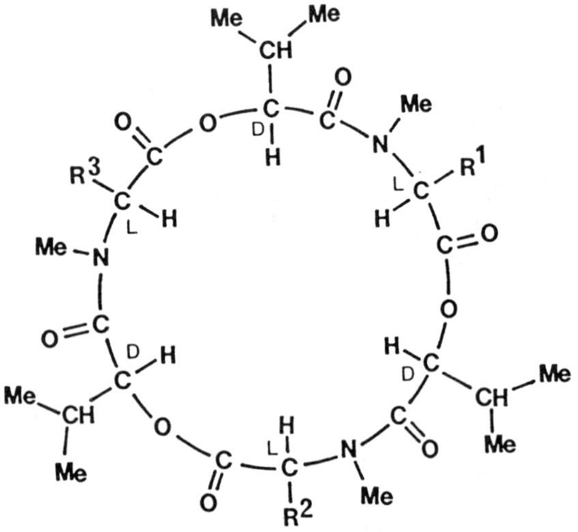

1

natural enniatin mixtures (15). Enniatin C, as well as other enniatins have been synthesized (20). Considering the difficulty of separating enniatin mixtures from natural sources into their individual components, the presence of N-methylleucine in the hydrolysates may suggest further the existence of a series of enniatins containing this amino acid in place of one or more N-methylisoleucine residues, (cf.15). Enniatin B has been obtained in a pure state from a species of Fusarium (13, 14), and a procedure was developed for separation of enniatin A from the complex produced by F. sambucinum (19). The resulting purified enniatin A was shown to contain both threo and erythro diastereomeric N-methylisoleucine residues (19).

When it is impractical or unnecessary to resolve the enniatin complex into its constituents, the approximate composition of the mixture may usually be estimated on the basis of mass spectrometry (21, 22) and hydrolysis experiments.

The mass spectrum of the enniatin complex from chromatography of the FA 120 extract showed peaks corresponding to the molecular ions of enniatins A, A_1, and B_1, with relative intensities of 3.90, 4.35, and 2.04%, respectively. A single crystallization from aqueous ethanol effected a significant change in the composition of the complex, manifested in the new values of 3.85, 1.44, and <1% for the corresponding relative intensities. The altered composition was also reflected in the relative intensities of the $C_6H_{14}N^+$ ion derived from N-methylisoleucine and N-methylleucine residues, and the $C_5H_{12}N^+$ ion from the N-methylvaline moiety.

The major peaks in the ^{13}C NMR spectrum of the crystallized material could be readily assigned to the carbon atoms of enniatin A, while smaller peaks signaled the presence, in lower concentration, of other enniatins, including one or more isomeric species containing the N-methylleucine residue (Strongman, D.W., Strunz, G.M., Giguère, P., Yu, C.M., and Calhoun, L., J. Chem. Ecol. In press, 1988).

Acid hydrolysis of the crystallized enniatin complex supported the spectroscopic data, affording as expected, D-2-hydroxyisovaleric acid and a mixture of N-methylisoleucine, N-methylleucine, and N-methylvaline, in which N-methylisoleucine was the predominant amino acid, and the other two were minor products.

It can be concluded from the combined data that enniatins A_1 and A are the principal components of the toxic fraction from FA 120, and that enniatin A becomes the predominant metabolite after crystallization of the complex. Among the metabolites isolated in previous studies of *F. avenaceum*, the only cyclohexadepsipeptides that appear to have been reported are enniatin B (23, 24) and 'avenacein' (16).

For the bioassays, the crystallized enniatin A-rich complex was uniformly incorporated into the insect diet at the desired concentrations, by dissolving it in methylene chloride, adding the powdered lyophilized diet, removing the solvent on a rotary vacuum evaporator, and reconstituting the diet as described above. At a concentration in the diet of 0.04% (wet weight), the material gave rise to toxic effects on spruce budworm larvae which were comparable to those produced by FA 120 hyphae incorporated at 1%. The mortality rate was 58%, whereas only 4% of the control insects died during the experiment. Surviving insects, like those in the bioassay of the hyphae, were debilitated and retarded in their development. Greatly reduced feeding by the test larvae was evident from the small amount of frass produced, and could reflect antifeedant effects in addition to the toxicity. The dose-response relationship for the enniatin A-rich complex is such that a tenfold reduction in concentration to 0.004% resulted in little evidence of toxicity, and at the conclusion of this assay, larval dry weight was not significantly different for test and control insects. The material at concentrations as low as 5 ppm did, however, cause damage to spruce budworm cells in culture (Clark, C., Univ. of New Brunswick, personal communication, 1986). The cell damage may result from the ability of these ionophores to affect ion transport across membranes (25, 26). The same property is believed to be responsible for their antibiotic activity (13, 25, 26). Enniatins also exhibit phytotoxicity, which could play a role in the plant injury associated with pathogenic *Fusarium* species (27).

Although some cyclodepsipeptides, such as beauvericin (28) and bassianolide (29) produced by *Beauveria bassiana* and other entomopathogenic fungi, are known to have insecticidal properties, only one previous report was found in which the effects of enniatins on insects were described (30). Grove and Pople (30) observed that enniatin A and enniatin complex from *F. lateritium* had moderate insectical activity when injected into adult *Calliophora erythrocephala* (blowflies) and *Aedes aegypti* (mosquito) larvae.

Toxic effects produced on a lepidopteran species by ingestion of enniatin complex do not appear to have been noted previously. With such properties, the enniatins could affect populations of spruce budworms in the field if produced in sufficient quantity by *F. avenaceum* on budworm-infested foliage. Definitive quantitative field data on enniatins and other fungal toxins associated with host-tree foliage, both insect-free and budworm infested, remain to be collected. Preliminary attempts to isolate enniatins from detached and autoclaved foliage of *Abies balsamea* inoculated with FA 120 were inconclusive.

Among the other metabolites present in the hyphal extract of FA 120, a minor product, identified as ergosterol peroxide, was also observed to produce toxic effects on spruce budworm cell cultures. This sterol can be formed in fungi from ergosterol by both photooxidative and enzymatic pathways (31, 32). Its wide distribution can be inferred from the common occurrence and abundance of ergosterol in filamentous fungi and yeasts (33). Ergosterol peroxide has recently been reported to be toxic to brine shrimp, bacteria, and chick embryos (34), and the possibility, suggested by the budworm cell culture assay, that it might also be toxic to insects was of considerable interest in the ecological context. However, notwithstanding the apparent manifestations of toxicity described above, no increased mortality was observed in spruce budworms, reared from the second to the sixth instar on a diet containing synthetic ergosterol peroxide (35) at concentrations substantially higher than those at which the enniatins showed insecticidal activity. It may be concluded that ergosterol peroxide does not make a major direct contribution to the observed toxicity of *F. avenaceum* hyphae to spruce budworm larvae, and that this property stems principally from the presence of the enniatin complex.

In summary, the concept that toxins produced by fungi associated with foliage can reduce predation by phytophagous insects and other herbivores (3-6) receives a preliminary measure of support from the properties of FA 120 and the enniatins observed to date. The presence on host tree foliage of *Fusarium avenaceum* and other fungi could contribute to spruce budworm mortality during collapse of an outbreak. The mycoflora on coniferous trees are, in principle, amenable to manipulation and it may be possible to exploit toxigenic fungi as control agents for the spruce budworm. Furthermore, the potential of microbial products, such as the avermectins (36) as insecticides is now widely accepted, and fungi toxic to the spruce budworm could provide a source of novel and useful insecticides. Future research will focus on collection of quantitative data on toxic metabolites from foliage in the field, and will include further screening of epiphytic and endophytic fungi from spruce budworm host trees.

Acknowledgments

We thank P. Giguère and C.M. Yu for technical assistance in the isolation and identification of the F. avenaceum metabolites. The ^{13}C NMR studies on the enniatins were expertly conducted by L. Calhoun.

Literature Cited

1. Greenbank, D.O. In The Dynamics of Epidemic Spruce Budworm Populations; Morris, R.F., Ed.; Ent. Soc. Can. Mem. 1963, 31, 19-23.
2. Royama, T., Ecol. Monogr. 1984, 54, 429.
3. Carroll, G.C. In Proceedings of the 4th International Symposium of the Phylloplane, Fokkema, N.J. and van den Heuvel, J., Eds.; Cambridge University Press, New York, N.Y, 1986.
4. Clay, K.; Hardy, T.N.; Hammond Jr, A.M.; Oecologia 1985, 66, 1.
5. Hinton, D.M.; Bacon, C.W. Can. J. Bot. 1985, 63, 36.
6. Siegal, M.R.; Latch, G.C.M.; Johnson, M.C. Phytopathology 1985, 69, 179.
7. Miller, J.D.; Strongman, D.; Whitney, N.J. Can. J. For. Res. 1985, 15, 896.
8. Booth, C. The genus Fusarium; Commonwealth Mycolog. Inst. Kew, England, 1971.
9. Marasas, W.F.O.; Nelson, P.E.; Toussoun, T.A. Toxigenic Fusarium Species: Identity and Mycotoxicology; Pennsylvania State University Press, University Park, 1984.
10. Davidson, A.G.; Prentice, R.M. Eds. Important Forest Insects and Diseases of Mutual Concern to Canada, the United States and Mexico, Department of Forestry and Rural Development, Canada (now Canadian Forestry Service) 1967.
11. Miczulski, B.; Machowicz - Stefaniak, Z. J. Invert. Path. 1977, 29, 386.
12. McMorran, A. Can. Ent. 1965, 97, 58.
13. Plattner, P.A.; Nager, U.; Boller, A. Helv. Chim. Acta. 1948, 31, 594.
14. Plattner, P.A.; Nager, U., ibid. 1948, 31, 665.
15. Plattner, P.A.; Nager, U., ibid. 1948, 31, 2203.
16. Cook, A.H.; Cox, S.F.; Farmer, T.H. J. Chem. Soc. 1948, 1022.
17. Shemyakin, M.M.; Ovchinnikov, Y.A.; Kiryushkin, A.A.; Ivanov, V.T. Izvest. Akad. Nauk SSSR., Ser. Khim. 1963, 1055.
18. Russell, D.W. Quart. Rev. Chem. Soc. 1966, 20, 559.
19. Audhya, T.K.; Russell, D.W. J. Chem. Soc. Perkin Trans. I 1974, 743.
20. Ovchinnikov, Y.A.; Ivanov, V.T.; Mikhaleva, I.I.; Shemyakin, M.M., Izvest. Akad, Nauk SSSR., Ser. Khim. 1964, 1823.
21. Wulfson, N.S.; Puchkov, V.A.; Bochkarev, V.N.; Rozinov, B.V.; Zyakoon, A.M.; Shemyakin, M.M.; Ovchinnikov, Y A; Ivanov, V.T.; Kiryushkin, A.A.; Vinogradova, E.I.; Feigina, M.Y.; Aldanova, N.A. Tetrahedron Lett 1964, 951.
22. Kiryushkin, A.A.; Rozynov, B.V.; Ovchinnikov, Y.A. Khim, Prir Soedin, 1968, 4, 182.

23. Tirunarayanan, M.O.; Sirsi, M. J. Indian Inst. Sci. 1957, 39, 1985.
24. Minasyan, A.E.; Chermenskii, D.N.; Ellanskaya, I.E. Mikrobiologiya 1978, 47, 67.
25. Shemyakin, M.M.; Vinogradova, E.I.; Feigina, M.Y.; Aldanova, N.A.; Loginova, N.F.; Ryabova, I.D.; Pavlenka, I.A. Experientia, 1965, 21, 548.
26. Shemyakin, M.M.; Ovchinnikov, Y.A.; Ivanov, V.T.; Kiryushkin, A.A.; Zhdanov, G.L.; Ryabova, I.D. Experientia, 1963, 19, 566.
27. Gaumann, E.; Naef-Roth, S.; Kern, H. Phytopathol. Zeit 1960, 40, 45.
28. Hamill, R.L.; Higgins, C.E.; Boaz, H.E.; Gorman, M. Tetrahedron Lett. 1969, 49, 4255.
29. Kanaoka, M.; Isogai, A.; Murakoshi, S.; Ichinoe, M.; Suzuki, A.; Tamura, S. Agric. Biol. Chem 1978, 42, 629.
30. Grove, J.F.; Pople, M. Mycopathologia, 1980, 70, 103.
31. Bates, M.L.; Reid, W.W.; White, J.D. J. Chem. Soc. Chem. Commun. 1976, 44.
32. Starratt, A.N. Phytochemistry, 1976, 15, 2002.
33. Turner, W.B.; Aldridge, D.C. Fungal metabolites II. Academic: New York, 1983.
34. Davis, N.D.; Cole, R.J.; Dorner, J.W.; Weete, J.D.; Backman, P.A.; Clark, E.M.; King, C.C.; Schmidt, S.P.; Diener, U.L. J. Agric. Food Chem., 1986, 34, 105.
35. Windaus, A.; Brunken, J. Justus Liebigs Ann. Chem. 1928, 460, 225.
36. Fisher, M.H.; Mrozik, H.; Abstr. Sixth International Congress of Pesticide Chemistry. IUPAC. Ottawa, Canada, 1986, 2F-06.

RECEIVED February 24, 1988

Chapter 9

Metabolites of Fungal Pathogens and Plant Resistance

J. David Miller and Roy Greenhalgh

Plant Research Centre, Agriculture Canada, Ottawa, Ontario K1A 0C6, Canada

The long term breeding of crops for yield and quality has substantially reduced their natural resistance to pests and pathogens, resulting in a dependency on the use of pesticides for the protection of crops. Biotechnology offers alternative strategies for crop protection. An approach thought to offer the best long term potential is one that improves disease resistance of crops. It is known that the production of phytotoxins by pathogenic bacteria and fungi are important determinants of disease and that an appreciable measure of resistance to a pathogen can be related to the ability of the plant to deal with phytotoxins. This general approach involving the characterization of the secondary metabolites produced by certain fungal pathogens and studying their phytotoxicity has been applied to the problem associated with the infection of cereals by *Fusarium graminearum*. A variety of secondary metabolites produced by this fungus have been isolated and their phytotoxicity determined with respect to a number of wheat cultivars. The results indicated a correlation between tolerance of cultivars to these secondary metabolites and the disease resistance. This work demonstrates the utility of this approach in terms of diagnosing resistant cultivars *in vitro* and determining the mechanisms involved.

This article will discuss the disease of wheat and corn (Fusarium head blight, maize ear rot) resulting from *Fusarium graminearum* infection. Recent discoveries have been made that are leading us to the identification of disease resistant germplasm in plants by the study of metabolites of this fungus and their effects on plant cells and *vice versa*. The integration of the disciplines of organic and analytical chemistry involving high resolution spectroscopic techniques with fermentation mycology and plant biotechnology has been very productive and lends itself to the

0097–6156/88/0379–0117$06.00/0

development of disease resistant crop varieties. In the past, the application of such sophisticated chemical techniques to plant disease-host interactions has rarely been done because plant pathologists and chemists have tended to ignore each others' efforts. In many cases, the traditional chemical solutions to crop protection problems (e.g. fungicides) have now become less acceptable to a number of groups in society and hence scientists responsible for developing resistance in plants may find it useful to apply their skills and resources in new ways.

Different strains of *Fusarium graminearum* are known to produce 30 to 40 secondary metabolites from different biogenic origins, including the trichothecenes deoxynivalenol, dihydroxycalonectrin and nivalenol, the sesquiterpenes sambucinol and culmorin (all mevalonate derived), zearalenone (polyketide derived), butenolide (amino acid derived), and fusarin C (unknown pathway). Most of these secondary metabolites are mycotoxins, some of which are found in wheat or corn, and are harmful to human and animal health (1,2). *Fusarium* disease and its associated toxins represent a worldwide problem, but especially in countries with north-temperate climates. In recent years, a large effort has been made to isolate and identify mycotoxins i.e. toxins harmful to humans and domestic animals (3,4,5) and to determine their effects principally on animal health (6). This work concludes that infections by *F. graminearum* must be minimized in order to produce crops that are sufficiently free of mycotoxins. Although several corn and wheat breeding programs exist in North America, China and other countries that are designed to produce very resistant cultivars or hybrids, however, to date none have succeeded.

FUNGAL PATHOGENS AND PLANT RESISTANCE

Wild plants, including ancestors of today's agricultural crops, generally possess good resistance to both pests and pathogens. It is generally assumed that pathogens evolved from a state of compatibility with the host rather than by acquiring incompatibility *de novo*. The mutation rate that drives this change resulting in a virulence condition, has been examined for a few species of fungal pathogens and it has been found to be modest. Plants have two types of genes that govern their responses to pathogens: (a) resistance genes and (b) so-called susceptibility genes. Resistance genes are identified using the virulence genes in the pathogen. Susceptibility genes are thought to be linked to, or are, useful genes for the health of the plant. Plant breeders realize that resistance genes should be sought in the geographic origin of the plant ("Vavilov's Rule"). Intensive breeding for yield normally increases the incidence of susceptibility genes (7). Another "biological fact of life" is that resistance genes require metabolic energy in terms of the synthesis of various gene products, the plant yield can be affected negatively. Similarly, the lack of susceptibility genes in wild plants contributes to 'yield' reduction. Evidence for these generalizations has come from intensive studies of a few host/pathogen combinations, especially the so-called biotrophic diseases (e.g. rusts) so that whatever models have been developed are largely based on a specific type of fungal host disease interaction. Thus, it is not known whether all the 'rules' about fungal disease of crop plants which were derived from the 'gene for gene' hypothesis (8) always apply.

With fungal diseases, it is clear that many genes govern the ability of a given strain to attack a plant. Vanderplank (7), from a breeding perspective noted, that "It is a common assumption in the literature to assume the existence of any two alleles for avirulence/virulence. A genotype of a pathogen is, on this assumption, either virulent for a particular resistance gene or it is not. This assumption is wrong. Within a pathogen, virulence is determined by multiple alleles having considerable variation". From a biochemical/plant pathology perspective, de Wit (9) stated that "Instances where resistances can be attributed solely to a single defense mechanism are very few; several responses may occur simultaneously or consecutively. Undoubtedly, many genes must be involved in the successful development of a parasite in its host and the relationships with host gene functions could be complex." Yoder and Turgeon (10) illustrate several examples where multiple pathogenecity factors are being analyzed from a molecular biology perspective. From a plant breeding point of view, efficacious resistance genes regulating biochemical events which control the attack of a given pathogen are the most desired.

Many, but not all aspects of the host pathogen interaction are amenable to further chemical study. These include: (A) elicitors - systems wherein molecules given off by the fungus induce the plant to produce new compounds or enzyme systems that restrict further growth of the pathogen e.g. phytoalexins, lignification, chitinases (9); chemistry is needed to purify and characterize the fungal products and the resulting plant products; (B) phytoalexin production by virulent strains of pathogens and degradation by resistant plants are known to occur in a variety of host/pathogen interactions (10,11). New research is needed to identify phytoalexins and their degradation products especially where the responsible pathogen genes are known (12). In both A and B, only the easy-to-identify molecules (by modern standards) have been adequately studied to date. (C) Phytotoxins.

PHYTOTOXINS REVISITED

The subject of phytotoxins is a contentious one in plant pathology. Only a generation ago, few scientists believed that phytotoxins played a role in pathogenesis. The discovery of a number of host specific phytotoxins has brought general acceptance to the notion that they are important in disease development but that non-specific phytotoxins are usually considered of secondary importance. Almost all the data we have on phytotoxins are from a few fungi and plant species. Several of the known phytotoxins are relatively complex and thus have proved to be difficult molecules to characterize. These problems among others, which are highlighted below, arguably conspire to devalue the importance of phytotoxins in plant diseases as well as the biochemical degradation of such compounds by resistant plant types. For the purposes of this discussion, it is useful to raise a few issues covered in Yoder's valuable review on toxins in pathogenesis (13).

1. HOST SPECIFICITY. Phytotoxin specificity for susceptible plants suggests a role in disease. Yoder argues that host specificity is not proof of their involvement in pathogenesis because a number of man-made chemicals as well as culture filtrates from saprophytic fungi also have apparent host specificity. Caution must be used here, in that as fungal metabolites share biogenic origins, the number of biochemical lesions possible in a plant is limited.

2. PHYTOTOXINS IN INFECTED PLANTS . Yoder notes that the presence or absence of toxins is insufficient evidence to determine if a fungal metabolite is a phytotoxin or not since "Inability to detect toxins in tissues may be due to their inactivation by host tissue, or their presence at concentrations below limits of detection". An example of the latter are the AM toxins I and II which induce necrosis in susceptible "Indo" apples at 0.1ppb (14,15). These points need to be considered in light of recent information on the physiology of toxin production by fungi *in vivo* (discussed below) and the power of analytical techniques.

3. INDUCTION OF TYPICAL DISEASE SYMPTOMS. Yoder notes that the production of visual or physiological (e.g. conductivity or respiration changes) symptoms are not reliable indicators of a compound's involvement in pathogenesis. It would be surprising if this were not so because fungi use multiple strategies to invade plant tissue. The application of a pure phytotoxin to a plant tissue cannot be considered evidence for involvement in pathogenesis, especially with respect to visual symptomology. Similarly, the absence of a plant tissue response to a putative phytotoxin is not necessarily evidence that a phytotoxin is not involved. It is necessary to examine both susceptible and resistant germplasm in order to discover the resistance genes.

Yoder (13) makes the additional comment that "virulence may not always be correlated with the quantity of toxin produced in culture" even when the toxin is known to play a role in pathogenesis. It is important to note that the production of either primary or secondary metabolites *in vitro* is a complicated problem as will be discussed later. Research to optimize metabolite production *in vitro* has seldom been carried out for phytotoxins. This renders any positive correlation between *in vitro* toxin production and virulence of a strain purely fortuitous. A poor correlation may simply indicate that suboptimal conditions for the production of the putative phytotoxins were used.

Reduced to the simplest case, fungal toxins are either primary or secondary metabolites, and all phytotoxins fall into one of these two categories in a physiological sense. Normally, wild type isolates of fungi do not produce large quantities of primary metabolites because - by definition - these compounds are required for the normal growth of the fungus. As a result almost all non-volatile toxins excreted by filamentous fungi are secondary metabolites. As noted earlier, there are few studies that suggest the type of experiments necessary to determine whether a phytotoxin is a true secondary metabolite. Some phytotoxins are definitely secondary metabolites but it is arguable that most are.

The strict definition of a fungal secondary metabolite is a compound that is produced as a result of a nutrient limitation. This means that a fungal cell, possessing the requisite genes, will excrete a secondary metabolite when it has grown, matured and is subjected to a limitation of a critical nutrient. In cultures grown in stirred jar fermentors, fungi have (A), first a rapid growth phase during which a synchronous population of cells are exposed to optimal nutrient and physical/chemical conditions, resulting in an increase in dry weight and the assembly of the necessary enzymes to produce one or more of the secondary metabolites controlled by the cell's genetic information followed by (B), exhaustion of some nutrient by the growth of the cells, which is required for primary metabolism. At this point the cells stop increasing in dry weight. During this time, when the cells are deprived of essential nutrients, e.g. nitrogen, the pools of primary metabolites such as acetate, mevalonolactone and/or amino acids accumulate. The cell's enzymes involved in secondary metabolism utilize these primary metabolites and the biosynthesis of the secondary metabolite commences and finally, at stage (C), the cells die (16, 17, 18).

The above steps in secondary metabolism have been known since the mid-1940's. It has also been known for a long time that certain fungal secondary metabolites end up in crop residues (e.g. *Fusarium* mycotoxins) although it has been difficult to correlate data from stirred jar fermentors studies with that from crop studies e.g. corn ears. Recently, however, Hale and Eaton (19) demonstrated that the characteristic cavities of wood soft rot fungi (mostly molds) are explained by the fact that filamentous fungi do not grow in a continuous fashion in solid substrates. Fungal growth was demonstrated to be oscillatory in nature and there are three stages of growth. A fine hyphal extension first grows out, growth stops, then the hyphae fatten and the cavity around the mycelium widens, during which time the nutrients around the hyphae become exhausted. The production of certain wood-degrading enzymes are known to be 'secondary' processes induced by nitrogen limitation (e.g. 20). Since hyphal growth is oscillatory in solids such as wood, this is comparable with our current understanding of the induction of 'secondary' biochemistries such as the production of certain degradative enzymes.

This growth pattern of fungi and the production of secondary metabolites in solid substrate appears to be correlated with the data from fermentor studies. In the 19th century, mycologists reported that growth and metabolic activity occured primarily in a few terminal cells of a hyphae (21). In molds such as *Fusarium*, nuclei occur most frequently in the terminal few cells. The production of the secondary metabolite zearalenone by *F. graminearum* in a solid fermentation has been shown to occur in localized segments of the mycelia which then stops (22) and the zearalenone metabolite diffuses into the substrate.

In our opinion, this information reconciles the *in vivo* and *in vitro* data concerning the production of secondary metabolites by filamentous fungi and suggests that a very serious look be taken at phytotoxins in plant disease. Yoder (13) notes that the problem of detecting a putative phytotoxin in infected plant tissue can be due to lability, inactivation of the compound or insensitive analytical procedures for detecting low concentrations. Additionally, the argument is made against a role of certain phytotoxins in plant disease is that their presence is merely coincidental with pathogenesis and is not responsible for pathogenicity. The likelihood that phytotoxins are produced as hyphal extension occurs raises serious issues regarding their detection and coincidence. In other words, the lability, detectability and turnover of phytotoxins in host tissue are real possibilities. Coincidence as an explanation for the presence of potent phytotoxins is not tenable since secondary metabolites, if they are produced at all, must be produced just after hyphal penetration of a new area of plant tissue.

FUSARIUM GRAMINEARUM METABOLITES IN CEREALS

Our studies on phytotoxins and plant disease started in 1982 when a 15-acetoxydeoxynivalenol producing strain of *F. graminearum* was used to inoculate corn. One parameter of the study was to monitor the occurrence of various mycotoxins in the infected ears weekly until harvest. Two remarkable observations were made: (A) very little 15-acetoxydeoxynivalenol was found, only high levels of deoxynivalenol and (B) when fungal growth stopped, there was a decline in the absolute amount of deoxynivalenol (23). Similar phenomena were observed in both experimental (24) and natural infections of wheat (25; A. Tiech, pers. com.). Analysis of experimentally infected spring cereals showed that susceptible wheat cultivars have a fungal biomass (ergosterol) to deoxynivalenol ratios of *ca.* 2:1; this ratio was also observed in heavy natural infections of Fusarium head blight of susceptible cultivars (26). By comparison, resistant cultivars experimentally inoculated with *F. graminearum* gave much higher ergosterol to deoxynivalenol ratios (27,28). These data imply that some cultivars of wheat can modify *Fusarium* toxins, a suggestion also made by Yoshizawa in 1975. This speculation has since been confirmed for a number of Fusarium head blight resistant cultivars while susceptible cultivars tested do not show this phenomena (29; Miller, unpublished).

Further experiments were initiated in which the effects of various metabolites produced by *F. graminearum* were evaluated using a coleoptile tissue assay devised by Cutler (30). These experiments demonstrated that the trichothecenes deoxynivalenol (XII) and 3-acetoxydeoxynivalenol (XI) are very phytotoxic to wheat tissue (28; Table I). Trichothecenes have been reported to affect plant tissue in various ways, especially increasing electrolyte loss (31,32). *Fusarium graminearum* is known to produce a large number of minor metabolites (Figure I) (3); some of these compounds such as dihydroxycalonectrin (VIII), butenolide (II) and sambucinol (XXII) were also tested using the coleoptile bioassay but were found to be less phytotoxic than the deoxynivalenol. The cyclic depsipeptides enniatin A and B, produced by related *Fusarium* species, were also found to be phytotoxic (see 33). For all the

Table I. Percentage inhibition of wheat coleoptile tissues by F. graminearum mycotoxins

Compound	Concentration (M)	Ankra	Belvedere	Casavant	Concorde	Dundas	Frontana	Su mei #3
deoxynivalenol (XI)	10^{-3}	83.3**	78.2**	93.7**	96.5**	61.9**	94.7**	93.6**
	10^{-4}	79.9**	72.0**	87.3**	84.7**	58.4**	86.7**	80.9**
	10^{-5}	64.6**	52.1**	62.9**	75.0**	42.0**	67.3**	70.0**
	10^{-6}	30.9**	19.8*	33.2**	19.0*	19.5*	36.7**	15.5
butenolide (II)	10^{-3}	37.2*	44.0**	80.9**	61.6**	76.0**	66.0**	51.1**
	10^{-4}	3.5	14.8	47.0**	48.0**	18.7	40.0**	29.3
	10^{-5}	3.5	12.1	44.4**	30.2**	5.4	26.0	23.7
	10^{-6}	15.8	17.0	24.4*	15.7**	8.6	21.3	25.2
culmorin (XIX)	10^{-3}	92.4**	94.5**	97.2**	98.4**	96.8**	96.9**	98.6**
	10^{-4}	6.2	2.8	57.8**	50.2**	7.5	13.9	32.2*
	10^{-5}	0	(5.5)	23.9	22.3**	1.1	15.5	16.8
	10^{-6}	(9.0)	2.8	11.3	25.2*	9.7	9.3	8.3
dihydroxycalonectrin	10^{-3}	55.9**	65.9**	80.3**	80.6**	40.1**	63.6**	67.8**
	10^{-4}	11.7	39.6**	47.9**	64.1**	50.3**	24.8	42.7**
	10^{-5}	(4.1)	2.2	40.9*	27.7**	19.8	11.6	16.0
	10^{-6}	2.7	17.0	12.7	28.9**	16.5	(20.1)	23.9
3-acetoxydeoxynivalenol	10^{-3}	98.7**	98.8**	95.0**	96.2**	97.1**	86.1**	87.6**
	10^{-4}	75.3**	77.3**	77.5**	86.2**	82.2**	41.6	49.7*
	10^{-5}	58.0**	37.7**	70.8**	60.0**	46.0**	(27.8)	21.5
	10^{-6}	6.7	11.4	26.4**	42.3**	40.2**	(5.6)	13.6
sambucinol (XXII)	10^{-3}	58.7**	56.3**	70.0**	58.5**	56.3**	38.1	57.8**
	10^{-4}	6.9	15.0	9.7	15.3	36.8*	4.8	10.7
	10^{-5}	10.6	4.2	2.8	16.2	14.9	(95.2)	0
	10^{-6}	(5.3)	3.5	11.1	9.2	22.4	(28.5)	(12.7)

1. Significantly different from control at p=0.05 (*) or p=0.01 (**) level.
2. Values in brackets refer to % increase in growth.

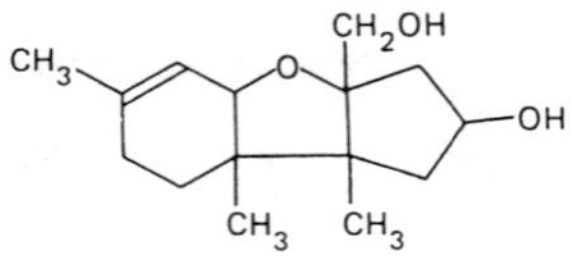

	R_1	R_2	R_3	R_4
III	H_2	H_2	OAc	OAc
IV	H_2	H_2	OAc	OH
V	H_2	H_2	OH	OAc
VI	H_2	OH, H	OAc	OAc
VII	OH, H	H_2	OAc	OAc
VIII	OH, H	OH, H	OAc	OAc
IX	H_2	H_2	OH	OH
XI	= O	OH, H	OH	OAc
XII	= O	OH, H	OH	OH
XIII	= O	OH, H	OAc	OAc
XIV	= O	H_2	OH	OAc
XV	H_2	H_2	H	OAc
XVI	H_2	OH, H	H	OAc
XVII	OH, H	H_2	H	OAc
XVIII	H_2	H_2	H	OH

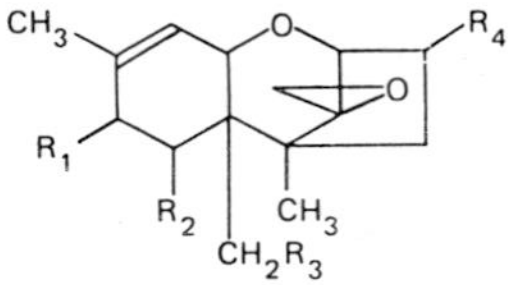

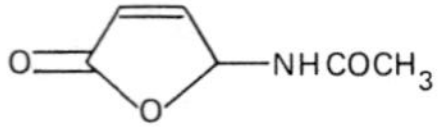

I

II

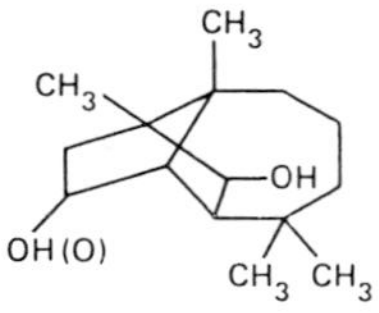

III – XVIII

XIX, XX

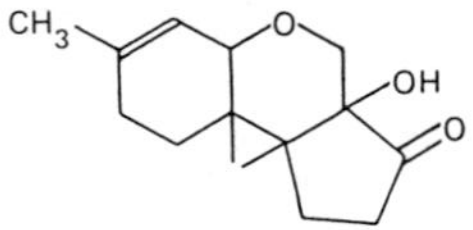

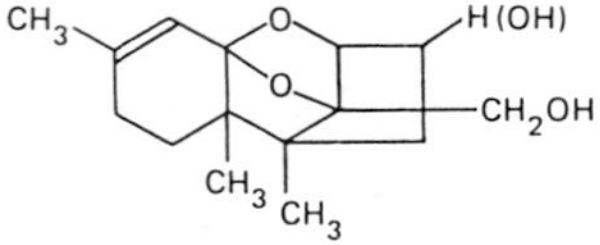

XXI, XXII

XXIII

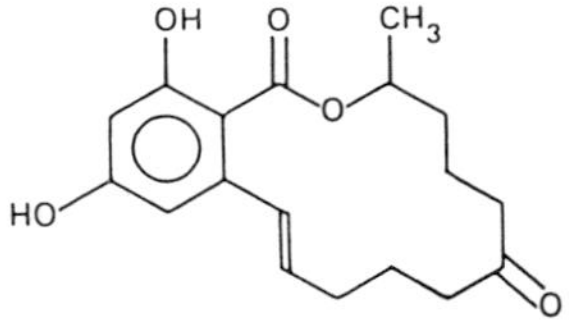

XXIV

Figure 1. Structures of some F. graminearum metabolites.

metabolites tested, cultivars known to be resistant to Fusarium head blight (e.g. Frontana, Su mei) were also more tolerant to the toxins compared to susceptible cultivars (Table II).

To understand the basis for this tolerance in resistant cultivars, a number of experiments have been carried out. *In vitro*, it has been possible to demonstrate conversion of trichothecenes (27) by a resistant cultivar. Cells of a resistant Chinese wheat cultivar were propagated in a 1.5 L Celligen fermentor (New Brunswick Scientific) and biosynthetically labelled ^{13}C 3-acetoxydeoxynivalenol (prepared after 4,18) was added to the culture. Analysis of the crude extract of the aqueous medium by ^{13}C-NMR showed 3-acetoxydeoxynivalenol together with another product postulated to be a glycoside (Figure 2). Approximately half of the labelled compound was recovered after a one week fermentation. The formation of other metabolites is currently being studied by the addition of ^{14}C-labelled deoxynivalenol to resistant and susceptible cultivars (prepared after 29).

Similar work has been reported with another trichothecene - producing plant pathogen, *Myrothecium roridum*, which is a pathogen of muskmelon. When the trichothecene roridin E was added to detached leaves of different melon varieties, the size of lesions produced were identical to leaves on plants inoculated with the pathogen; varieties resistant to the toxin were resistant to the pathogen. This trichothecene also significantly increased sporulation and lesion size on inoculated leaves. Closely related trichothecenes from non-pathogenic *Myrothecium* strains did not affect sporulation, lesion size or conductivity of leaf tissue after application to leaf tissue. The authors stated that the data does not totally resolve the issue of whether the trichothecene roridin E is involved in pathogenesis but suggests that it may play a role in predisposing tissue to infection (34,35). Cutler (loc. cit.) discusses other aspects of the possible involvement of trichothecenes in plant disease and their site of action. The invasion of corn kernels by *F. graminearum* viewed by histology demonstrates that the corn cells are killed in advance of the invading mycelium (D. Wicklow, pers. com.).

The tests required to discern absolutely the role of trichothecenes in virulence or pathogenesis have yet to be carried out. The strictest test involving mutants with or without a gene that controls production of these toxins will be difficult to do since it would also depend on the existence of the incompatibility system in the plant. Our studies have shown that in wheat, all susceptible cultivars tested are equally susceptible to *F. graminearum* trichothecenes. These cultivars would probably be equally affected by a trichothecene non-producing strain because a number of incompatibility/ compatibility interactions are known (36). Cultivars, very tolerant to trichothecenes by virtue of degradation and other responses, should also to be used to test the hypothesis that toxins are involved in pathogenecity.

Strains of *F. graminearum* from Canada, China, Europe, Japan, South Africa and the USA produce 3- or 15-acetoxydeoxynivalenol and usually deoxynivalenol. It may be adaptive for 'generalist pathogens' like *F. graminearum*, which attacks many plant species, to employ potent general phytotoxins such as the trichothecenes

Table II Tests for Fusarium head blight resistance of spring wheats

Cultivar	FHB[1] Index %	Ergosterol (ppm) / kernel	Deoxynivalenol (ppm) / kernel	*in vitro*[2] Test Score
Frontana (Ste.Hy)	17.5	15.8	2.7	15
Su mei #3	12.5	13.4	2.1	13
Ankra	100.0	157.6	84.9	5
Belvedere	100.0	194.3	32.3	4
Dundas	100.0	191.8	61.9	-2
Casavant	92.5	169.7	114.4	-6
Concorde	100.0	229.0	90.4	-9

[1] Plots inoculated by spraying conidia on the heads at anthesis.

[2] Wheat coleoptile tissue test; values assigned based on the minimum concentrations of *F. graminearum* metabolites that resulted in significant decreases (P=0.05 or 0.01) in growth of the tissue after 20h exposure. Values assigned as follows: deoxynivalenol and 3 acetyldeoxynivalenol (10^{-3}M=15, 10^{-4}M=10, 10^{-5}M=5, 10^{-6}M=0) and dihydroxycalonectrin, butenolide, culmorin and sambucinol (10^{-3}M=1, 10^{-3}M=0, 10^{-4}M=-1, 10^{-5}M=-2, 10^{-6}M=-3).

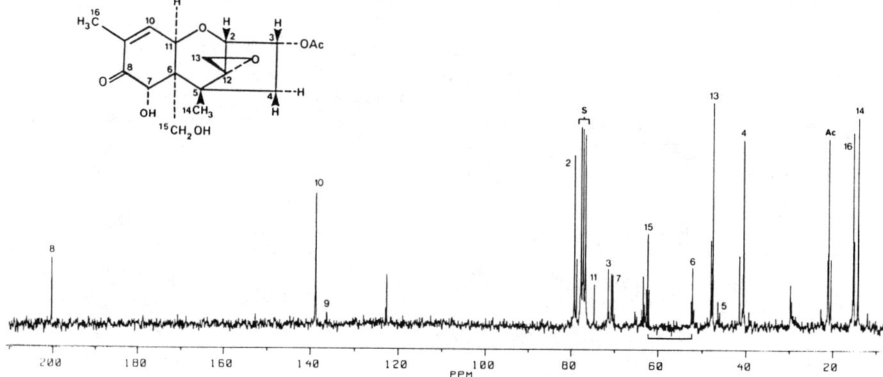

Figure 2. ^{13}C NMR spectrum of crude fungal extract showing the presence of 3-acetyldeoxynivalenol and putative glycoside. Pure ^{13}C labelled 3-acetoxydeoxynivalenol (20 mg) was added to 1L of a cell suspension of wheat (Su mei) in an N.B. Scientific Celligen fermentor for 14 days. The medium was filtered and the filtrate extracted with ethyl acetate. The solvent was removed and the residues dissolved in $CDCl_3$.

rather than evolve a whole series of host specific toxins. Plant types that can tolerate these potent inhibitors of protein synthesis that are produced during fungal invasion should be expected to better resist the pathogen. New proteins can be synthesized as the fungus invades the plant that result in the synthesis of phytoalexins, chitinases, etc.

SUMMARY

These experiments have identified a number of approaches involving the techniques of chemistry and biotechnology (tissue culture, molecular biology) that are helping to identify disease resistant germplasm. Rapid screening techniques based upon plant tissue or suspension culture response to the toxins from *Fusarium* are proving to be a useful way to verify the resistance of advanced breeding material (37). The enzymatic basis for the deoxynivalenol degradation phenomenon by cell culture is also being explored. This approach should lead to the identification of the genetic basis of response, with a view to both the selection for and addition to the development agronomically of improved cultivars (see 37).

ACKNOWLEDGMENT

The authors wish to thank Dr. B.A. Blackwell for the ^{13}C-NMR spectral analysis, and Drs. George Fedak and John Simmonds for assistance with wheat tissue culture.

LITERATURE CITED

1. Foster, B.C.; Trenholm, H.L.; Friend, D.W.; Thompson, B.K.; Hartin, K.E. *Can. J. Anim. Sci.* 1986, *66*, 1149.
2. Hesseltine, C.W. In P.S. Steyn and R. Vleggar (Eds), *Mycotoxins and Phycotoxins*. Elsevier Science Publishers, Amsterdam. 1986, 1.
3. Greenhalgh, R.; Meier, R.M.; Blackwell, B.A.; Miller, J.D.; Taylor, A.; ApSimon, J.W.. *J. Agric. Fd. Chem.* 1984, *32*, 1261.
4. Greenhalgh, R.; Levandier, D.; Adams, W.; Miller, J.D.; Blackwell, B.A.; McAlees, A.; Taylor, A. *J. Agric. Fd. Chem.* 1986. *34*, 98.
5. Greenhalgh, R.; Meier, R.M.; Blackwell, B.A.; Miller, J.D.; Taylor, A.; ApSimon, J.W. *J. Agric. Fd. Chem.* 1986. *34*, 115.
6. Trenholm, H.L.; Hamilton, R.M.G.; Friend, D.W.; Thompson, B.K.; Hartin, K.W. *J. Am. Vet. Med. Assoc.* 1984. *185*, 527.
7. Vanderplank, J.E. *Disease resistance in plants*. Academic Press, New York. 1984.
8. Flor, H.H. *Ann. Rev. Phytopathol.* 1971, *9*, 275.
9. deWit, P.J.G.M. In J. Bailey (Ed), *Biology and Molecular Biology of plant pathogen interactions*. Springer-Verlag Berlin, 1986, 149.
10. Yoder, O.C., Turgeon, D.G. *Gene manipulations in fungi*. Academic Press, New York, 1985, 417.
11. Jost, J.D. *Phytopathol. Z.* 1965, *54*, 338.

12. Yoder, O.E.; Turgeon, B.G. In T.W.E. Timberlake (Ed), Molecular genetics of filamentous fungi. A.R. Liss, Inc. New York, 1985, 383.
13. Yoder, O.C. Ann. Rev. Phytopathol. 1980, 18, 103.
14. Veno, T.; Nakashima, T.; Hayashi, Y.; Fukami, H. Agri. Biol. Chem. 1975. 39, 1115.
15. Veno, T.; Nakashima, T.; Hayashi, Y.; Fukami, H. Agric. Biol. Chem. 1975. 39, 2081.
16. Anon. Protection against trichothecene mycotoxins, National Academy Press, Washington, D.C. 1983, 26.
17. BuLock, J.D. In J.E. Smith and D.A. Berry (Eds), The filamentous fungi, Vol. 1, Academic Press, New York, 1975, 33.
18. Miller, J.D.; Blackwell, B.A. Can. J. Bot., 1986, 64, 1.
19. Hale, M.D.; Eaton, R.A. Trans. Bro. Mycol. Soc. 1985, 84, 277.
20. Keyser, P.; Kirk, T.K.; Zeikus, J.G. J. Bact. 1978, 135, 790.
21. Burnett; J.H. Fundamentals of mycology, Edward Arnold, London, 1976, 61.
22. Morita, H.; Singh, J.; Fulcher, R.G. Can. J. Plant Pathol. 1984, 6, 179.
23. Miller, J.D.; Young, J.C.; Trenholm, H.L. Can. J. Bot. 1983, 61, 3080.
24. Miller, J.D.; Young, J.C. Can. J. Plant Pathol. 1985, 7, 132.
25. Scott, P.M.; Nelson, K.; Kanhere, S.R.; Karpinski, K.F.; Hayward, S.; Neish, G.A.; Teich, A.H. Appl. Environ. Microbiol. 1984, 48, 884.
26. Love, G.R.; Seitz, L.M. Cereal Chem. 1987, 62, 124.
27. Miller, J.D.; Young, J.C.; Sampson, D.R. Phytopathol. Z. 1985, 113, 354.
28. Wang, Y.Z.; Miller, J.D. J. Phytopathol. 1987, in press.
29. Miller, J.D.; Arnison, P.G. Can. J. Plant Pathol. 1986, 8, 147.
30. Cutler, H.G.; Jarvis, B.B. Environ. Expt. Bot. 1985, 25, 115.
31. Brian, P.W.; Dawkins, W.; Grave, J.F.; Aemming, H.G.; Lowe, D.; Norvis, G.L.F. J. Exp. Bot. 1961, 12, 1.
32. Jacobellis, N.S.; Bottalico, A. Phytopath. Medit. 1981, 20, 129.
33. Drysdale, R.B. In M.O. Moss and J.E. Smith (Eds), The applied mycology of Fusarium. Cambridge University Press. 1984, 95.
34. Kuti, J.O.; Ng, T.J.; Bean, G.A. 1987. Hort Science. 1987, 22, 635-637.
35. Kuti, J.O. Genetic and biochemical studies of resistance in muksmelon (Cucumis melo L.) to Myrothecium roridum Tode. ex. Fries and its trichothecene mycotoxin roridin E. Ph.D. thesis, University of Maryland, 1987.
36. Schroeder, H.W.; Christensen, J.J. Phytopathology, 1963, 53, 831.
37. Wang, Y.Z.; Miller, J.D. Proc. CIMMYT conference on tropical wheat production, Jan. 1987, Thailand, CIMMYT, Mexico.
38. Hall, R. Can. J. Plant Pathol. 1987, 9, 152.

RECEIVED March 8, 1988

CONTROL OF INSECTS

Advances in Insect Control

Biotechnological advances in the crop sciences have a strong counterpart in research on insects. The metazoan most intensely studied from a genetic standpoint is the fly *Drosophila melanogaster* (Meigan). Molecular genetics research has been very productive with this insect, and builds on many decades of work dealing with classical genetics. Most spectacular is the ability to insert genes into the germline of *D. melanogaster*. This potential, if realized with pest species, will make it possible to prepare sterile, single-sex populations for sterile-release programs, or to enhance the effectiveness of beneficial insects. Additionally, the potential for altering insect-specific viruses genetically is at hand, and may make it possible to engineer more effective viral insecticides. The techniques of molecular biology also make it possible to isolate and manipulate behaviorally, structurally, and developmentally significant proteins and peptides and to determine the potential of these molecules (or their antagonists) for biotechnological approaches to insect control.

In recent years the fly *Drosophila melanogaster* (Meigan), the organism of choice for studies in classical genetics for over half a century, also has become the eukaryote of choice for many investigations in molecular biology. This remarkable insect has provided generations of geneticists with seemingly endless possibilities for research on genetic regulation. It should have come as no surprise then, that a few years ago it became possible in *Drosophila* to inject a gene directly into an embryo and produce a genetically altered fly. Rubin and Spradling (1), in their 1982 paper in Science, revolutionized the approach to research on insect molecular genetics. They injected a DNA fragment containing the *rosy* gene and a transposable genetic element, the P element, into fly embryos and achieved stable gene integration and effective gene therapy by yielding flies with wild-type *rosy* eye color from mutant flies lacking the gene. The success of this process requires the efficient integration of the injected DNA, without rearrangement, into the germline cells so that successive generations of the transformed flies retain the injected gene.

Interest in *Drosophila* for molecular genetic studies has increased substantially because of research on transposable genetic elements, as well as other advances in understanding the nature of genetic control and gene structure. A *Drosophila* conference held in May, 1987 attracted *ca*. 900 scientists, a very large turn-out indeed for specialists interested in a single species of insect.

The question addressed in this symposium is whether the exciting genetic mainpulations that have appended advances in molecular biology, in general, and molecular genetics of *Drosophila* in particular, can be applied to insects of economic importance. Perspectives on this issue can be found in articles published in 1984 by Schneiderman (2) and by Cockburn et al. (3), and in Kirschbaum's (4) review in 1985. Schneiderman points out the possibilities for genetic-engineering of crop plants for insect resistance. Kirschbaum (4) focuses particularly on genetic alteration of anti-insect microbes and discusses *Bacillus thuringiensis* as a significant first target for this approach. In fact, his suggestion of integration of the *B. thuringiensis* endotoxin gene into plants by genetic transformation was later described in detail by Marvel (5). In this system a natural soil bacterium, *Agrobacterium tumefaciens*, contains a plasmid which successfully transmits a foreign gene into a host plant. Although plant genetic engineering is a very important approach, it is the task of this series of papers to focus on applications of biotechnology to insects *per se*.

The potential for propagating genetically altered pest insects and releasing them into the environment is discussed in an article by Cockburn et al. (3). Sterile insect release strategies have been used for many years in connection with screwworm fly eradication and control of fruit flies. The new techniques of molecular genetics may make it possible to increase effectiveness of the released insects, as well as to cut insect production costs in half through sex selection schemes. Potentially, genetically engineered sterilization could replace or augment radiation induced sterility, and may produce sterile flies that are more competitive in the field than irradiated flies. In the present symposium, the use of molecular genetics to obtain sex-specific selection will be considered by Shirk et al. (this volume).

Molecular techniques may be used to investigate neuropeptides that control diverse processes in insects. Ultimately, as pointed out by Keeley (this volume) in his paper, inhibitors of neuropeptide-controlled processes, or perhaps genetically enhanced microbial insecticides, could take advantage of this knowledge for new options in pest control.

In the third paper, Kramer (this volume) addresses some new developments in the biochemistry of insect cuticle. The cuticle, of course, remains a critical insect-specific system that should provide a vulnerable target for biotechnological pest control strategies. The inhibition of chitin synthesis is only one possibility for disrupting the formation of insect cuticle. An enhanced understanding of the structure and biochemistry of insect cuticle is needed to uncover other potential approaches to insect control.

In his 1985 article, Kirschbaum (4) pointed out the importance of understanding insect defenses against microbial infection so that more effective microbial insecticides can be designed. In this connection Postlethwait (this volume) discusses exciting new developments concerning insect anti-microbial defense systems. He shows that in fruit flies bacterial inoculation induces the production of proteins in the insect hemolymph which may kill the bacteria. Thus, it may be possible to manipulate the insect's immune system so that more effective means may be devloped to utilize microbial infections to control insects.

In the final chapter in this section Knipple (this volume) discusses the use of molecular genetics to understand resistance by insects to important chemical insecticides. Collectively, these chapters point out both recent progress as well as future directions in biotechnology research with insect pests. Manifestly, the interactions between biotechnology and entomology will be synergistic and will lead to new concepts and practices in the management of insect pests.

Literature Cited

1. Rubin, G. M. and A. C. Spradling. 1982. Genetic Transformation of Drosophila with Transposable Element Vectors. Science 218: 348-353.
2. Schneiderman, H. A. 1984. What entomology has in store for biotechnology. Bull. Entomol. Soc. Am. 30: 55-61.
3. Cockburn, A. F., A. J. Howells, and M. J. Whitten. 1984. Recombinant DNA technology and genetic control of pest insects. Biotech. and Genet. Engr. Rev. 2: 69-99.
4. Kirschbaum, J. B. 1985. Potential implications of genetic engineering and other biotechnologies to insect control. Ann. Rev. Entomol. 30: 51-70.
5. Marvel, J. T. 1985. Biotechnology in crop improvement. In P. A. Hedin [ed.], Bioregulators for Pest Control. ACS Symp. Series #276, 34: 477-510.

H. Oberlander
Insect Attractants, Behavior, and Basic Biology Laboratory
Agricultural Research Service
U.S. Department of Agriculture
Gainesville, FL 32604

RECEIVED February 4, 1988

Chapter 10

Sex-Specific Selection Using Chimeric Genes

Applications to Sterile Insect Release

Paul D. Shirk, David A. O'Brochta, P. Elaine Roberts, and Alfred M. Handler

Insect Attractants, Behavior, and Basic Biology Research Laboratory, Agricultural Research Service, U.S. Department of Agriculture, Gainesville, FL 32604

Application of recombinant DNA technology to isolate, manipulate, and transfer genetic material to a pest insect may offer an efficient and cost effective means of achieving sex-specific selection of males for use in sterile insect release programs. Central to the implementation of genetic modification using recombinant DNA technology is the ability to efficiently transfer and integrate genes into the genome of a pest insect. An embryonic excision assay was developed to assess the potential of utilizing a P-element transposable vector for genetic transfer in insects other than _Drosophila melanogaster_ (Meigen). When tested in various _Drosophila_ species, the excision assay indicated normal P-element function in embryos of _D. melanogaster_, _D. simulans_, and _D. grimshawi_. However, excision events were not observed in embryos of the Caribbean fruit fly, _Anastrepha suspensa_ (Loew), or the Indianmeal moth, _Plodia interpunctella_ (Hübner). Once gene transfer techniques have been established for an insect, conditional sex-specific selection of males can be achieved by transforming the insects with chimeric genes that impart sex-specific sensitivity. The proposed structures of chimeric genes that may be useful in genetic sexing schemes will be presented. The chimeric genes will consist of promoter sequences from sex-specific genes such as yolk protein genes and will either express structural genes that impart chemical sensitivities to the females or produce antisense sequences to sex determining genes to disrupt development of females.

The combination of restrictions on the use of agrochemicals, the costs of developing and registering new chemicals, and the development of chemical resistance by pest insects has placed farmers and agribusiness in a position whereby they must rely to an ever increasing extent on biorational methods as supplements to chemical measures to achieve effective insect pest management. The greatest successes in achieving biological control of pest insects have appeared in the use of autocidal programs. The sterile insect technique has been an effective means of limiting dipteran populations for nearly 20 years and has been examined as a means of controlling all pest insects (1). Although the sterile insect technique can be used very effectively to control pest insects, as evidenced by the screwworm eradication program (1), sterile insect release (SIR) in its present form is not applicable to all pest insects. Because lepidopterous insects require greater doses of radiation than dipterans to achieve sterility, the use of substerilizing dosages has been considered. Although release of insects receiving substerilizing doses has an impact on a population, larger releases of insects would be necessary to achieve effective control of a pest and would add considerably to the cost of the program (1).

A major limitation to SIR programs is the requirement for the availability of large numbers of reproductively competent sterile males (2). Because there is no efficient mechanism for the specific selection of males, the current SIR programs rear, sterilize, and release both sexes. The presence of the females results in added costs to the programs for rearing and in reduced efficiency of sterile male matings after release. Implementation of improved procedures for the selection of males or removal of females early in development would offer measures necessary to reduce the costs and increase the efficiency of SIR programs.

A number of classical genetic manipulations theoretically applicable to any insect have been developed in the fruit fly *Drosophila melanogaster* (Meigen) and several other insects that result in breeding populations that produces only one viable sex. Typically, the selection relies on the disruption of normal sex ratios by induced or naturally occurring mutants of the sex determination genes, chromosomal translocations, maternal effect lethals, or aberrant chromosomes (i.e. compound-X). However, these methods have not been generally applicable to agriculturally important insects because there is a paucity of significant genetic information on these insects. Ultimately, it has been the lack of basic genetic information that has precluded the development of genetic selection schemes applicable to SIR programs.

One scheme utilizing conventional genetic techniques, which has been developed for use in the control of the Mediterranean fruit fly, *Ceratitis capitata* Wiedemann, has not been completely successful (3). A portion of the chromosome carrying the gene for wild type pupal color was attached to the Y chromosome, which is carried only in males. The wild type color-attached Y chromosome was then mated into a white strain of flies. This composite mutant strain is being tested to determine the efficacy of utilizing a color marker for selecting males under large scale rearing conditions. While similar schemes may eventually prove useful, the preliminary experiments with the pigmentation marker revealed a problem with the breakdown of the marker chromosome by recombination (3). In addition to the breakdown of the marker chromosome, similar genetic manipulations in *Drosophila* routinely result in decreased viability of the mutant strain when compared with wild type strains (4).

Although most recent proposals for utilizing recombinant DNA technology in insects have focused on measures that would allow for direct genetic control of pest insects (5, 6), the potential for developing effective genetic-sexing procedures for SIR programs through application of recombinant DNA technology is being explored as a more immediate solution to the problem of SIR efficiency. The technology of molecular biology offers the ability to construct chimeric genes that encode for a readily selectable gene-product that is expressed in a specific sex, stage, or tissue (7). Once introduced into the host genome, a sex-specifically expressed chimeric gene would offer a means of selecting for a desired sex. For example, combining a gene for a chemical or drug resistance with a male-specific promoter would confer resistance to males and females could be selected against when using a chemical treatment. Alternatively, females could be selected against by utilizing a chimeric gene containing a structural gene that imparts chemical sensitivity with a female-specific promoter that again would make the females lethally sensitive to chemical treatment. Chimeric gene constructs that result in chemical sensitivity in females are being produced and tested in D. melanogaster. However, application of these genetic selection systems to insects other than Drosophila requires the ability to introduce these genes efficiently into the genomes of pest insects, something that has not been achieved.

P-Element Excision Assay in Insect Embryos

The only efficient means of introducing genes stably into the genome of an insect is the modified transposable P-element vector system from D. melanogaster developed by Spradling and Rubin (8,9). The P-elements were found in P strains of D. melanogaster and were lacking in M strains. The presence of the transposable element could be demonstrated by the development of hybrid dysgenesis due to the movement of the P-element when males of a P strain were mated to females of an M strain (10). The DNA sequence of the P-element was cloned and inserted into a bacterial plasmid (8). When the P-element plasmid is injected into embryos of M strain flies, i.e. those flies that do not contain P-elements, the P-element transposes from the plasmid and stably integrates into the genome of the host fly. The ability to identify genetically transformed Drosophila initially depended upon "rescuing" a mutant host fly with a copy of a wild type gene carried by a P-element vector (9). Although genetic rescue can be useful in a well defined genetic organism such as D. melanogaster, the P-element vector required a more generally selectable element to identify transformed flies. The inclusion of the neomycin phosphotransferase gene linked with a heat shock promoter (hspneo) in the P-element vector supplies a dominant marker for selection of germline transformants (11). Other laboratories have utilized the hspneo P-element vector to test the ability of P-elements to effect germline transformation in a variety of insects without success. However, failure to impart neomycin resistance to an insect does not give an adequate assessment of P-element transposability in an insect or a direct test of the utility of the P-element as a transformation vector in that insect. A simple more direct assessment of P-element functionality in heterologous germline transformation systems therefore is essential.

P-elements have been used as gene vectors in the closely related species D. simulans (12). In addition, the P-elements were found to undergo transposition in the more distantly related species D. hawaiiensis, although the capacity to act as a gene vector could not be demonstrated in this species (13). However, the transposable functionality of P-elements has not been demonstrated for insects outside the genus Drosophila.

Insertion of a P-element into a gene often results in the partial or complete inactivation of the gene and precise or reading-frame-conserved excision of the P-element restores gene function. In D. melanogaster, P-element excision appears to rely on the same enzymatic activity of the P-element transposase as does P-element insertion, and the excision activity is not genomic site dependent but extends as well to P-element sequences in plasmids (14). The excision of a P-element from a gene that results in restoration of gene activity is a means of monitoring the functionality of the P-element vector system, and forms the basis of an assay to test P-element activity in preblastoderm insect embryos.

The P-element excision assay was a modification of the assay used by Rio et al. (14) to assess P-element activity in tissue culture cells. The plasmid pISP (14) (Fig. 1) contains an internally deleted P-element that interrupts the lacZ alpha peptide coding region of the plasmid pUC8 and was used to monitor P-element excision. The pISP plasmid was co-injected into preblastoderm embryos of each species with the plasmid pUChsπΔ2-3, which was used as a helper plasmid to provide a source of P-element transposase by heat shock (15). Precise or reading-frame-conserved excision of the P-element from a pISP plasmid restores lacZ alpha peptide complementing ability to the plasmid, which can be determined easily by transforming the recovered plasmids into a lacZ$^-$ E. coli.

The excision assay plasmid and the helper plasmid were introduced into preblastoderm embryos of the various flies using the well established embryo injection procedures developed for D. melanogaster (8). The fly eggs were dechorionated in dilute bleach, immersed under oil, injected at 22°C, and incubated under O_2 for 15-18 hr. However, the injection procedures were modified for the embryos of the Indianmeal moth because they did not survive dechorionation or immersion in oil. The Indianmeal moth embryos, 1-2 hr old, were injected without dechorionation close to the micropyle and sealed with two layers of Krazy® glue. The embryos were incubated under O_2 for either 20 or 44 hr at 30°C. After incubating all embryos in O_2, both dipteran and lepidopteran embryos were heat shocked at 37°C for 1 hr to induce the P-element transposase activity. The flies were allowed to recover for 1 hr at 22°C and the moths for 2 hr at 30°C, under O_2. The embryos were collected and low molecular weight DNA,

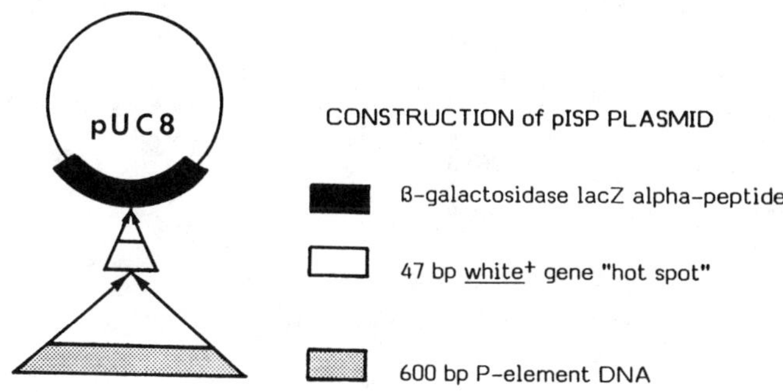

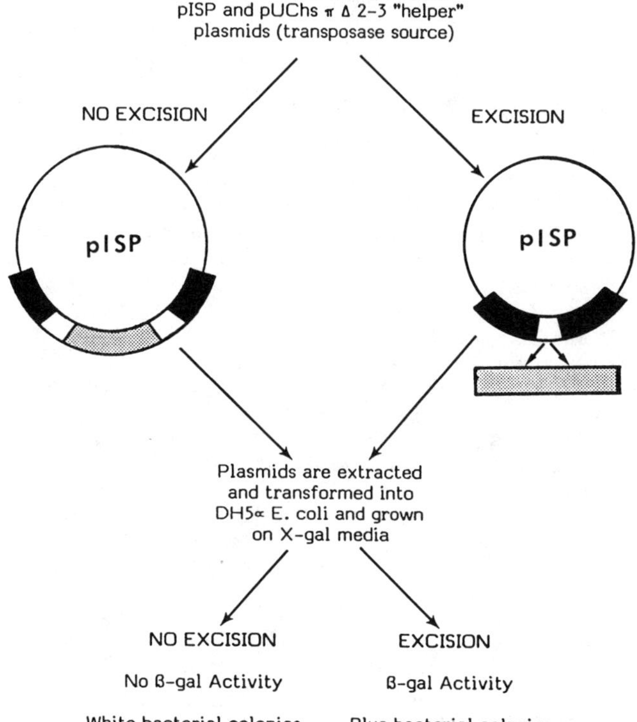

Figure 1. Diagrammatic representation of the P-element excision assay in embryonic soma.

which contained the injected excision assay plasmids, was isolated using the method of Hirt (16). The low molecular weight DNA recovered from injected embryos was used to transform the *E. coli* strain DH5α made competent according to the methods of Hanahan (17). *E. coli* transformants were plated on LB plates containing ampicillin (70 μg/ml) and X-gal (50 μg/ml). Transformants containing pISP plasmids that had lost the resident P-element through excision appeared blue on this medium.

Table 1. P-element excision in insect embryos

Insect species	pISP2 Plasmids recovered	$lacZ^+/amp^R$ Colonies	Frequency
D. melanogaster (M strain)	6.0×10^4	109	1.8×10^{-3}
D. melanogaster (P strain)	6.6×10^4	41	0.6×10^{-3}
D. simulans	4.0×10^4	40	1.0×10^{-3}
D. grimshawii	5.0×10^3	6	1.2×10^{-3}
A. suspensa	1.4×10^5	0	0.0
P. interpunctella	3.8×10^5	0	0.0

The functionality of the excision assay was initially tested in preblastoderm embryos of the M and P strains of *D. melanogaster*. The isolation of $lacZ^+/amp^R$ transformants, as indicated by blue colonies (see Fig. 1), showed that the excision assay was functional in the embryos as it was in cell lines as reported previously (14). The rates of excision for both the M and P strain embryos (Table 1) were equivalent to those observed for genomic P-elements and about ten times higher than the rates observed in cell lines. The latter result is somewhat surprising because P-element transposition is absent or greatly reduced in P strains. Restriction endonuclease mapping of plasmids recovered from these transformants indicated that the plasmids were pISP plasmids that had lost the resident P-element (data not shown). P-element excision in the embryos was observed only when the pISP plasmid was co-injected with the helper plasmid, i.e. when P-element transposase was present at high levels.

The excision assay was tested also in two other *Drosophila* species to determine the functionality of the assay in other insects. Excision occurred in both *D. simulans* and *D. grimshawii* at nearly the same rate as observed in *D. melanogaster* (Table 1). The transposition activity of the P-element in the embryos of these species demonstrates that the excision assay can be employed as a reliable indicator of P-element behavior in other insects.

When the excision plasmid was injected into the embryos of *A. suspensa* and *P. interpunctella*, no excision events were observed (Table 1). We conclude that these two species lack the cellular machinery necessary for normal P-element transposition that is present in D. melanogaster and the other Drosophila species tested.

Testing P-element functionality using an excision assay as described here has a number of advantages over testing for germline transformation directly using dominant selectable markers or measuring P-element activity by conducting the excision assay in cell lines. First, because the assay is conducted only in embryos and without having to achieve correct insertion into the host genome, the procedure does not require extensive rearing and selection of injected animals and their progeny, nor does it depend upon achieving genetic transformation. Second, the plasmid excision assay is extremely sensitive to P-element activity. As the assay is currently conducted, we can recover approximately 10^3 pISP plasmids per injected embryo allowing as many as 10^6 plasmids to be screened easily from any species. Finally, since the assay is conducted in insect embryos, the problem of differences in transposon activity observed between cell lines and embryos is obviated. When transformed into cell lines, the Drosophila transposon, copia, undergoes extensive transposition whereas it rarely undergoes transposition in the whole animal (18). When all of these factors are considered, the P-element plasmid excision assay will be useful for determining P-element functionality in heterologous systems, and in addition will permit extensive analysis of the mechanics of P-element excision in D. melanogaster.

The failure to observe P-element excision in A. suspensa and P. interpunctella embryos suggests that P-elements may not be functional in all insects. Although the reason for the lack of functionality has not been identified, the excision assay is designed with adequate latitude for modification so that it is amenable to experimental analysis. With sufficient effort, the biochemical basis for the lack of P-element activity can be identified and potentially be corrected. Preliminary results indicate that the transposase gene carried by the helper plasmid is being transcribed in A. suspensa, and efforts are now underway to determine if all of the posttranscriptional modifications necessary to generate a functional transposase mRNA occur (15).

The rapid and simple assay of P-element functionality described here can be used readily to assess P-element function in a wide variety of economically important insects. The assay also will be used in the development of transformation vectors that are phylogenetically unrestricted, which in the end is the absolute prerequisite for the application of recombinant DNA technology to the genetic modification of pest insects and their integration into insect pest management programs.

Chimeric Gene Constructs for Genetic-Sexing Schemes

Once reliable methods for genetic transformation of pest insects have been developed, several chimeric gene constructs may be useful in genetic-sexing schemes. The schemes presented here employ existing molecular genetic technologies and use promoters and structural genes that already have been isolated and characterized. The constructs and rationales can be tested in Drosophila species while methods for transformation of other insects are being developed. Thus, when gene transfer is more universally available, the constructs for genetic-sexing can be implemented with greater efficiency.

One general approach that can be considered for genetic-sexing is to impart a chemical sensitivity to the females and then select for the males by chemical treatment. The use of male-limited alcohol dehydrogenase (Adh) activity has been presented as one possible scheme (19). Development of this technique could be accomplished in insects that have a functional alcohol dehydrogenase even using classical genetic techniques but would require genetic transformation in those that lack Adh activity. The rationale is to develop a strain that has a functional Adh gene attached to a Y-chromosome, and then introduce the Adh-bearing Y chromosome into a strain that lacks Adh activity. This would result in a breeding population in which the males would have Adh activity and the females would not. Thus, the females would be lethally sensitive to low ethanol concentrations and could be selected out of the cultures. The male selection scheme described here is of course applicable to those species where the males are heterogametic. For insects such as lepidopterans, where the female is heterogametic, the genetic-sexing scheme would have to be modified to reverse the selection (see below). As pointed out above, the attached Y chromosomes are of limited use because they are subject to breakdown by recombination. In addition, the translocation of autosomal genes onto the Y chromosome can cause decreased viability of the strain. Therefore a more efficient and stable means of establishing the mechanics for genetic selection is required.

A selection scheme for males can be developed without classical genetics by employing recombinant DNA technology to construct a chimeric gene between the structural gene for Adh and a female-specific promoter for introduction into an Adh^- insect. This scheme would permit a stronger selection program to be used against the females. Since Adh would be expressed only in females, chemical treatment with 1-pentyn-3-ol or related compounds which are metabolized by Adh to lethal products (20) would result in the elimination of all females in the culture and survival of all males which would lack Adh activity. The yolk protein genes have been cloned from several insect species including D. melanogaster (21), Locusta migratoria (22, 23), Aedes aegypti (24), A. suspensa (A. Handler, unpublished), and P. interpunctella (P. Shirk, unpublished), and offer an easily obtainable female-specific promoter that could be used in the chimeric gene constructs. Unfortunately, it may be necessary to use a homologous cloned promoter for each insect as not all female-specific promoters are regulated correctly in heterologous insects (25).

A second general scheme for developing genetic-sexing, which perhaps has more potential, is the use of antisense RNA to disrupt normal gene activity. Antisense RNA is a transcript generated from the complementary DNA strand of a gene and therefore codes for an RNA that is complementary to the normal RNA transcript produced during transcription of a gene (26). A chimeric gene that produces antisense RNA to a gene can be made by cutting the original gene from the promoter sequences and resplicing the gene with the promoter at the opposite end thus reversing the polarity of the structural gene relative to the transcriptional signals. By reversing the polarity of the DNA strands, the opposite strand will be transcribed and a complementary or antisense RNA will be produced. When antisense RNA was introduced either in vitro (27) or in vivo (28), the presence of the antisense RNA inhibited the expression of the normal RNA transcript produced by the resident gene. Suppression of RNA translation by antisense RNA may be due to the formation of an RNA heteroduplex between the two complementary RNA strands thus blocking the attachment of ribosomes. An example of this application in the context of the Adh example would be the construction of a chimeric gene containing the antisense sequence for Adh linked with a female-specific promoter. Thus, Adh activity would be eliminated from females and make them lethally sensitive to treatment with ethanol.

Another prospective application of antisense RNA to genetic-sexing being tested in our laboratories involves the use of antisense RNA to control sexual differentiation. In Drosophila, sex determination is controlled by the interaction of several autosomal genes (29). The activity of two of sex determination genes, transformer (tra) and transformer-2 (tra-2), is required to maintain a female state of differentiation. If either of the two genes is mutant or nonfunctional in a chromosomal female (XX vs. XY in males), the individual will develop phenotypically as a sterile male. Because of the required activity of the transformer genes, the phenotype of the insect can be changed from female to male during development if a transformer gene is switched off (30); temperature sensitive tra-2 females are phenotypically female when reared at the permissive temperature but become males biochemically when switched to the restrictive temperature as adults. Chromosomal XY males mutant for tra or tra-2 are not affected phenotypically except that tra-2 males are germline sterile; tra males are fertile. The rationale for utilizing the sex determining genes is to construct a chimeric gene containing an antisense sequence for a transformer gene linked with a promoter that can be controlled conditionally. This would allow XX females to develop as phenotypic females under the nonrestrictive conditions. However, when reared under restrictive conditions, the XX females would develop as phenotypic sterile males as would the XY males. This technique in effect would obviate genetic-sexing by creating a population of sterile males only. For this scheme to be useful, transformed genetic females as well as mutant carrying males would have to mate successfully with wild type females and be competitive in the field. Laboratory testing of Drosophila sex-determination mutants indicate that mutant phenotypic males do court and mate (31).

At present, the *tra* gene has been isolated from *Drosophila* (32, 33) and our laboratory is in the process of constructing a chimeric gene with antisense *tra* linked with a heat shock promoter. This construct will be introduced into *D. melanogaster* and its ability to inhibit normal *tra* activity in females will be tested morphologically and biochemically. Since a *tra* mutation does not cause male sterility, the complete scheme cannot be tested and awaits the isolation of the *tra-2* gene. While this scheme may prove functional in *Drosophila*, the genetics of sex-determination has not been examined extensively in other insects. Possibly, the sex-determination genes in other insects can be identified by hybridization with a *tra* or *tra-2* probe. Although nothing is known about the similarity of structure and function of the sex-determining genes in other species, analogous if not homologous genes are expected to be functioning in other insects. Should experimental disruption of sex-determination prove successful in *Drosophila*, the high efficiency of this scheme would justify the research necessary to identify the genes and implement similar programs in agriculturally important pest insects.

Acknowledgments -- This work was funded in part by USDA Grant 86-CRCR-1-2012 to AMH and PDS. D.A.O'B. was supported by a USDA Research Associate Fellowship. P.E.R., Department of Entomology, Colorado State University, Fort Collins, CO 80523.

Literature Cited

1. Knipling, E. F. (1982) "Present status and future trends of the SIT approach to the control of arthropod pests." In *Sterile Insect Technique and Radiation in Insect Control*, pp. 3-23. IAEA/FAO, Vienna.
2. Knipling, E. F. (1979) *The Basic Principles of Insect Population Suppression and Management*." USDA Agric. Handbook No. 512, Washington, DC.
3. Hooper, G. H. S., A. S. Robinson and R. P. Marchand (1986) "Behavior of a genetic-sexing strain of Mediterranean fruit fly, *Ceratitis capitata*, during large-scale rearing. FAO/IAEA Report on Research Co-ordination Meeting on the Development of Radiation-induced Conditional lethals and Other Genetic Measures, pp. 34-49. Colymbari, Crete, Greece.
4. Lindsley, D. L., and E. H. Grell. (1968) Genetic variations of *Drosophila melanogaster*, pp. 471. Carnegie Inst., Washington.
5. Cockburn, A. f., A. J. Howells, and M. J. Whitten (1984) "Recombinant DNA technology and genetic control of pest insects." Biotechnol. Genet. Engr. Rev. 2: 69-99.
6. Courtright, J. B., and A. Krishna Kumaran. (1986) "A genetic engineering methodology for insect pest control: Female sterilizing genes." In *Beltsville Symposia in Agricultural Research (10) Biotechnology for Solving*

Agricultural Problems, (eds. Augustine, P. C., H. D. Danforth, and M. R. Bast), pp. 326-336. Martinus Nijhoff Publ., Dordrecht, The Netherlands.

7. Kelly, J. H., and G. J. Darlington. (1985) "Hybrid genes: Molecular approaches to tissue-specific gene regulation." Ann. Rev. Genet. 19: 273-296.
8. Spradling, A. C. and G. M. Rubin. (1982) "Transposition of cloned P elements into Drosophila germ line chromosomes." Science 218: 341-347.
9. Rubin, G. M. and A. C. Spradling. (1982) "Genetic transformation of Drosophila with transposable element vectors." Science 218: 348-353.
10. Rubin, G. M., M. G. Kindwell and P. M. Bingham. (1982) "The molecular basis of P-M hybrid dysgenesis: the nature of induced mutations." Cell 29: 987-994.
11. Steller, H. and V. Pirrottta. (1985) "A transposable P vector that confers selectable G418 resistance to Drosophila larvae." EMBO J. 4: 167-171.
12. Scavarda, N. J. and D. L. Hartl. (1984) "Interspecific DNA transformation in Drosophila." Proc. Natl. Acad. Sci. USA 81: 7515-7519.
13. Brennan, M. D., R. G. Rowan and W. J. Dickinson. (1984) "Introduction of a functional P element into the germ-line of Drosophila hawaiiensis." Cell 38: 147-151.
14. Rio, D. C., F. A. Laski and G. M. Rubin. (1986) "Identification and immunocytochemical analysis of biologically active Drosophila P element transposase." Cell 44: 21-32.
15. Laski, F. A., D. C. Rio and G. M. Rubin. (1986) "Tissue specificity in Drosophila P element transposition is regulated at the level of mRNA splicing." Cell 44: 7-19.
16. Hirt, B. (1967) "Selective extraction of polyoma DNA from infected mouse cultures." J. Molec. Biol. 26: 365-369.
17. Hanahan, D. (1983) "Studies of transformation of Escherichia coli with plasmids." J. Molec. Biol. 166: 557-580.
18. Rubin, G. M. (1983) "Dispersed repetitious DNAs in Drosophila," pp. 329-361. In: Mobile Genetic Elements, (ed. J. A. Shapiro). Academic Press, New York.
19. Riva, M. E. and A. S. Robinson. (1986) "Induction of alcohol dehydrogenase null mutants in the Mediterranean fruit fly, Ceratitis capitata." Biochem. Genet. 24: 765-774.
20. O'Donnell, J., L. Gerace, F. Leister and W. Soffer. (1975) "Chemical selection of mutants that affect alcohol dehydrogenase in Drosophila. II Use of 1-pentyn-3-ol." Genetics 79: 73-83.
21. Barnett, T., C. Pachl, J. P. Gergen and P. C. Wensink. (1980) "The isolation and characterization of Drosophila yolk protein genes." Cell 21: 729-738.
22. Wyatt, G. R., J. Locke, J. Y. Bradfield, B. N. White and R. G. Deeley. (1981) "Molecular cloning of vitellogenin gene sequences from Locusta migratoria," pp. 299-307. In Juvenile Hormone Biochemistry, (eds. G. E. Pratt and G. T. Brookes). Elsevier/North Holland, Amsterdam.

23. Locke, J. (1985) Molecular Cloning and Characterization of Two Vitellogenin Genes from Locusta migratoria. PhD Thesis, Queen's University, Kingston, Ontario.
24. Racioppi, J. V., R. M. Gemmill, P. H. Kogan, J. M. Calvo and H. H. Hagedorn. (1986) "Expression and regulation of vitellogenin messenger RNA in the mosquito, Aedes aegypti." Insect Biochem. 16: 255-262.
25. Wyatt, G. R., M. R. Kanost, J. Locke and V. K. Walker. (1986) "Juvenile hormone-regulated locust vitellogenin genes: Lack of expression after transfer into Drosophila." Arch. Insect Biochem. Physiol. Suppl. 1: 35-46.
26. Green, P. J., O. Pines and M. Inouya. (1986) "The role of antisense RNA in gene regulation." Ann. Rev. Biochem. 55: 569-597.
27. Izant, J. G. and H. Weintraub. (1985) "Constitutive and conditional suppression of exogenous and endogenous genes by antisense RNA. Science 229: 345-352.
28. Rosenberg, U. B., A. Preiss, E. Seifert, H. Jackle and D. C. Knipple. (1985) "Production of phenocopies by Krupple antisense RNA injection into Drosophila embryos." Nature 312: 703-709.
29. Baker, B. S. and J. M. Belote. (1983) "Sex determination and dosage compensation in Drosophila melanogaster." Ann. Rev. Genet. 17: 345-397.
30. Belote, J. M., A. M. Handler, M. F. Wolfner, K. J. Livak and B. S. Baker. (1985) "Sex-specific regulation of yolk protein gene expression in Drosophila." Cell 40: 339-348.
31. McRoberts, S. B. and L. Tompkins. (1985) "The effect of transformer, doublesex, and intersex mutations on the sexual behavior of Drosophila melanogaster." Genetics 111: 89-96.
32. McKeown, M., J. M. Belote and B. S. Baker. (1987) "A molecular analysis of transformer, a gene in Drosophila melanogaster that controls female sexual differentiation." Cell 48: 489-499.
33. Butler, B., V. Pirrotta, I. Irminger-Finger and R. Nothiger. (1986) "The sex-determining gene tra of Drosophila: molecular cloning and transformation studies." EMBO J. 5: 3607-3613.

RECEIVED February 4, 1988

Chapter 11

Potential Applications of Neuroendocrine Research to Insect Control

Larry L. Keeley

Laboratories for Invertebrate Neuroendocrine Research, Department of Entomology, Texas Agricultural Experiment Station, Texas A&M University, College Station, TX 77843

Advances in biotechnology that result from neuroendocrine research may have applications for the control of insect pests. Neurohormones are the master regulatory hormones of insects and affect critical physiological processes that include: molting, metamorphosis, reproduction, and general homeostasis. Manipulation of neurohormone titers would disrupt sensitive physiological processes and kill or debilitate treated insects. Sensitive neuroendocrine events include: hormone synthesis, hormone secretion and degradation; hormone-target cell-receptor interactions and target cell responses. Neuroendocrine events and titers may be altered by: (1) genetic engineering of neurohormone genes into suitable baculovirus cloning-expression vectors; (2) disruption of neurosecretory activity by exogenous chemicals that affect aminergic neurons; (3) development of analogs that are antagonists or superagonists of neurohormones at receptor proteins; and (4) inhibition of enzymes instrumental in hormone synthesis or degradation.

The role of the animal brain as a source for endocrine factors was discovered in insects with the pioneering work of Kopec in 1917 (1). He found that ligation of the head prevented molting and metamorphosis in gypsy moth larvae. In 1941, the phenomenon of neurosecretion was described for insects by Berta Scharrer (2) using the cockroach, Leucophaea maderae. Despite this early use of insects as models for studying neuroendocrine processes, the topic of insect neuroendocrinology has lagged behind its vertebrate counterpart because of the scarcity of identified neurohormones. Over the four decades since Scharrer first described the neurosecretory role for cockroach brain neurons, about two dozen physiological processes of insects have been identified as being neuroendocrine dependent (3). However, it was not until 1975-76 that the first insect neurohormone structures were defined (4,5). No further structures were reported until 1984 (6-8). At present,

0097-6156/88/0379-0147$06.00/0

there are about two dozen neurohormonal structures reported in the literature, but many of these are natural analogs (bioanalogs) of each other and affect the same physiological processes.

Neurohormones are the master regulators for physiological processes in both vertebrate and invertebrate animals. In insects, neurohormones regulate general homeostatic activities such as: behavior; the synthesis of circulating carbohydrates, lipids and proteins; water and salt balance; heartbeat rate; involuntary muscle contractions and basal metabolism. Neurohormones also regulate the synthesis and secretion of the ecdysteroids and juvenile hormones that directly control the processes of molting, metamorphosis and reproduction. Furthermore, neurohormones affect specific events of growth and reproduction such as eclosion, cuticle tanning, ovulation and oviposition.

The potential for using endocrine imbalance as a means of insect pest control was suggested by Williams (9,10). He proposed that exposing immature insects to juvenile hormone (JH) at the time of metamorphosis, when JH is normally absent, would cause abnormal development and individuals incapable of survival. Since insect metamorphosis is unique, JH disruption would affect only insects. This would result in an environmentally safe approach to insect control as compared to current chemical pesticides which are less insect specific and more biocidal. The JH approach to pest insect control is most effective when adults are the destructive stage, and commercial preparations of JH mimics are available for use in the control of adult flies, mosquitoes, and fleas and, recently, for cockroach reproduction. However, many pest insects are destructive as larvae.

The use of neurohormones for pest control remains speculative, but considerable interest is developing concerning the possibility, and several articles have proposed approaches for applying neuroendocrine information (11-14). Examples exist in the literature which suggest that neuroendocrine imbalance affects insects adversely. Molting ceases after destruction of the neurosecretory cells that secrete the prothoracicotropic hormone (PTTH) (15), and removal of the corpora cardiaca as a neuroendocrine source affects the physiological performance of active insects. For example, the corpora cardiaca contain a neurohormone that mobilizes trehalose, the major insect blood sugar (16,17). Normal *Calliphora erythrocephala* blowflies fly for more than 45 min without exhaustion or a measurable decline in hemolymph trehalose, but cardiacectomized blowflies have an 85% reduction in trehalose and are exhausted after 45 min of flight (18). Similarly, removal of the corpora cardiaca reduces flight speed by 40% in locusts (19), probably because of a deficiency in the adipokinetic hormone (AKH) that mobilizes fat body lipids as an energy source for the flight muscles. Alternatively, excessive amounts of neurohormones appear debilitating, as well. Large doses of proctolin cause temporary stupor and immobility in cockroaches (20).

It is a great leap from observing experimental responses to endocrine imbalance in the laboratory to controlling insects in the field by neuroendocrine disruption. Nevertheless, laboratory experiments indicate that sustained neuroendocrine disruption would have profound effects on growth, development and physiological functions. Whereas experimental procedures such as injection of a natural hormone do not exhibit extreme responses because the hormone is degraded by proteases, agents designed to disrupt the neuroendocrine balance for control purposes will have to be stable in both the environment and the insect. Such agents will cause a persistent endocrine imbalance that results in the target tissue receiving either a sustained, uncontrolled hyperstimulation or suppression of response. This persistent, uncontrolled response by the target tissue will cause serious debilitation and, ultimately, death of the treated insect.

Objections to Using Neuroendocrine-Based Pest Insect Control

Peptide Nature of the Neurohormones. Neurohormones are proteins and, as such, are intrinsically unstable and unsuited for application in the environment or directly to insects for pest control purposes. Peptides are easily degraded in the environment by light, heat, and microorganisms. Also, it is unlikely that natural neurohormones could gain entry into exposed insects. Most proteins cannot penetrate the insect cuticle, and they would be digested to their constituent amino acids by gut proteases if consumed orally. For example, topically applied proctolin is not absorbed by *M. sexta* larvae and ingested proctolin is degraded rapidly (21). Therefore, it is highly unlikely that natural neurohormones could be used to affect the neuroendocrine balance of target insects.

However, it may be possible to overcome some of the disadvantages of peptides. Peptide analogs might be stabilized so as to resist digestive proteases and be permeable to the insect gut. Analogs might be designed to be stable or permeable to the cuticle in the same manner as present insecticides. An example exists of a series of peptides that were synthesized with isoprene- or terpenoid-like structures and exhibited juvenile hormone activity when applied topically to larval and pupal insects (22). Finally, the use of insect viruses as highly-efficient cloning-expression vectors for neurohormone genes might provide a vehicle for carrying the genes directly into the insect where the hormone could be produced in uncontrolled superabundance.

Universality of Neurohormones Among Animals. If both insects and vertebrates share common peptide structures as neurohormones, then strategies for manipulating insect neurohormones could be detrimental to nontarget animals and present serious environmental concerns. However, present findings do not support the contention that insects and vertebrates share common hormones for the same functions. Antibodies to vertebrate peptide hormones react with peptides present in insects (23). But, in those cases where an immunoreactive vertebrate-like peptide has been obtained, the peptide does not have the physiological action found in its vertebrate counterpart. Conversely, in those cases where a physiological process was used as the basis for isolation of an insect neurohormone, the resulting peptide has had a unique structure. Similarities are sometimes noted between the peptide structures of insect and vertebrate neurohormones. For example, leucosulfakinin is an insect myotropin that has a 50% identity with vertebrate gastrin-cholecystokinin (24). Intestinal functions are stimulated in vertebrates by gastrin and cholecystokinin and in insects by leucosulfakinin; however, neither gastrin nor cholecystokinin affect intestinal contractions in the insect-gut bioassay. The 4 kDa PTTH from silkworms also has an approximately 50% identity with vertebrate insulin (25), yet vertebrate insulin does not activate the prothoracic glands at doses that exceed normal PTTH titers by 10,000 times.

Vertebrate-like neuropeptides are present in insects as demonstrated immunologically, and probably vice versa, but it is unclear what the function is for these peptides in their heterologous animal system. The structural similarities between these molecules suggest that common, biologically-active ancestral molecules may have existed and evolved to perform different functions depending on the physiological diversity and needs of the animals involved.

It is still possible to make use of the neuroendocrine approach even if the procedures used affect both insect and vertebrate neuroendocrine systems. For example, most current insecticides are general nerve poisons of higher animals. The broad-based toxicity of these chemicals has been minimized by directing the insecticides at target organisms through specific application and formulation procedures. Such directed application procedures will surely be part of the strategies developed for disrupting neuroendocrine processes in insects.

Neurohormone Distribution Among Insect Species. It is important to consider the distribution of the various neurohormones among the insect species when developing a neuroendocrine-based control strategy. Some hormones such as the neuropeptides that regulate molting must be universal since all insects grow by molting. On the other hand, some hormones may be restricted in their distribution. It is likely that aquatic insects and stored grain insects may have little need for antidiuretic and diuretic factors, respectively. In addition, hormones with the same function may differ structurally between species, and families of hormones exist. The adipokinetic/hypertrehalosemic (AKH/HTH) hormones constitute a peptide family that, now, has eight representative structures from eight species of insects. One other AKH-related structure (red pigment-concentrating hormone) is found in crustaceans and crossreacts in insects. In some cases, the same hormone is found in two distinct insect species; in other cases two AKH/HTH peptides with different structures and functions are found in the same species. In all cases, the AKH/HTH bioanalogs affect lipid or carbohydrate metabolism by the fat body, although the different hormones have differing potencies among the various species for their metabolic effects.

Advantages to Using Neuroendocrine-Based Pest Insect Control

There are a number of advantages to a neuroendocrine approach to insect pest management.

1. Neurohormones regulate numerous critical physiological processes susceptible to alteration.

2. Neurohormones are amenable to genetic engineering technology because of their peptidic nature.
3. Several neuroendocrine events are susceptible to alteration.
4. Neurohormones are structurally diverse so that it is possible to synthesize a wide array of agonistic or antagonistic analogs.

Endocrine Regulated Physiological Processes. The variety of physiological processes affected by neurohormones suggests that disruptions in neuroendocrine-dependent processes may be used as the =basis for a new approach to insect pest control. Since neurohormones such as PTTH and allatotropin and allatostatin regulate the secretion of the ecdysteroids and JH, manipulation of these neurohormones could influence molting and metamorphosis in larvae and reproduction in adults just as would manipulation of the ecdysteroids or the JHs, themselves. Disruption of eclosion hormone could prevent the onset of molting behavior and ecdysis, and disruption of bursicon could prevent tanning of the soft cuticle of newly-molted individuals. Disruption of either ecdysis or tanning would result in the death of affected insects.

Enhancement of diuretic hormone actions could be fatal for those insects that have critical problems of maintaining body water, esp. soft-bodied larvae. Soft-bodied larvae rely on hemolymph volume to maintain turgor for movement, and dehydration would result in immobility followed by death. Inhibition of antidiuretic hormones could have serious effects on insects that reside in extremely dry environments, e.g. stored grain. Antidiuretic hormone antagonists in combination with diuretic hormone agonists would be especially potent.

Disruption of neuropeptides that affect muscle contraction might result in immobility or block intestinal peristalsis to inhibit feeding and digestion. Proctolin-induced immobility in cockroaches illustrates this situation (20).

Disruption of AKH or HTH could prevent the release of lipids and carbohydrates from the fat body and their use as energy fuels for mobility. Flight speed, wingbeat frequency and endurance decreased in insects rendered deficient in AKH or HTH by corpora cardiaca removal (18,19). Although lowering blood metabolites might have a relatively benign effect on survival, the resulting decrease in mobility would make treated insects more susceptible to environmental mortality factors, such as parasitization, predation and adverse weather and prevent migration or movement for long distances in search of food, shelter, or mates.

Several steps in the reproductive process are neurohormone regulated. Pheromone synthesis-release, ovulation, oviposition, and gonad maturation might also be considered as neuroendocrine-dependent processes susceptible to manipulation for reproductive disruption.

Recent studies have demonstrated that a given hormone may have multiple effects. For example, for over 20 years we have known that factors existed in the corpora cardiaca of cockroaches that affected the heartbeat rate and trehalose levels of the hemolymph. Only with the isolation-characterization of the two AKH/HTH factors from the corpora cardiaca of *P. americana* did we find that the heartbeat and carbohydrate regulations are performed by identical factors (6-8). Finally, we have found in my laboratory that the HTH of the cockroach *Blaberus discoidalis* affects not only trehalose synthesis and heartbeat rate but also heme synthesis for cytochrome production in the fat body (unpublished data).

Endocrine Events Susceptible to Manipulation. There are a series of events which are common to the production and action of any hormone. These events include: hormone synthesis, secretion and transport; hormone-target cell receptor interaction; target cell response; and hormone degradation. Several of these events are susceptible to manipulation by external influences that might be devised into control strategies.

Neurohormones are proteins and are, therefore, ultimately gene products. It seems questionable that the transcriptional activity of neurohormone genes could be disrupted by a means that would be specific to insects and safe for general use. However, if we were to find that the neurohormone genes are regulated by insect hormones such as the ecdysteroids or JHs, then it might be possible to suppress insect neurohormone gene activity selectively.

The enzymes that process prohormones into active hormones are candidates for artificial manipulation. Neuroendocrine research on vertebrates and noninsect invertebrates indicates that the primary translation products from insect neurohormone genes can be expected to be large prohormones that are post-translationally processed into the active hormone by cleavage at pairs of basic amino acids. No evidence is published on the nature of the insect prohormones, although several laboratories are examining this topic. Nothing is known about the prohormone processing enzymes of insects but hormone processing enzymes are reported for vertebrates (26,27), and it is probable that the enzymes are similar for the two groups. Again, if the processing enzymes are similar, then these enzymes may not be the ideal site for directing a control procedure since such an approach might also affect nontarget organisms.

Neurosecretion is susceptible to manipulation to alter the neuroendocrine balance. Neurohormones are secreted from their source neurons under the direction of other endocrine agents or in response to nervous input from regulatory neurons. The regulatory neurons affect neurosecretory activity via specific biogenic amine neurotransmitters. Vertebrate studies show that biogenic amine neurotransmitters stimulate specific neurohormone secretion (28) Peptidic releasing factors also promote specific neurosecretion (29) in vertebrates, but such releasing factors are unknown in insects. Disruption of neurohormone secretion patterns in insects, if performed at critical times in the life cycle, could disturb sensitive physiological events such as molting, metamorphosis or reproduction.

Little is known about the transport of insect neurohormones from their source to target tissues. Presumably, the hormones are transported via the hemolymph, although there are indications of transport by nerves extending from the endocrine source to the target tissue (30) or by hemocytes (31). Hemolymph-borne hormones should be susceptible to degradation by circulating proteases, but this may not be the case. The hormones may be bound to a protecting carrier protein (neurophysin), or specific degradative proteases may be associated with receptor sites on target cell membranes so that the circulating hormone is not affected. The presence of the amide and pyro-glutamate groups at the C- and N-termini of the AKH/HTH hormones may serve a protective role to prevent degradation during transport.

Since neurohormones are proteinaceous, it is unlikely that they can be applied directly to an insect with the hope of implementing an endocrine imbalance; however, it is feasible to design and develop stable, nonpeptide analogs to the active conformation of the hormone. The use of stable analogs that either mimic or block the neurohormone receptor sites of target tissues constitute one of the most promising means for manipulating endocrine-dependent processes in an insect control strategy. Alternatively, inhibition of the active site of degradative protease enzymes by neurohormone analogs could result in the accumulation of natural hormones and the hyperstimulation of sensitive processes.

Neuroendocrine Manipulation

Four ways that neurohormones and neurohormone-dependent physiological processes might be affected are:

1. genetic engineering and recombinant DNA biotechnology;
2. altered neurosecretion;
3. inhibition of enzymes for hormone synthesis or degradation; and
4. neurohormone analogs.

Recombinant DNA Biotechnology. Since neurohormones are proteins or peptides, they are gene translation products and are amenable to recombinant DNA technology and genetic engineering. The most obvious appproach to using recombinant DNA technology is to engineer a neurohormone gene directly into the genome of the host plant for a pest insect. In this manner, the host plant would theoretically produce the hormone and pests that feed on the genetically-transformed plants would receive excessive doses of the hormone resulting in hyperstimulation of sensitive processes. Other approaches for delivering neurohormones by genetic engineering are to transform microorganisms such as bacteria, protozoa or yeasts with neurohormone genes. In all of the above cases, the hormone is available to the insect by consuming the genetically-engineered vector. However, digestive degradation of the natural hormone still remains as a problem since the hormone must pass the alimentary system of the insect. Furthermore, since neurohormones are synthesized as prohormones and are post-translationally processed into active hormones, it remains to be proven whether the products of neurohormone genes can be processed appropriately in the cells of plants, bacteria, or protozoa. It might be possible to synthesize the short DNA sequence that transcribes specifically for the amino acid sequence of the active hormone, and engineer this short "gene" into a plant or cloning-expression vector so that the DNA is transcribed and the hormone translated. Whether short segments of DNA that encode directly for the active hormone can be expressed remains to be demonstrated.

Insect baculoviruses are an alternative to plants or single-celled organisms as cloning-expression vectors for neurohormone genes. Baculoviruses are frequent disease agents of lepidoptera larvae and were considered at one time as having a high potential as biocontrol agents for pest insects. Recently, the baculovirus of Autographa californica has been genetically engineered to serve as a cloning-expression vector for foreign genes (32). Transformation of the baculovirus expression vector (BEV) with the human beta-interferon gene followed by infection of insect cell cultures resulted in the production and secretion by infected cells of large amounts of biologically-active interferon. The ability of BEVs to produce foreign gene products in vivo was confirmed using the baculovirus of B. mori silkworms transformed with the human alpha-interferon gene. In this case, alpha-interferon was found in the hemolymph of silkworm larvae infected with the recombinant BEV (33). These findings demonstrate that nuclear polyhedrosis viruses can carry foreign genes into infected larvae, and the genes are expressed and the translation products processed into biologically-active molecules that are released from infected cells into the host's circulatory system. Based on this, it is theoretically possible that BEV genetically-engineered with neurohormone genes would carry the genes into host insects and express the hormonal peptide products in uncontrolled excess. The hormones would by-pass the digestive system, be released directly into the circulatory system and would cause hyperstimulation of sensitive physiological processes, for example water excretion which would cause rapid dehydration and death of the infected larva.

Recombination with appropriate foreign genes could make the virus a more effective biocontrol agent. BEV pathology is slow and often requires nearly a week to kill infected insects. Also, there is a dose-dependency between the number of ingested virus capsules and larval size. A neuroendocrine product produced by a transformed virus might improve the virus as a biocontrol agent by inhibiting feeding and killing an infected larva early, before the onset of the viral pathology, or by converting a sublethal infection into a fatal infection due to physiological weakening and increased susceptibility. A BEV transformed with a neurohormone gene has the advantage that it is an insect-specific biological agent combined with a natural, insect biorational factor. Such a combination should be of minimal environmental concern.

It is also feasible to transform BEVs with natural or synthetic genes for proteinaceous agents that would modulate neurohormone balance. For example, it is reported that the antisense strand of DNA carries a genetic code that is complementary to the sense strand (34). Where the sense strand codes for hydrophobic amino acids, the antisense strand codes for hydrophilic amino acids, and vice versa. Neutral amino acids pair generally with neutral amino acids. The amino acid sequence was synthesized for a peptide complementary to ACTH as based on the antisense code for ACTH-mRNA (35). The resulting anti-ACTH mimicked a receptor protein and bound selectively to ACTH. If this pattern of peptide complementarity is valid, then once insect neurohormone genes are isolated and their base sequences determined, it should be possible to synthesize anti-hormone genes, clone them into a BEV and produce an antihormone that will tie-up the natural hormone at a critical time in the life of the insect (e.g. cuticle tanning, eclosion). This technology would be most useful with the large molecular weight neurohormones such as PTTH, eclosion hormone or bursicon.

Several other alternatives may be possible using genetic engineering. For example, it may be possible to design artificial proteins and their associated genes that will act as hormone antagonists or mimics at hormone acceptor sites (target cell, degradative proteases). Finally, it may be possible to produce monoclonal antibodies to a given hormone, isolate the gene for the antibody, insert the antibody gene into the BEV and express an antibody that would bind and functionally remove specific hormones from BEV-infected insects.

Altered Neurosecretion. Biogenic amines affect specific hormone secretion in the neuroendocrine system of vertebrates. The biogenic amines may either inhibit or stimulate neurosecretion directly (36), or they may regulate neurosecretion indirectly through controls over hormone-specific releasing factors (28). Biogenic amines also affect neurosecretion in insects. For example, reserpine depletes biogenic amines from the nervous system, and treatment of *P. americana* with reserpine stimulates the secretion of hyperglycemic hormones (37). This result suggests that the biogenic amines inhibit hyperglycemic hormone secretion. Locusts poisoned with insecticides that act on the cholinergic nervous system show enhanced AKH secretion (38). However, AKH secretion is suppressed if the test insects are pretreated with agents that inhibit the aminergic nervous system before they are exposed to the insecticides (39,40). These findings suggest that the cholinergic nervous system influences the aminergic nervous system which, in turn, regulates specific neurosecretion.

Of the insect biogenic amines, octopamine appears to exert the most effect on neurosecretion (41,42). Octopamine has little significance in the vertebrate nervous system where it is either degraded rapidly or may serve as a cotransmitter with norepinephrine. Exposure of corpora cardiaca to octopamine *in vitro* promotes AKH secretion (41), and chlordimeform, an insecticidal agent and octopamine mimic, also stimulates AKH secretion from isolated corpora cardiaca (43). This latter finding illustrates that chemicals can be applied to insects with the expectation that they will affect or mimic the aminergic nervous system of insects to regulate specific neurosecretion. More fundamental information about the regulation of neurosecretory processes in insects may suggest ways to manipulate neurosecretion for pragmatic purposes.

Inhibition of Enzymes for Hormone Synthesis or Degradation. The question of inhibiting enzymes that are instrumental in either the synthesis or degradation of insect neurohormones is difficult to address at this time. Little is known about the synthetic enzymes of vertebrate neurohormones and nothing is known about the insect enzymes. This area must await more research in the field of both animal groups. The yeast KEX2 protease has the properties of a prohormone-converting enzyme (44) and may provide a model for studying prohormone processing. Captopril, illustrates that it is possible to manipulate peptide hormone titers by inhibiting appropriate synthetic or activating enzymes. Captopril is a synthetic oral antihypertensive that was designed specifically to inhibit the angiotensin-converting enzyme that converts inactive angiotensin I to active angiotensin II (45). Alternatively, little is also known about the degradative enzymes for insect neurohormones. Aminopeptidase was identified as being important for *in vivo* inactivation of proctolin in cockroaches (46,47). The ability to alter hormone titers by altering the synthesis and degradation of insect neurohormones awaits more basic research, specifically: chemical definition of insect neurohormones; isolation of the genes and prohormone forms involved in the synthesis of the defined hormones; and preparation of labeled hormones for use in degradation studies.

Neurohormone Analogs. Finally, analogs of the neurohormones probably have the greatest potential for manipulating neuroendocrine-regulated processes. There are several reasons why analogs are the most feasible approach to a neuroendocrine-based strategy for pest insect control. First, there is a long history of experience and success concerned with the use of chemicals to control insect pests. Second, analogs have a nearly infinite potential for modifications to improve stability, penetrability, toxicity and specificity. We know that there are numerous insect neurohormones and that several forms may exist for the same hormone, both between and within insect species. Therefore, neurohormones present the synthetic peptide chemist with a multitude of possibilities for use in designing insecticidal agents.

It is argued that the commercial manufacture of peptides is too expensive for agricultural use. However, I do not propose that the final analog will be a peptide. Rather, the practical agent will probably be a nonpeptidic derivative of a neurohormone that is locked into a structure that mimics the active conformation of the hormone. Historically, if an agent is sufficiently effective and an adequate market exists, then industry will solve the problems of economical production and application.

A useful analog for a peptide may no longer be peptidic. There is not sufficient space to describe the details of how an analog can be designed for a given neurohormone. Furthermore, each peptide possesses unique properties that require individual approaches for analog design, and more details related to analog design were presented in a separate paper (14). What is modeled in designing an analog is the biologically-active conformation assumed by the peptide as it interacts with its target-cell receptor. A peptide hormone carries two types of information: an address and a message. The address portion of the peptide is that region that fits into the correct receptor. The message is either part of the address or it can be a separate region, and it activates the receptor to initiate the cellular response. Antagonistic analogs contain only an address, and antagonists fit into the receptor without conveying a message. Alternatively, agonistic analogs contain both a functional address and message and produce a cellular response. The desire of the analog approach is to develop molecules that are stable and locked into the active conformation so that the analog either binds the receptor and prevents the natural hormone from exerting its action (an antagonist), or the analog binds the receptor and causes hyperstimulation of the physiological response (a superagonist). The analog is either stabilized to resist enzymic degradation or the enzyme does not act on the analog because of its nonpeptidic nature.

These properties permit the analog to maintain a continuously high titer and to sustain its effect.

Several physiological processes are susceptible to superagonists. For example, a superagonist to the diuretic hormone could cause severe dehydration, immobility and death in larval insects, especially those in a dry environment such as stored grain. Hyperstimulation of JH production by an allatotropin agonist might result in abnormally high levels of JH at the time for epidermal cell recommitment, hence metamorphosis would not occur. Conversely, an allatostatin agonist would inhibit JH production and promote premature metamorphosis. Antagonistic analogs to allatotropin and allatostatin could be used in a reciprocal manner to agonists for regulating JH production. An antagonist to allatotropin would inhibit JH synthesis and promote premature metamorphosis; whereas, an allatostatin inhibitor would promote JH synthesis and prevent metamorphosis. Finally, analogs might be useful for disrupting the action of myokinins. Antagonistic analogs to myokinins that affect gut contractility might suppress feeding activity, and analogs to neuropeptides that affect skeletal muscles might result in paralysis and immobility. If the analogs were stable to enzymic degradation, they might remain active for an extended time and cause sustained paralysis and death.

Summary

It is my intent that this discussion should stimulate consideration of the practical potential for insect neuroendocrine research. It is unlikely that we will develop practical neurohormone-based strategies for insect pest management in the near future. Conservatively, I would project that any pragmatic use of current neuroendocrine research would be five years away at the earliest and more on the order of ten or more years. At present, we need more fundamental information on the chemical nature of the insect neurohormones, their endocrinology and their physiological significance within the insects. It is essential that we structurally define more neurohormones and develop specific assays for these hormones in order to study their endocrinology. Finally, we need to determine what significance the disruption of a physiological process would have on survival. Without this base of fundamental information, it will not be possible to develop the types of pragmatic strategies proposed here.

Acknowledgments

Personal research related to this publication was supported by NSF grant no. DCB 8511058 and by the Robert J. Kleberg, Jr. and Helen C. Kleberg Foundation.

Literature Cited

1. Kopec, S. Bull. Int. Acad. Cracovie 1917, B, 57-60.
2. Scharrer, B. J. Comp. Neurol. 1941, 74, 93-108.
3. Raabe, M. In Insect Neurohormones; Plenum: NY, 1982.
4. Starratt, A.N.; Brown, B.E. Life Sci. 1975, 17, 1253-1256.

5. Stone, J.V.; Mordue, W.; Batley, K.E.; Morris, H.R. Nature 1976, 263, 207-211.
6. Scarborough, R.M.; Jamieson, G.C.; Kalish, F.; Kramer, S.J; McEnroe, G.A.; Miller, C.A.; Schooley, D.A. Proc. Nat. Acad. Sci. USA 1984, 81, 5575-5579.
7. Baumann, E.; Penzlin, H. Biomed. Biochim. Acta 1984, 43, 13-16.
8. Witten, J.L.; Schaffer M.H.; O'Shea, M.; Cook, J.C.; Hemling, M.E.; Rinehart, K.L., Jr. Biochem. Biophys. Res. Comm. 1984, 124, 350-358.
9. Williams, C.M. Nature 1956, 178, 212-13.
10. Williams, C.M. Sci. Amer. 1967, 217, 13-7.
11. Menn, J.J. J. Pestic. Sci. 1985, 10, 372-376.
12. Menn, J.J.; Henrick, C.A. In Agricultural Chemicals of the Future; Hilton, J.L., Ed.; Rowman & Allanheld: Totowa, 1984; p. 247.
13. O'Shea M. In Approaches to New Leads for Insecticides; Keyserlingk, H.C.Jager, A.; Szczepanski, Ch. Eds. Springer-Verlag: New York, 1985,; p. 133.
14. Keeley L.L.; Hayes, T.K. Insect Biochem. 1987, 17, 639-51.
15. Steel, C.G.H. Gen. Comp. Endocr. 1978, 34, 219-28.
16. Steele, J.E. Nature 1961, 192, 680-681.
17. Friedman, S. J. Insect Physiol. 1967, 13, 397-405.
18. Vejbjerg, K.; Normann, T.C. J. Insect Physiol. 1974, 20, 1189-92.
19. Goldsworthy, G.J.; Coupland, A.J. J. Comp. Physiol. 1974, 89, 359-68.
20. Starratt, A.N.; Steele, R.W. Insect Biochem. 1984, 14, 97-102.
21. Quistad, G.B.; Adams, M.E.; Scarborough, R.M.; Carney R.L.; Schooley, D.A. Life Sci. 1984, 34, 569-76.
22. Zaoral, M.; Slama, K. Science 1970, 170, 92-3.
23. Duve, H.; Thorpe, A. In Insect Neurochemistry and Neurophysiology; Borkovec, A.J.; Kelly T.J. Ed.; Plenum: New York, 1984, p. 171.
24. Nachman, R.J.; Holman, G.M.; Haddon, W.F.; Ling, N. Science 1986, 234, 71-3.
25. Nagasawa, H.; Kataoka, H.; Isogai, A.; Tamura, S.; Suzuki, A.: Ishizaki, H.; Mizoguchi, A.; Fujiwara, Y.; Suzuki, S. Science 1984, 226, 1344-5.
26. Douglass, J.; Civelli, O.; Herbert, E. Ann. Rev. Biochem. 1984, 53, 665-715.
27. Fricker, L.D. Trends in NeuroSci. 1985, 8, 210-14.
28. Anton-Tay, F.; Wurtman, R.J. In Frontiers in Neuroendocrinology; Martini,L; Ganong, W.F. Ed.; Oxford Univ.: London, 1971, p 45.
29. McCann,S.M. In Neurosecretion Molecules, Cells, Systems; Farner, D.S.; Lederis, K. Ed.; Plenum: NY, 1981, p 139.
30. Johnson, B.; Bowers, B. Science 1963, 141, 264-6.
31. Takeda, N. J. Insect Physiol. 1977, 23, 1245-54.
32. Smith, G.E.; Summers, M.D.; Fraser M.J. Molec. Cell Biol. 1983, 3, 2156-65.
33. Maeda, S.; Kawai, T.; Obinata, M.; Fujiwara, H.; Horiuchi, T.; Saeki, Y.; Sato, Y.; Furusawa, M. Nature 1985, 315, 592-94.
34. Blalock, J. E.; Smith, E.M. Biochem. Biophys. Res. Commun. 1984, 121, 203-7.
35. Bost, K.L.; Smith, E.M.; Blalock, J.E. Proc. Nat. Acad. Sci. USA, 1985, 82: 1372-5.

36. Weiner, R.I.; Ganong W.F. Physiol. Rev. 1978, 58, 905-76.
37. Gersch, M. J. Insect Physiol. 1972, 18, 2425-39.
38. Samaranayaka, M. Gen. Comp. Endocr. 1974, 24, 424-36.
39. Samaranayaka, M. J. Exp. Biol. 1976, 65, 415-25.
40. Samaranayaka, M. Pestic. Biochem. Physiol. 1977, 7, 283-8.
41. Orchard, I.; Loughton, B.G. J. Neurobiol. 1981, 12, 143-53.
42. Orchard, I. Can. J. Zool. 1982, 60, 659-69.
43. Singh, G.J.P.; Orchard. I.; Loughton, B.G. Pestic. Biochem. Physiol. 1981, 16, 249-55.
44. Bathurst, I.C.; Brennan, S.O.; Carrell, R. W.; Cousens, L.S.; Brake, A.J.; Barr, P.J. Science 1987, 235, 348-50.
45. Cushman, D.W.; Cheung, H.S.; Sabo E.F.; Ondetti, M.A. Biochemistry 1977, 16, 5484-91.
46. Starratt, A.N.; Steele, R.W. Insect Biochem. 1984, 14, 97-102.
47. Steele, R.W.; Starratt, A.N. Insect Biochem. 1985, 15, 511-9.

RECEIVED February 4, 1988

Chapter 12

Insect Cuticle Structure and Metabolism

Karl J. Kramer[1,2], Theodore L. Hopkins[3], and Jacob Schaefer[4]

[1]U.S. Grain Marketing Research Laboratory, Agricultural Research Service, U.S. Department of Agriculture, Manhattan, KS 66502
[2]Department of Biochemistry, Kansas State University, Manhattan, KS 66506
[3]Department of Entomology, Kansas State University, Manhattan, KS 66506
[4]Department of Chemistry, Washington University, St. Louis, MO 63130

Insects have become the most diverse and numerous animal group partly because of evolution of a multifunctional integument, which provides for growth, mobility, protection and communication. The cuticle, which is secreted by the epidermis, is a composite of materials, primarily a polymeric structure of protein and chitin chains with lesser amounts of phenolics, lipids and minerals. The organization and interactions of these components, which are only partially understood, confer stability to the exoskeleton and provide for its varied functional properties. For several years, we have been investigating the identity and metabolism of phenolic compounds that permeate the exoskeleton and serve as precursors for agents that sclerotize and pigment the cuticle. More recently, we have used solid state nuclear magnetic resonance spectroscopy to determine cuticle composition and aromatic cross-link structure between protein amino acids and chitin. The chemical and spectroscopic data support a general scheme for assembly of cuticle with increasing amounts of protein and chitin, as well as a gradual accumulation of diphenolic compounds primarily into the outer parts of the cuticle during sclerotization. Aromatic cross-links derived from quinonoid derivatives stabilize the protein-chitin matrix against chemical and physical degradation, and confer stiffness and other essential mechanical properties. The degradation of cuticle periodically occurs as part of the molting process and is catalyzed by enzymes, which digest the less stabilized portion of the exoskeleton, leaving the highly sclerotized exocuticle and outer epicuticle to be shed as the exuviae. In order to understand how the cuticle is degraded naturally, hydrolytic enzymes from insect epidermis and entomopathogens have been characterized. The primary hydrolases are proteolytic and chitinolytic enzymes, which facilitate digestion and recycling of

0097–6156/88/0379–0160$07.50/0

cuticular components and the penetration of entomopathogens through the cuticle. The insect-specific metabolism that occurs in the integument is vital for growth and, therefore, is a good target for development of highly selective insecticides. Some of the physical, chemical and biological means of attacking the integument and disrupting cuticle formation and function are described in this article.

A continuing demand for more innovative and selective ways to control insect pests is necessitated by the capacity of insect populations to develop resistance to insecticides and by the need to minimize the impact of these toxic substances on the environment. One approach takes advantage of the evolutionary differences in biochemistry and physiology of insects and other types of organisms or between diverse groups of insects. This approach requires knowledge of specialized or unique life processes vital to the development, reproduction or survival of the pest insect, definition of the physical and chemical characteristics of the systems involved, and delineation of the pertinent biochemical steps. With over 300 million years separating the evolution of insects and higher animals, evolutionary design has generated several kinds of insect-specific metabolism that can be targeted by selective toxicants or growth regulators. Because the integument that makes up the exoskeleton is so important to the maintenance and protection of the insect's internal environment, respiration, sensory reception, locomotion and a multitude of other functions, it is a tissue system deserving detailed investigation, and researchers have studied its diverse features and functions for several decades (1-5). Although great progress has been made in understanding the secretion and stabilization of the insect exoskeleton and its multifunctional properties, many important questions remain to be answered that could provide new selective approaches to insect pest control.

The advantages of a cuticle that contains protein and chitin as biological structural materials are its lightness of weight, strength and flexibility. These properties have allowed insects to develop flight as a means of locomotion and dispersion and to inhabit diverse ecological niches. The major disadvantage of the exoskeleton is its inability to expand beyond certain physical limits, thereby constraining growth and requiring insects to molt periodically in order to grow and mature to the adult reproductive stage. During each molting cycle, a new cuticle must be secreted and stabilized, and the old one must be partially digested to make possible escape by splitting the latter along predetermined lines of weakness. The cuticle along these ecdysial lines consists mainly of epicuticle after digestion of the underlying endocuticle. The new cuticle then briefly becomes extensible to accommodate growth, followed by sclerotization and pigmentation to complete the functional exoskeleton. The cuticle of some soft bodied larval insects however, continues to grow during the intermolt period to accommodate growth during each stage. The patterns of cuticle that are stiffened or remain flexible are very specific for each stage of development and are determined by the underlying epidermal cells by mechanisms little understood.

The formation and molting of the insect exoskeleton require a myriad of enzymatic and endocrine processes that center principally in the epidermal cells. The chitinous and proteinaceous cuticle is a tempting target for control strategies, because it is chemically distinct from the integument of vertebrates, which is made up of keratinaceous and collagenous proteins. Although many aspects of the structure and chemistry of insect cuticle are known, much still remains to be learned about how protein and chitin are secreted and assembled into a cuticular structure and about the physical and chemical changes these macromolecules undergo as they become sclerotized and pigmented. Some of the research on cuticle structure and metabolism will be reviewed here, including discussion about composition, metabolites, enzymes, insect growth regulators and entomopathogens that affect the cuticle.

One aspect of biotechnology that may not be sufficiently appreciated by biologists is the application of high technological analytical instrumentation, such as nuclear magnetic resonance, to biology. For several years, we have been using solid state nuclear magnetic resonance spectroscopy and chemical analysis to study insect cuticle structure. The former technique is especially promising, because it allows study of intractable materials in a noninvasive manner. Our data support the cuticle model originally proposed by Pryor (6) in the 1940's, which depicts protein chains cross-linked by quinonoid derivatives of diphenolic compounds. We have also used chemical and kinetic procedures to study how molting fluid digests the unsclerotized layers of the old cuticle into component amino acids and amino sugars for recycling and the construction of a new one. For the sake of this discussion, cuticle will be classified into two general types that differ in mechanical properties. First, there are soft cuticles like those of most larvae, which are flexible and extensible and therefore, hydrostatic in nature. Second, there are hard cuticles like those of dipteran puparia, lepidopteran pupae, and coleopteran adults, which are highly sclerotized or mineralized, stiff and self-supporting. All of these structures contain protein, chitin, diphenols, lipids, water and mineral salts. Depending on functional demands, the relative levels of individual components can vary. Flexible cuticles tend to be more hydrated than stiffer cuticles and also contain fewer phenolic or inorganic stabilizing agents. Conversely, stiffer cuticles are more dehydrated and contain higher levels of phenolics or minerals. The phenolic compounds are associated with aromatic cross-links between macromolecular substituents of cuticle or serve as dehydrating or protein denaturing agents (1).

Insect cuticle is a heterogeneous structure that varies in chemical composition according to species, stage of development, appendage, location and physical property (1,7). In general, the thinner the cuticle, the weaker it becomes, but the blend of components, degree of compaction and hydration, as well as the number of cross-links, can render even a very thin cuticle quite tough. Epicuticle and the underlying exocuticle are the principle barriers to environmental challenges and, therefore, are stabilized by a greater abundance of aromatic cross-links, phenolics and lipids. For example, the endocuticular-rich intersegmental membranes and unions between different types of locust cuticle appear to be preferential entry sites for insecticidal compounds and fungal pathogens (8).

Insect supportive structures generally consist of a macromolecular assembly of protein, which is stabilized and dehydrated to various degrees either by aromatic cross-links or by deposition of diphenols or mineral salts (1). One of the simplest non-cuticular structures in chemical terms that undergoes some form of stabilization is moth cocoon silk, such as that from Bombyx mori, which is primarily made up of two proteins, the thread-like protein, fibroin, and the glue-like protein, sericin (9). Some cocoon silks may be cross-linked by diphenolic tanning agents (10) or by tryptophan metabolites (11). Chorion and egg capsules or cases from cockroaches, mantids, grasshoppers and other insects are other examples of proteinaceous structures that are tanned by cross-linking reactions involving diphenolic metabolites (6,12-15). Insect cuticle, however, is a composite of not only protein and diphenolic compounds, but also chitin as a major component, which results in a laminated framework and perhaps renders the cuticle flexible, strong and recyclable. Lipids, which are relatively minor components, occur primarily at or near the surface of the cuticle and in the waxy layer of the epicuticle. They may provide the major barrier against water loss, but they probably contribute little to the strength and shape of the cuticle. Mineral content is low in most cuticles (less than a few percent), except in those dipteran species that impregnate substantial amounts of minerals into the puparial cuticle to harden it (7,16-20).

Cuticle Sclerotization Versus Mineralization.

Why is sclerotization instead of mineralization the preponderant mechanism used by insects to stiffen their exoskeleton? Sclerotization utilizes relatively high levels of organic components such as proteins, chitin and diphenolic compounds for structural support and form whereas inorganic salts such as calcium carbonate and phosphate are incorporated into an organic matrix for the same purpose during biomineralization. The former mechanism is generally capable of producing a much lighter cuticle with requisite properties of strength and flexibility necessary for rapid terrestrial locomotion and flight. During sclerotization, diphenols are oxidized to reactive metabolites that cross-link and form covalent adducts with macromolecular components of the cuticle. During biomineralization, inorganic salts precipitate in a protein-chitin matrix, and in the process, a hard rigid cuticle is formed. Salts such as colloidal calcium phosphate have been reported to stabilize protein aggregates in the presence of chaotropic solvents, probably by forming intermolecular cross-links (21). However, the structure of the mineral cross-link has not been determined. Whether inorganic types of cross-linkages occur in mineralized insect cuticle is unknown.

During postembryonic development, the mechanical properties of cuticle are altered to accommodate changing functional demands. A classic example of this phenomenon is dipteran pupariation, in which the soft larval cuticle becomes stiffened to support and protect the developing adult (22,23). In the house fly, Musca domestica, the puparial cuticle is stabilized primarily by sclerotization, in which diphenolic compounds accumulate in the protein and chitin-rich larval cuticle (19). In the face fly, Musca autumnalis, the puparium is hardened primarily by the deposition of calcium and magnesium phosphate into the larval cuticle. By examining the chemical and physical properties of puparial exuviae of the house fly and face fly, the superiority of an organic strengthening material over an inorganic one is demonstrated. Table I lists the major organic and inorganic elements found in these two types of cuticles. Sclerotized house fly cuticle is composed of approximately five times more organic elements and tenfold fewer inorganic elements than mineralized face fly cuticle (19). Physically, the sclerotized cuticle is about 80% as dense and twice as flexible as the mineralized cuticle, and the former cuticle requires about three times more force to fracture it than the mineralized cuticle (Table II, 20). Approximately threefold more inorganic mass is required to make a cuticle of comparable strength to an organic one. Thus, organic constituents appear to assemble and stabilize a cuticle more effectively than inorganic elements. Except in environments where a heavy body weight and lack of mobility and flexibility are not detrimental to survival (for example, aquatic habitats), sclerotization is the mechanism of choice for cuticle stiffening. One apparent disadvantage of sclerotization is the amount of energy required to synthesize the diverse kinds of building materials. Assuming that minerals are readily available, less energy is probably needed to deposit minerals into a biological structure than to synthesize and deposit both low and high molecular weight organic components.

Table I. Major Elements in Sclerotized and Mineralized Insect Cuticles[a]

Element	% Dry Weight		Ratio (S:M)
	Sclerotized	Mineralized	
Carbon	45.3	9.7	4.7
Nitrogen	9.5	1.5	6.3
Hydrogen	6.8	3.3	2.1
Oxygen	29.7	23.7	1.3
Calcium	0.9	18.5	0.05
Magnesium	0.3	3.0	0.10
Phosphorus	0.3	9.9	0.03

[a]Data from Roseland et al. (19). Sclerotized and mineralized cuticles are puparial exuviae from house fly and face fly, respectively.

What are the major biochemical constituents of insect cuticles? We have used solid state cross polarization-magic angle spinning (CPMAS) NMR to identify and compare the levels of constituents in sclerotized and mineralized cuticles and to look for covalent cross-links (24). To estimate the chemical composition, we integrated resonances in ^{13}C-spectra with chemical shifts at 144 ppm for diphenol content, 104 ppm for chitin, 60 and 55 ppm for protein (after subtracting the contributions from chitin), and 33 ppm for lipid (after subtracting contributions from diphenol and protein carbons, see Fig. 1 and Table III). Water and mineral contents were determined by gravimetric and ash analyses, respectively. Whereas mineral salts comprise more than 60% of the face fly puparial exuvium, they make up only 3% of the house fly exuvium (Table IV). Sclerotized cuticle contains substantially more protein, chitin and diphenolic compounds than does mineralized cuticle. Note in particular the high abundance of diphenolic carbon resonances at 116 and 144 ppm in the spectrum of house fly puparia but not in that of face fly puparia (Fig. 1). Approximately 90% of the wet weight of house fly puparial cuticle is protein, chitin and diphenolic compounds, whereas those constituents account for less than 30% of face fly cuticle. Similar amounts of lipid and water are present in both types of cuticle.

Table II. Physical and Mechanical Properties of Sclerotized and Mineralized Insect Cuticles[a]

	Type of Cuticle		
Property	Sclerotized	Mineralized	Ratio (S:M)
Puparial diameter (mm)	2.6	2.7	1.0
Cuticular thickness (μm)	26.8	41.7	0.6
Density (g mm^{-2})	2.2	2.6	0.8
Stress modulus of fracture (kg mm^{-2})	23.6	8.1	2.9
Elastic modulus (kg mm^{-2})	711.5	346.4	2.1

[a]From Grodowitz et al. (20). Sclerotized and mineralized cuticles are puparial exuviae from house fly and face fly, respectively.

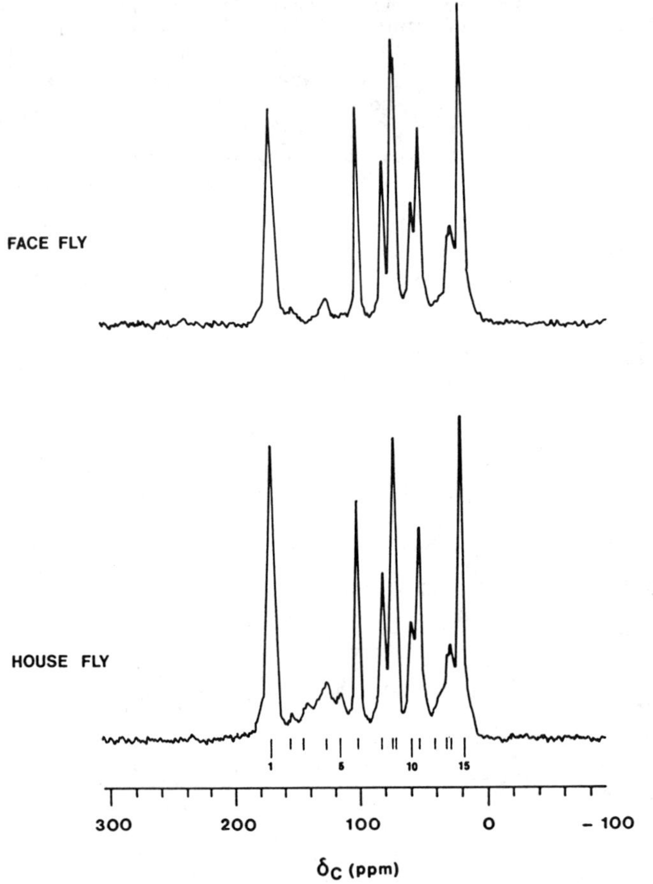

Figure 1 ^{13}C-CPMAS NMR spectra of face fly and house fly puparial exuviae (Kramer et al. unpublished data). See Table III for resonance assignments.

Table III. Chemical Assignments of Resonances in the CPMAS-^{13}C-NMR-Spectra of Insect Cuticles[a]

Resonance	δ-Values (ppm)	Assignment
1	172	Carbonyl carbon in chitin, protein, lipid and diphenol acyl groups
2	155	Phenoxy carbon in tyrosine, guanidino carbon in arginine
3	144	Phenoxy carbon in diphenols
4	129	Aromatic carbons
5	116	Tyrosine carbons 3 and 5, imidazole carbon 4, diphenol carbons 2 and 5
6	104	GlcNAc carbon 1
7	82	GlcNAc carbon 4
8	75	GlcNAc carbon 5
9	74	GlcNAc carbon 3
10	60	GlcNAc carbon 6, amino acid α-carbon
11	55	GlcNAc carbon 2, amino acid α-carbon
12	44	Amino acid, diphenol and lipid aliphatic carbons
13	33	Amino acid, diphenol and lipid aliphatic carbons
14	23	Methyl carbons in chitin, protein, lipid and diphenol acetyl groups, amino acid methyne carbons
15	19	Amino acid and lipid methyl carbons

[a]From Schaefer et al. (24). δ-Values relative to external TMS reference.

Table IV. Major Components of Sclerotized and Mineralized Insect Cuticles[a]

Component	Types of Cuticle		Ratio (S:M)
	Sclerotized	Mineralized	
Protein	31	10	3.1
Chitin	45	19	2.4
Diphenol	13	1	13
Lipid	2	2	1.2
Water	6	5	1.2
Mineral	3	63	0.05

[a]Data from Kramer et al., unpublished. Unit = percentage of wet weight. Sclerotized and mineralized cuticles are puparial exuviae from house fly and face fly, respectively. Protein, chitin, diphenol and lipid determined by ^{13}C-NMR; water by gravimetric analysis; and mineral by ash content.

In terms of amino acid and carbohydrate content liberated by acid hydrolysis, sclerotized cuticle contains more than twofold more total amino acids and also 2-acetamido-2-deoxy-D-glucopyranoside (as chitin) than does mineralized cuticle (Table V). As previously shown, we have confirmed that the amino acid β-alanine is virtually absent in the mineralized puparium of the face fly, whereas it is the second most abundant amino acid in the house fly puparium (19,25). We have also determined that much of the β-alanine is conjugated with the catecholamines N-β-alanyldopamine (NBAD), N-β-alanylnorepinephrine (NBANE) or their quinonoid adducts linked to macromolecules (26-28). Diphenol content of the two types of cuticle is even more dissimilar (Table VI). Whereas only dopamine at relatively low levels is present in mineralized cuticle, about 100-fold more diphenols occur in sclerotized cuticle, with NBAD being the major compound (19). Overall, sclerotized cuticles have higher concentrations of proteins, diphenols and chitin, whereas mineralized cuticles have much higher levels of inorganic salts, primarily calcium and magnesium phosphates and carbonates. Thus, the chemical composition of a particular cuticle reflects the relative contribution of either sclerotization or mineralization to its physical properties. The evolutionary choice of using an organic matrix of protein and chitin stabilized by intra- and intermolecular interactions and cross-links, together with lesser amounts of diphenols, lipid, minerals, pigments, water and other components for cuticle construction has provided the insect with many options in regard to cuticle properties and structure. By varying the kind and amount of components as well as their interactions, the epidermis may assemble cuticles with various degrees of hardness, flexibility or stiffness and pigmentation.

Table V. Amino Acid Composition (μmole g^{-1}) of Sclerotized and Mineralized Insect Cuticles[a]

Amino Acid	Sclerotized	Mineralized
ASP	208.4 ± 26.7	72.8 ± 2.9
THR	88.6 ± 10.1	41.3 ± 1.5
SER	78.5 ± 8.0	36.6 ± 0.2
GLU	214.5 ± 25.0	86.3 ± 3.1
PRO	115.7 ± 12.2	46.3 ± 0.9
GLY	319.5 ± 57.8	303.8 ± 9.9
α-ALA	123.4 ± 16.7	57.2 ± 0.8
VAL	134.4 ± 12.2	53.8 ± 1.9
ILU	58.3 ± 7.5	25.2 ± 1.2
LEU	76.1 ± 10.2	28.7 ± 1.6
TYR	53.6 ± 8.7	22.9 ± 0.8
PHE	47.8 ± 7.9	17.8 ± 1.2
β-ALA	299.2 ± 43.1	<5
HIS	227.4 ± 37.6	41.7 ± 4.7
LYS	85.8 ± 16.3	27.5 ± 3.3
ARG	41.2 ± 16.1	17.1 ± 2.9
Total amino acid relative ratio	1.0	0.4

[a]Data from Roseland et al. (19). Sclerotized and mineralized cuticles are puparial exuviae from house fly and face fly, respectively.

Table VI. Diphenol Composition (nmole g^{-1}) of Sclerotized and Mineralized Insect Cuticles[a]

Diphenol	Sclerotized	Mineralized
NBAD	312.0 ± 56.7	<1
NADA	27.2 ± 10.9	<1
DA	21.4 ± 1.5	5.0 ± 0.8
Total diphenol relative ratio	1.0	0.01

[a]Data from Roseland et al. (19). Sclerotized and mineralized cuticles are puparial exuviae from house fly and face fly, respectively.

In addition to studying the chemical composition of fly cuticles, we have also used solid state ^{13}C-NMR to measure the relative abundance of four major carbonaceous components, gravimetric analysis to determine moisture content, and ash analysis for mineral content of cuticles from three developmental stages of a single species, the tobacco hornworm, Manduca sexta (Fig. 2). There are substantial differences in composition between the larval, pupal

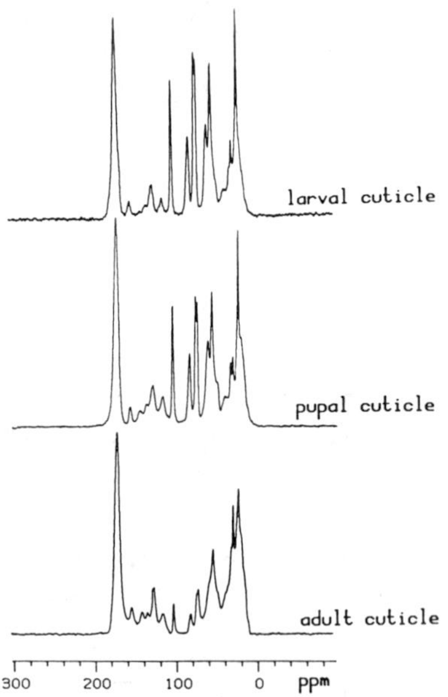

Figure 2 ^{13}C-CPMAS NMR spectra of M. sexta larval, pupal and adult cuticles. The larval and adult samples are devoid of cuticle from the head and appendages. Scales were removed from adult cuticle (from Kramer et al. 43).

and adult body cuticles (Table VII). The larval cuticle is very flexible and largely unpigmented, the pupal cuticle is stiff and brown in coloration, and the adult cuticle excluding scales is a thin and lightly colored stiff cuticle adapted for flight behavior. The major component in all of the Manduca body cuticles is water, which ranges from about 50 to 75% of the wet weight (Table VII). Chitin is the major organic component in larval and pupal cuticles, whereas protein is major in the adult. Diphenols and lipids range from 1 to 7%.

Table VII. Major Components of Manduca sexta Cuticles During Development[a]

Component	Stage of Development		
	Larvae	Pupa	Adult
Water	73	49	52
Protein	10	19	28
Chitin	14	25	7
Diphenol	1	4	7
Lipid	1	2	5
Mineral	1	1	1

[a]Data from Kramer et al. unpublished. Unit = percentage of wet weight. Protein, chitin, diphenol and lipid determined by ^{13}C-NMR; water by gravimetric analysis; and mineral by ash content. Cuticles from 3-day old fifth instar, pupae and adult, respectively.

The major chemical components present in some other hornworm cuticular structures are shown in Table VIII. The composition of individual sclerites or structures in each stage can vary greatly from the composition of the general body cuticle (Table VII). Digestion of the larval or pupal endocuticle by molting fluid hydrolytic enzymes results in an epicuticle-exocuticle complex or exuviae. Larval exuviae are similar in composition to the larval cuticle, except for a three-fold increase in the level of diphenolic compounds. Two of the hardest larval structures are the exuviae of head capsule and mandible. These are characterized by a high abundance of diphenols as are the pupal body exuviae and the adult wing, which is one of the thinnest types of cuticle. The diphenols in the hard dark mandibles are primarily N-β-alanyl derivatives of dopamine and norepinephrine, particularly the former, whereas the colorless head capsule cuticle contains mostly N-acetyldopamine (27,28). The mandibles also contain zinc or manganese at their cutting edges in order to increase density and fracture toughness (29,30). Of all the cuticle structures that we have analyzed by solid state NMR, the one with the highest diphenol content is the elytron or wing cover of the red flour beetle, Tribolium castaneum. It is composed of 26% diphenols together with 33% chitin, 29% protein, 4% lipid and 8% water. Most of the diphenol content is

3,4-dihydroxyphenylacetic acid (28,31). The red flour beetle elytron and Manduca pupal exuviae are remarkably similar in appearance and also chemical composition, with the exception of diphenol composition. All of our data on cuticle composition suggest that the lightest and strongest cuticles are characterized by substantial amounts of protein and chitin, together with 5 to 25% diphenols. The diphenols primarily act as cross-linking or adduct-forming precursors, cementing the cuticular protein and chitin macromolecular chains together. Other functions suggested for the high levels of unreacted diphenols easily extracted from cuticle are dehydrating agents, lipid antioxidants and antimicrobial substances (32,33).

Chemical Changes During Sclerotization.

We have used solid state ^{13}C-NMR to follow the changes in chemical composition that occur when M. sexta pupal cuticle becomes sclerotized (Table IX). The soft, newly ecdysed pupal cuticle is mostly water (80%) and protein (14%), with minor amounts of chitin, diphenol and lipid (<3%). The sclerotized pupal exuviae contain sixfold less water, twofold more protein and 17-fold more of both chitin and diphenol. Thus, the process of pupal cuticle sclerotization consists of dehydration together with substantial enrichment of chitin, protein and diphenolic compounds.

Table VIII. Major Components of Manduca sexta Cuticular Structures[a]

Component	Larval			Pupal	Adult
	Exuvium	Head Capsule	Mandibles	Exuvium	Wings
Water	69	59	32	13	41
Protein	15	14	24	31	34
Chitin	11	12	18	34	12
Diphenol	3	13	21	17	8
Lipid	1	1	3	4	4
Mineral	1	1	2	1	1

[a]Data from Kramer et al. unpublished. Unit = percentage of wet weight. Protein, chitin, diphenol and lipid determined by ^{13}C-NMR; water by gravimetric analysis; and mineral by ash content.

Table IX. Major Components of *Manduca sexta* Unsclerotized and Sclerotized Pupal Cuticles[a]

Component	Unsclerotized	Sclerotized	Ratio (S:U)
Water	80	13	0.16
Protein	14	31	2.2
Chitin	2	34	17
Diphenol	1	17	17
Lipid	3	4	1.3
Mineral	1	1	1

[a]Data from Kramer et al. unpublished. Unit = percentage of wet weight. Water determined by gravimetric analysis; protein, chitin, diphenol and lipid by ^{13}C-NMR; and mineral by ash content. Unsclerotized cuticle from newly ecdysed pupa; sclerotized cuticle is pupal exuvium.

A determination of the chemical composition of the outer forewing pupal cuticle of *M. sexta* reveals that a major difference between the sclerotized and unsclerotized layers is the concentration of diphenols. Three-day-old pupal outer forewing cuticle was separated into outer tanned and inner untanned portions and subjected to diphenol and amino acid analyses. There is greater than 50 times more NBAD, NBANE, dopamine (DA) and 3,4-dihydroxyphenylethanol-4-O-sulfate (DOPET-SO_4) in the sclerotized layer than in the unsclerotized portion of the cuticle (Table X). However, the amino acid compositions are very similar in both portions, except for β-alanine, the bulk of which is conjugated with DA and norepinephrine (NE) (Table XI). Thus, the accumulation of substantial amounts of diphenols into the outermost portion of the cuticle correlates with its high degree of sclerotization.

Table X. Diphenol Composition of Outer Forewing Cuticle of *M. sexta* Pupa[a]

Diphenol	Concentration (nmole g^{-1}) Endo	Exo	Ratio (Exo:Endo)
NBAD	4	2440	610
NBANE	5	450	90
DA	1	57	57
NADA	3	13	4
DOPET-SO_4	10	1200	120

[a]Data from Morgan et al. unpublished. Cuticle from 3-day old pupal outer forewing.

Table XI. Amino Acid Composition of Outer Forewing Cuticle of *M. sexta* Pupa[a]

Amino Acid	Mole % Endo	Mole % Exo	Ratio (Exo:Endo)
ASP	5.8	5.1	0.9
THR	4.0	3.4	0.9
SER	7.3	7.4	1.0
GLU	5.9	5.0	0.9
GLY	5.4	5.7	1.1
ALA	19.5	17.8	0.9
VAL	4.2	4.2	1.0
ILU	2.2	2.2	1.0
LEU	5.4	4.6	0.9
TYR+PHE+GlcN	25.8	24.3	0.9
β-ALA	<1	6.3	>6
HIS	10.2	10.2	1.0
LYS	2.0	2.0	1.0
ARG	1.8	1.8	1.0

[a]Data from Kramer et al. unpublished. Cuticle from 3-day old pupal outer forewing. Hydrolysis in 6 N HCl containing 5% phenol at 110°C for 22 h *in vacuo*.

Covalent bonds between protein nucleophilic groups and the aromatic or aliphatic carbons of diphenols have been proposed to occur in sclerotized cuticles (2,34-37). We have used solid state NMR to probe for covalent bond formation between cuticular components in M. sexta pupal cuticle (24). NMR analyses of cuticle containing protein labeled with either 1,3-[$^{15}N_2$]-histidine or ε-[^{15}N]-lysine demonstrated that a side chain histidyl or lysyl nitrogen becomes attached to a carbon atom (N-aryl or N-alkyl) during sclerotization. After the pupal cuticle was doubly labeled with both 1,3-[$^{15}N_2$]-histidine and ring-[$^{13}C_6$]dopamine and subjected to NMR analysis, the double cross polarization spectrum revealed that one of the aromatic catecholamine carbons is covalently bonded to a ring nitrogen of histidine (Fig. 3). This aromatic carbon-nitrogen cross-link structure is consistent with an imidazoyl nitrogen attacking a phenyl carbon of an o-quinone derivative of the diphenolic compound.

Also it has been proposed that bonds exist between cuticular protein and chitin (38). Solid state ^{15}N-NMR analysis of chitin prepared by alkali extraction of 1,3-[$^{15}N_2$]-histidine labeled *M. sexta* pupal exuviae revealed an ^{15}N chemical shift expected for the substituted imidazole nitrogen cross-link structure depicted in Fig. 3 (24). Apparently, the chitin is not coupled directly to protein but, instead, to a diphenolic carbon, which serves as a part of the bridge between protein and chitin macromolecules.

Figure 4 shows the proposed mechanism for the metabolism of phenolic and diphenolic compounds to quinonoid cross-linking agents for macromolecules in the outer portions of dark brown cuticle from

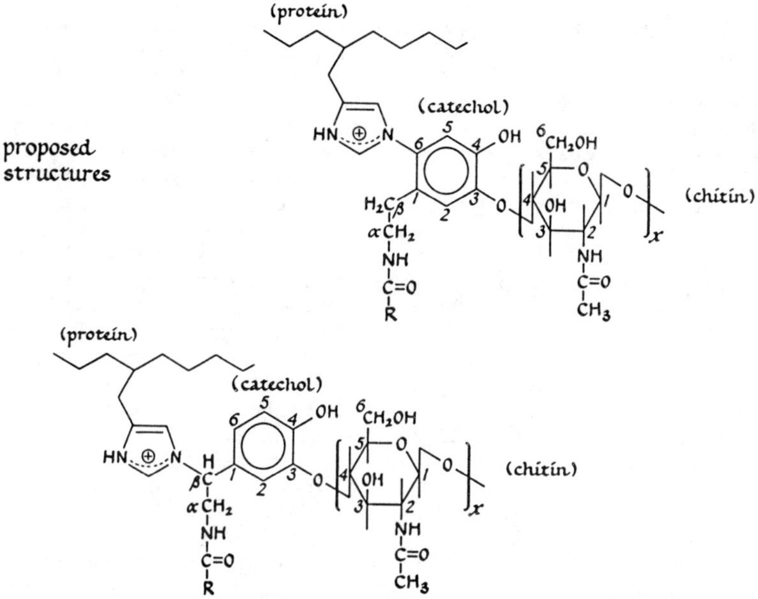

proposed structures

Figure 3 Proposed structure for diphenol mediated cross-link between protein and chitin in *M. sexta* pupal cuticle (from Schaefer et al. 24).

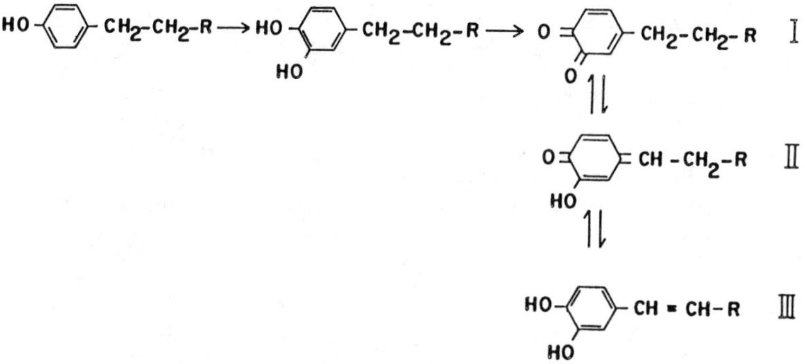

Figure 4 Proposed pathway of phenol metabolism for sclerotization of cuticle including monophenol, diphenol, o-quinone (I), *p*-quinone methide (II) and α,β-dehydrocatechol (III) derivatives. Reactions catalyzed by phenoloxidase or tyrosinase followed by keto-enol tautomerization.

structures such as tobacco hornworm pupae, house fly puparia and flour beetle elytra. N-Acylcatecholamines, such as NBAD and NBANE, are converted by oxidative enzymes, such as tyrosinases, phenoloxidases, peroxidases or laccases, to o-quinone, _p_-quinone methide, semiquinone or free radical electrophilic intermediates, which then cross-link cuticular proteins and chitin (33,39-41). The N-β-alanylcatecholamines are generally more associated with stiff brown cuticles and the N-acetylcatecholamines with stiff colorless cuticle, whereas excessive dopamine polymerizes into black pigment, such as melanin, in the outer parts of the cuticle (26,27,31,42). N-Acylation of dopamine with β-alanine or acetate may facilitate electron delocalization from the aromatic ring carbons of the o-quinone to the aliphatic side chain α- or β-carbon and favor production of the tautomeric _p_-quinone methide or α,β-dehydro diphenol. These tautomers provide multiple carbon sites for attack by nucleophilic groups, which ultimately lead to the formation of adducts or cuticular cross-links (24,36,41,43).

Noncuticular structures have been proposed as model cuticles in which similar hardening reactions occur, but these materials are generally less complex in terms of chemical composition, even though they undergo chemical transformations similar to those that occur during cuticular tanning. Structures such as cocoons and egg cases of insects are composed mostly of protein but may also contain some diphenolic compounds, which tan the protein (Table XII). They consist of little or no chitin or lipid. The proteinaceous _Drosophila_ chorion is cross-linked by an endogeneous peroxidase that causes formation of interpolypeptide di- and tri-tyrosyl adducts (44). The silkmoth, _Bombyx mori_, makes a cocoon that is untanned and consists of the fibrous protein fibroin and the protein glue sericin. Protein but no diphenols are evident in the ^{13}C-NMR spectrum of _B. mori_ cocoons (Table XII). On the other hand, the cocoons of _Anthereae polyphemus_, _A. mylitta_ and _Hyalophora cecropia_ are tanned and contain not only protein but also 2-3% diphenols. The egg case of the praying mantis, _Tenodera sinensis_, is composed of 95% protein and 5% diphenol. In these structures, which provide protective housings for developing pupae or eggs, the diphenolic compounds probably provide strength by cementing (cross-linking) the protein chains together. Thus, diphenolic compounds or their oxidized metabolites may sclerotize structures that contain primarily both protein and chitin or only protein.

Table XII. Major Organic Components of Noncuticular Insect Structures[a]

Component	Cocoons				Egg case
	B. mori	_A. polyphemus_	_A. mylitta_	_H. cecropia_	_T. sinensis_
Protein	100	98	98	97	95
Diphenol	<1	2	2	3	5

[a]From Kramer et al. unpublished. Unit = percentage of total carbon. Determined by ^{13}C-NMR. Lipid and chitin were <1%.

The presence of substantial amounts of diphenolic compounds in insect cuticular structures does not necessarily mean that these structures will be sclerotized. Only when the diphenol is oxidized to a cross-linking agent will sclerotization take place. The soft integument of *M. sexta* fifth larval instars accumulates large amounts of NBAD during the feeding period but loses that catecholamine as pupation nears (Fig. 5, Krueger et al., unpublished data). In vitro culture of integument supplemented with dopamine revealed that NBAD is synthesized by the integument (epidermis) throughout the instar but at the highest rate on both the last day of larval feeding and 6 days later when the pharate pupal cuticle is being deposited (Fig. 6). Fat body synthesizes substantially lesser quantities of N-acylated catecholamines than integument. Other diphenols occur at levels 10-fold or more lower. During the latter part of the fifth instar, the NBAD titer in the hemolymph is inversely related to that in the integument (Fig. 5). Thus, NBAD is prevalent in the larval integument during the intermolt period, but at the end of the fifth instar, hemolymph is the major source of NBAD. Apparently, the larval integument and hemolymph of *M. sexta* are storage sites for NBAD during the intermolt period. Almost 90% of the integumental diphenol is localized in the cuticle and only about 10% in the epidermis (Krueger et al. unpublished data). The reciprocal nature of NBAD in integument and in hemolymph suggests that NBAD translocation occurs during the intermolt, perhaps as part of a cuticle recycling process. Cuticle may also serve as a storage site for protein and carbohydrate. For example, in the tsetse fly, cuticular protein and chitin are turned over in the female to supplement ingested blood meals, which are used to support rapidly growing larvae in utero (45).

How Is the Cuticle Attacked in Nature? The cuticle acts as a barrier to environmental hazards, such as predators, pathogens and toxic substances. Predators may mechanically penetrate the cuticle while the insect, when molting, and also some microorganisms use hydrolytic enzymes to degrade it. Entomopathogenic fungi, such as *Metarhizium anisopliae*, *Beauveria bassiana* and *Verticilium lecanii*, penetrate through the insect cuticular barrier by secreting extracellular proteolytic, chitinolytic and lipolytic enzymes, which hydrolyze some of the major components (46-53). The higher the degree of sclerotization the lower the susceptibility of cuticle to enzymatic attack by fungal or molting enzymes. Penetration of pathogens is likely to be greater in soft bodied or unsclerotized endocuticular portions than in hard bodied or sclerotized exocuticular portions. If all other things were equal, we could predict that larval stages of Coleoptera and Lepidoptera would be more susceptible to fungal pathogens than adult stages because of differences in cuticular mechanical properties. To our knowledge, this hypothesis has not been tested.

Results of *in vitro* and model experiments indicate that proteases and lipases attack the cuticle earlier than the chitinases and that, together, those enzymes act in a synergistic manner to solubilize the cuticle. When grown in cultures containing comminuted cuticle, fungi produce endoproteases, exoproteases, lipases, esterases, chitinases and β-N-acetylglucosaminidases, with

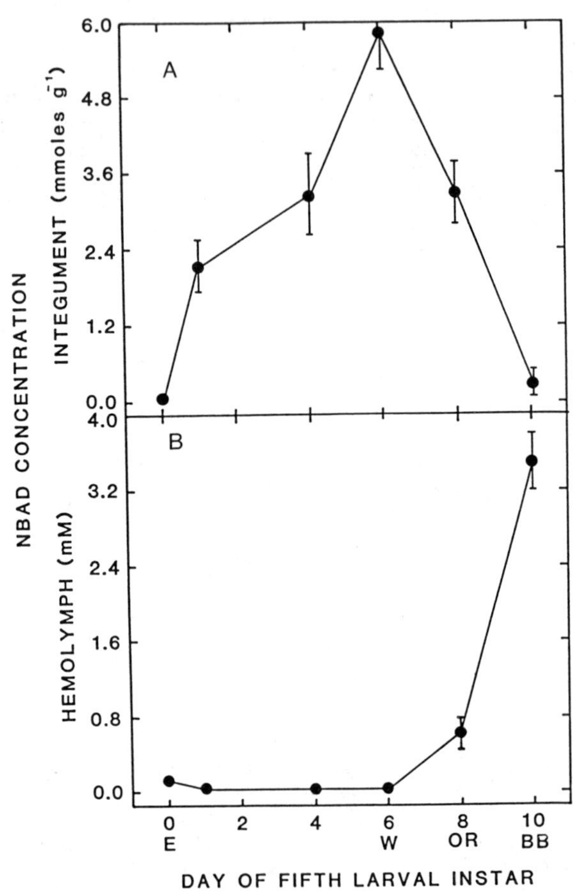

Figure 5 NBAD in integument (A) and hemolymph (B) of M. sexta during fifth larval instar (from Krueger et al. unpublished). E = ecdysis, W = wandering, OR = ocellar retraction during apolysis, BB = brown metathoracic bar stage of pharate pupa.

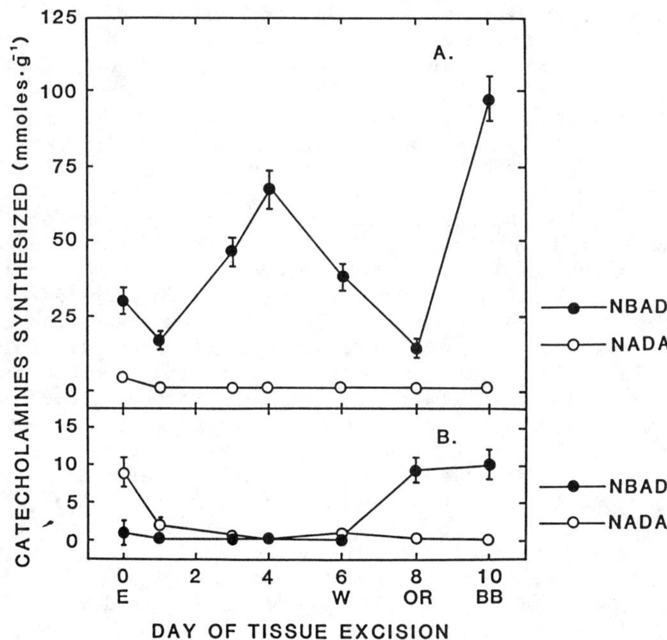

Figure 6 Synthesis of NBAD and NADA by integument (A) and fat body (B) in vitro from M. sexta fifth larval instar (from Krueger et al. unpublished). E = ecdysis, W = wandering, OR = ocellar retraction during apolysis, BB = brown metathoracic bar stage of pharate pupa.

the proteolytic enzymes appearing early and chitinolytic enzymes later (51-53). Smith _et al._ (54) found that chitin-protein skeletal "ghosts" extracted from corn earworm cuticle are digested more effectively by a combination of enzymes (proteases followed by chitinases). Most of the cuticular chitin is apparently shielded or masked by proteins, since pretreatment of cuticle with proteases enhances chitin hydrolysis by chitinases (49).

The Cuticle As a Target For Selective Control of Insect Pests. Biotechnology offers us the opportunity of using gene products that have insecticidal properties for insect control (55). For an anticuticle genetic application to be developed, it is necessary to identify and characterize genes that control the biosynthesis and degradation of cuticular components, including the chitin-protein procuticle matrix and the diphenolic sclerotizing agents. The potential for manipulating the genetic control of cuticle degradation is promising, especially considering that insects and some entomopathogens already produce enzymes that break down the cuticle during their life cycles. Genes that code for enzymes having the capacity to degrade integral protective structures outside or inside the insect can be engineered into vectors such as bacteria, fungi or viruses or into plants. Genes that code for anticuticle enzymes include those coding for cuticular chitinases, proteases and lipases. It has already been demonstrated that plant chitinases are potent inhibitors of fungal growth, but those enzymes have not yet been tested for inhibition of insect growth (56). In order to digest the cuticle most effectively, a super vector or several vectors that elaborate the best anticuticle proteolytic, chitinolytic and lipolytic enzymes might be constructed. At present, several laboratories are characterizing molting fluid and microbial enzymes and their genes, with a long-term goal of manipulating enzyme levels using chemical and biotechnological procedures for the purpose of insect control. The characterization of promoter and structural genes for anticuticle enzymes may facilitate the introduction of those genes into microbes and plants.

The most common and effective anticuticle devices that have been used by humans to control insects are mechanical ones, such as flyswatters that physically crush the exoskeleton. However, these devices are not adaptable to widescale pest control. Sorptive or abrasive dusts including clays, silicas, diatomaceous earths and carborundums, which damage the cuticle primarily by destroying the lipid water proofing layers, have also been used for insect control but with limited success (57). There are naturally occurring and synthetic compounds whose insecticidal activities have been at least partially attributed to a disruption of cuticle formation. One class of anticuticle compounds are the juvenile hormone (JH) analogues which may not only prevent maturation but also probably target JH receptors in tissues such as epidermal cells, affecting cuticle morphology and structure. Commercially developed compounds, such as methoprene, hydroprene and fenoxycarb, apparently alter cuticular composition of protein and diphenol levels (58,59). Another class of anticuticle compounds are the so-called chitin synthesis inhibitors or benzoylphenylureas, which alter chitin levels but whose mode of action at the molecular level remains obscure (57). Originally proposed to inhibit the epidermal enzyme, chitin synthase, their exact site of action is not a direct interaction with that enzyme, since cell free preparations are uninhibited. Mitsui et al. (60,61) suggested that the benzoylphenylureas prevent chitin synthesis by blocking transport of uridine diphospho-N-acetylglucosamine across biomembranes. Another antichitin inhibitor is the natural product, allosamidin, a basic pseudotrisaccharide consisting of two molecules of 2-acetamido-2-deoxy-D-allose and an aminocyclitol derivative (62). It competitively inhibits B. mori chitinase in vitro at least 500-fold better than it inhibits bacterial or plant chitinases, lysozyme or β-N-acetylglucosaminidase (63). It also prevents ecdysis of Leucania separata fifth larval instars to pupae (Sakuda, S., unpublished data).

Cyromazine is a substituted triazine insecticide that produces necrotic lesions in the cuticle of the sheep blow fly, Lucilia cuprina (64,65), and both stiffens larval cuticle and causes histological damage in the flesh fly, Sarcophaga bullata (A. Iseki, unpublished data). Although cuticle appears to be the primary target of cyromazine, its chemical mode of action remains unknown. Several substituted benzimidazole derivatives exhibit various insect growth regulating effects, including the inhibition of silkworm molting (66). It has been proposed that this class of compounds arrests cuticle formation by inhibiting mitochondrial respiration in a manner like some respiratory chain inhibitors and uncoupling agents (67). Whether oxidative enzymes that tan the cuticle are also inhibited is unknown. A number of 1,5-disubstituted imidazoles exhibit antijuvenile hormone activity, which can be counteracted by juvenile hormone (68,69). One of these apparently inhibits ecdysone synthesis by the prothoracic glands of B. mori in vitro with an ID =1 nM (71). The structure of some of these imidazole compounds resembles that of the carbon-nitrogen cross-link structure in M. sexta pupal cuticle, which involves a histidine residue (24). Certain hindered phenols are compounds that are highly toxic to Diptera (70,72,73). One of the di-tert-butylphenolic derivatives, MON-0585, apparently interferes with sclerotization of mosquito cuticle by inhibiting arylation of cuticular proteins with quinonoid derivatives which results in malfunctions of the breathing tubes of the pupae (74,75). There are other anticuticle insecticides including dopa decarboxylase inhibitors, that have been reviewed previously by Chen and Mayer (57) and Kramer and Koga (38).

Concluding Remarks

Insect cuticle serves both as an exoskeleton and skin. It is a highly evolved, multifunctional, supramolecular system that is in large part responsible for the great success of insects in achieving flight over 300 million years ago and dominating the terrestrial environments of the earth in both diversity and numbers. This unique structure also offers a challenging diversity of research problems for understanding how insects periodically synthesize the component molecules and assemble and stabilize these into the finished exoskeleton during each molt. Although aspects of cuticle synthesis and degradation have been extensively researched, little is known about the regulation and genetics of macromolecular metabolism, orientation and interactions. Many difficulties have been encountered in the study of cuticle metabolism, chief among them being the lability of cuticular oxidative enzymes and reactive metabolic intermediates, as well as the intractable architecture of the cuticle. The role of molting fluid in recycling cuticular components is also poorly understood. Significant progress has been made recently on elucidating the nature of carbon-nitrogen cross-links and reactants with the newer procedures of solid state NMR, electrochemistry and high performance liquid chromatography. The importance of N-acylated catecholamines, such as NADA and NBAD, in cuticle morphogenesis is now better understood (37) and their detection in neural tissue suggests additional roles in neurophysiology (76,77, Krueger et al. unpublished). We know very little about the nature of carbon-carbon and carbon-oxygen cross-links in cuticle. Much more needs to be learned about how insects precisely regulate synthesis and secretion of cuticular components in synchrony with the molting cycle and also the role of the nervous and hormonal systems in this regulation. The cuticle and its overall regulation offer a variety of targets for selective control of insect pests and a number of these approaches have already been exploited with some success. The potential of developing better microbial or chemical products as anticuticle insecticides appears to be substantial. As academic, governmental and industrial researchers continue to elucidate the novel features of cuticle structure and function, many new opportunities will undoubtedly become available for selective disruption of insect development.

Acknowledgments

This article is dedicated to Carroll Williams who has made enormous contributions to insect developmental biology and continues to inspire researchers in insect science. This research was conducted by U.S. Grain Marketing Research Laboratory, Agricultural Research Service, U.S. Department of Agriculture in cooperation with the Kansas Agricultural Experiment Station, Manhattan, KS 66506. This is contribution no. 88-271-J from the Kansas Agricultural Experiment Station, Manhattan, KS 66506. Supported in part by U.S. Department of Agriculture competitive research grant 85-CRCR-1-1667 and by National Science Foundation grant DCB 86-09717. We are grateful to Thomas D. Morgan, Renee Krueger, Roy D. Speirs, Alberto B. Broce, Michael Grodowitz, Craig R. Roseland, Vincent Bork, Joel Garbow, Gary Jacob and Ed Stejskel for discussion and assistance with these studies.

Literature Cited

1. Hepburn, H. R. In Compre. Insect Physiol. Biochem. Pharamcol.; Pergamon: New York, 1985; Vol. 3, p 1.
2. Andersen, S. O. In Compre. Insect Physiol. Biochem. Pharmacol.; Pergamon: New York, 1985; Vol. 3, p 59.
3. Kramer, K. J.; Dziadik-Turner, C., Koga, D. In Compre. Insect Physiol. Biochem. Pharamcol.; Pergamon: New York, 1985; Vol. 3, p 75.

4. Blomquist, G. J.; Dillwith, J. W. In Compre. Insect Physiol. Biochem. Pharamcol.; Pergamon: New York, 1985; Vol. 3, p 117.
5. Hadley, N. F. Sci. Amer., 1986, 254, 104.
6. Pryor, M. G. M. Proc. Roy. Loc. B., 1940, 128, p 378.
7. Hackman, R. H. In The Physiology of Insecta; Academic Press: NY, 1974; Vol. 6, p 215.
8. Fontan, A.; Zerba, W. N. Arch. Insect Biochem. Physiol. 1987, 4, 313.
9. Seifter, S.; Gallop, P. M. In The Proteins; Academic Press: New York, 1966; Vol. 4, p 201.
10. Kawasaki, H.; Sato, H. Insect Biochem. 1985, 6, 681.
11. Brunet, P. C. J.; Coles, B. C. Proc. Roy. Soc. B., 1974, 187, p 133.
12. Bodine, J. H.; Allen, T. H. J. Cell Comp. Physiol. 1941, 18, 151.
13. Kawasaki, H.; Yago, M. Insect Biochem. 1983, 13, 267.
14. Yago, M.; Sato, H.; Kawasaki, H. Insect Biochem. 1984, 14, 7.
15. Yago, M.; Kawasaki, H. Insect Biochem. 1984, 14, 487.
16. Hackman, R. H. In Physiol. Insecta; Academic Press: New York, 1964; Vol. 3, p 471.
17. Gilby, A. R; McKellar, J. W. J. Insect Physiol. 1976, 22, 1465.
18. Darlington, M. V.; Meyer, H. S.; Graf, A.; Freeman, T. P. J. Insect Physiol. 1983, 29, 157.
19. Roseland, C. R.; Grodowitz, M. J.; Kramer, K. J.; Hopkins, T. L.; Broce, A. B. Insect Biochem. 1985, 15, 521.
20. Grodowitz, M. J.; Roseland, C. R.; Hu, K. K.; Broce, A. B.; Kramer, K. J. J. Exp. Zool. 1987, 243, 201.
21. Aoki, T; Kawahara, A; Kako, Y.; Imamura, T. Agric. Biol. Chem. 1987, 51, 817.
22. Fraenkel, G.; Hsiao, C. J. Insect Physiol. 1967, 13, 1387.
23. Zdarek, J; Fraenkel, G. J. Exp. Zool, 1972, 179, 315.
24. Schaefer, J.; Kramer, K. J.; Garbow, J. R.; Jacob, G. S.; Stejskol, E. O.; Hopkins, T. L.; Speirs, R. D. Science 1987, 235, 1200.
25. Bodnaryk, R. P. Insect Biochem. 1972, 2, 119.
26. Hopkins, T. L.; Morgan, T. D.; Aso, Y.; Kramer, K. J. Science 1982, 217, 364.
27. Hopkins, T. L.; Morgan, T. D.; Kramer, K. J. Insect Biochem. 1984, 14, 533.
28. Morgan, T. D.; Hopkins, T. L.; Kramer, K. J.; Roseland, C. R.; Czapla, T. H.; Tomer, K. B.; Crow, F. W. Insect Biochem. 1987, 17, 255.
29. Hillerton, J. E.; Vincent, J. F. V. J. Exp. Biol. 1982, 101, 333.
30. Hillerton, J. E.; Robertson, B.; Vincent, J. F. V. J. Stored Prod. Res. 1984, 20, 133.
31. Kramer, K. J.; Morgan, T. D.; Hopkins, T. L.; Roseland, C. R.; Aso, Y.; Beeman, R. W.; Lookhart, G. L. Insect Biochem. 1984, 14, 293.
32. Vincent, J. F. V.; Hillerton, J. E. J. Insect Physiol. 1979, 25, 653.
33. Atkinson, P. W.; Brown, W. V.; Gilby, A. R. Insect Biochem. 1973, 3, 309.
34. Andersen, S. O. Ann. Rev. Entomol. 1979, 24, 29.

35. Brunet, P. C. J. Insect Biochem. 1980, 10, 467.
36. Lipke, H.; Sugumaran, M.; Henzel, W. Adv. Insect Physiol. 1983, 17, 1.
37. Kramer, K. J.; Hopkins, T. L. Arch. Insect Biochem. Physiol. 1987, 4, in press.
38. Kramer, K. J.; Koga, D. Insect Biochem. 1986, 16, 851.
39. Kramer, K. J.; Nuntnarumit, C.; Aso, Y.; Hawley, M. D.; Hopkins, T. L. Insect Biochem. 1983, 13, 475.
40. Hasson, C.; Sugumaran, M. Arch. Biochem. Biophys. 1987, 5, 13.
41. Sugumaran, M. In Molecular Entomology; UCLA Symposia; A. R. Liss: New York, 1987; Vol. 49, p 357.
42. Roseland, C. R.; Kramer, K. J.; Hopkins, T. L. Insect Biochem. 1987, 17, 21.
43. Kramer, K. J.; Hopkins, T. L.; Schaefer, J.; Morgan, T. D.; Garbow, J. R.; Jacob, G. S.; Stejskal, E. O.; Speirs, R. D. In Molecular Entomology; UCLA Symposia; A. R. Liss: New York, 1987; Vol. 49, p 331.
44. Mindrinos, M. N.; Petri, W. H.; Galanopoulos, V. K; Lombard, M. F.; Margaritis, L. H. Wilhelm Roux's Arch. 1980, 189, 187.
45. Solowiej, S.; Davey, K. G. Arch. Insect Biochem. Physiol. 1987, 4, 287.
46. Gabriel, B. P. J. Invert. Path. 1968, 11, 70.
47. Leopold, J.; Samsinakova, A. J. Invert. Path. 1970, 15, 34.
48. Samsinakova, A.; Misikova, S.; Leopold, J. J. Invert. Path. 1971, 18, 322.
49. St. Leger, R. J.; Cooper, R. M.; Charnley, A. K. J. Invert. Path. 1986a, 47, 167.
50. St. Leger, R. J.; Charnley, A. K.; Cooper, R. M. J. Invert. Path. 1986b, 47, 295.
51. St. Leger, R. J.; Charnley, A. K.; Cooper, R. M. J. Invert. Path. 1986c, 48, 85.
52. St. Leger, R. J.; Charnley, A. K.; Cooper, R. M. Arch. Biochem. Biophys. 1987, 253, 221.
53. St. Leger, R. J.; Cooper, R. M.; Charmley, A. K. J. Gen. Microbial. 1987, 133, 1971.
54. Smith, R. J.; Pekrul, S.; Grula, E. A. J. Insect. Path. 1981, 38, 335.
55. Kirschbaum, J. B. In Molecular Entomology; UCLA Symposia; A. R. Liss: New York, 1987; Vol. 49, p 465.
56. Schlumbaum, A.; Mauch, F.; Vogeli, U.; Boller, T. Nature 1986, 324, 365.
57. Chen, A.C.; Mayer, R. T. In Compre. Insect Physiol. Biochem. Pharmacol.; Insect Control, Pergamon: New York, 1985; Vol. 12, p 57.
58. Mitsui, T.; Riddiford, L. M. Devel. Biol. 1978, 62, 193.
59. Brooks, G. T. J. Pesticide Sci. 1987, 12, 113.
60. Mitsui, T.; Nobusawa, C.; Fukami, J. J. Pest. Sci. 1984, 9, 19.
61. Mitsui, T.; Tada, M.; Nobusawa, C.; Yamaguchi, I. J. Pest. Sci. 1985, 10, 55.
62. Sakuda, S.; Isogai, A.; Matsumoto, S.; Suzuki, A.; Koseki, K. Tetrahedran Lett. 1986, 27, 2475.
63. Koga, D.; Isogai, A.; Sakuda, S.; Matsumoto, S.; Suzuki, A.; Kimura, S.; Ide, A. Agric. Biol. Chem. 1987, 51, 471.

64. Binnington, K. C. Tissue Cell 1985, 17, 131.
65. Binnington, K. C.; Retnakaran, A.; Stone, S.; Skelly, P. Pest. Biochem. Physiol. 1987, 27, 201.
66. Kuwano, E.; Sato, N.; Eto, M. Agric. Biol. Chem. 1982, 46, 1715.
67. Nakagawa, Y.; Kuwano, E.; Eto, M.; Fujita, T. Agric. Biol. Chem. 1985, 49, 3569.
68. Asano, S.; Kuwano, E.; Eto, M. Appl. Ent. Zool. 1986, 21, 63.
69. Kuwano, E.; Eto, M. Agric. Biol. Chem. 1986, 50, 2919.
70. Walton, B. T.; Sanborn, J. R.; Metcalf, R. L. Pest. Biochem. Physiol. 1979, 12, 23.
71. Yamashita, O.; Kadono-Okuda, K.; Kuwano, E.; Eto, M. Agric. Biol. Chem. 1987, 51, 2295.
72. Jurd, L.; Fye, R. L.; Morgan, J. J. Agric. Food Chem. 1979, 27, 1007.
73. Van Mellaert, H.; DeLoof, A.; Jurd, L. Ent. Exp. Appl. 1983, 33, 83.
74. Zomer, E.; Lipke, H. Pest. Biochem. Physiol. 1981, 16, 28.
75. Semensi, V.; Sugumaran, M. Pest. Biochem. Physiol. 1986, 26, 220.
76. Murdock, L. L.; Omar, D. Insect Biochem. 1981, 11, 161.
77. Brown, C. S.; Nestler, C. In Compre. Insect Physiol. Biochem. Pharmacol.; Pergamon: New York, 1985; Vol. 11, p 435.

RECEIVED February 29, 1988

Chapter 13

Mechanisms of Immunity in *Drosophila* and Mediterranean Fruit Flies

John H. Postlethwait[1], Stephen Saul[2], Mark Robertson[1], Dale Grace[1], Kristie Abrams[1], and Ellie Brandenburg[1]

[1]Department of Biology and Institute of Molecular Biology, University of Oregon, Eugene, OR 97403
[2]Department of Entomology, University of Hawaii, Honolulu, HI 96827

Fly blood does not normally contain substances that kill bacteria, but flies inoculated with bacteria rapidly accumulate antibacterial proteins (ABs) in their blood. Wild type Drosophila have at least three different antibacterial proteins based on isoelectric points. Genetic variants identify structural genes for these antibacterial proteins. A DNA sequence that can encode a conserved portion of moth and fleshfly antibacterial proteins has been used to synthesize a complementary oligonucleotide probe. This probe recognizes a messenger RNA that appears in the fat body of Drosophila and Medflies only after they have been inoculated with bacteria. Bacteria-sensitive lethal mutations were induced to identify genes necessary for flies to survive a bacterial infection. These studies lay the foundation for developing strategies to control pest insect populations by interfering with the functioning of their immune systems.

There are more species of insects than all other animal groups combined. While many factors contribute to this enormous success, it would not have happened if insects could not vigorously combat pathogens. The immune system of insects can recognize and eliminate bacteria, fungi, and metazoan parasites. From a small number of laboratories, substantial information has been generated to help understand how the immune system works to protect insects. Published work shows that although the immune systems of insects lack important features of vertebrate immune systems, such as lymphocytes and immunoglobulins, both cellular and humoral factors play a role in the insect defense system, as they do in vertebrates (1).

Insect cellular defense reactions involve phagocytosis of bacteria or fungi (2,3) and encapsulation of foreign objects by a layer of blood cells to seal the invader from the hemolymph (4,5). Humoral immunity in some Diptera involves a phenoloxidase-based humoral encapsulation reaction (6,7), and in many types of

0097–6156/88/0379–0186$06.00/0

insects inducible antibacterial factors that are generated by bacterial infection (see for review ref. 8-10).

This communication reports studies on the humoral antibacterial response in Drosophila and *Ceratitis capitata*, the Mediterranean fruit fly. The goal of our work is to understand the molecular mechanisms that protect flies from bacterial infection. Here we examine three questions: 1) What genes encode antibacterial proteins? 2) Does the antibacterial response involve the accumulation of new messenger RNAs? And 3), what genes are necessary to survive a bacterial attack? Answers to these questions will help us to determine the potential of blocking the immune system to control populations of insect pests.

Materials and Methods

Drosophila melanogaster (Oregon R), *Drosophila simulans* (*st pe*), *Drosophila mauritiana* (wild type), *Ceratitis capitata* (Mediterranean fruit fly or Medfly), *Dacus cucurbitae* (melon fly), and *Dacus dorsalis* (Oriental fruit fly) were cultured by standard methods. The *Cy/Pm;D/Sb* stock of *Drosophila melanogaster* we call here *Cu*; *At* and *Am* are the wild type stocks At18 and Amherst.

Animals were inoculated with *Enterobacter cloacae* and hemolymph was collected as described (11,12). Hemolymph proteins were separated by isoelectric focusing (IEF) on LKB PAG plates at pH 3.5-9.5 according to the manufacturers instructions. To detect antibacterial activity after IEF, the electrophoretic plates were overlaid with phosphate buffered LB agar seeded with *E. coli* strain D31. The antibacterial factors diffused from the IEF plate into the bacteria-containing overlay thus blocking bacterial growth and revealing bands of activity (13).

Total RNA was collected by the lithium chloride/urea method (14), separated on agarose gels and transferred to nitrocellulose filters (15). The probe was end-labeled (16) and hybridized to the filters.

Mutants were screened by mutagenizing males with 4000 R of gamma-rays and mating them to *y v* attached-X females. The G1 males were mated again to the attached-X females to avoid mosaics and the G2 males were individually mated to attached-X females. Five G3 sons of each individual G2 male were inoculated with 0.1 microliters of an overnight culture of *Escherichia coli* A-585. Three days later lines were scored for survival. Lines in which 4 or 5 of the G3 males had died were retested to confirm that they were bacteria-sensitive lethal mutations.

Results

Genes Specifying Antibacterial Protein Structure. Adult flies and larvae of Drosophila and Ceratitis do not normally contain in their blood proteins that specifically kill bacteria. However, when animals are inoculated with *Enterobacter cloacae*, potent antibacterial activity appears in the blood (11,12,17,18). Antibacterial activity is detected in the blood of adult *Drosophila melanogaster* flies within two hours after inoculation, and is still detectable sixty days later (11). Investigation of this activity by isoelectric focusing reveals several blood proteins with antibacterial activity

(ABs) including one neutral and several basic forms (Fig. 1). ABs in Drosophila, Ceratitis, and two species of Dacus display different patterns of isoelectric points. These results show that the immune systems of these flies share some basic similarities.

To identify genes affecting the structure of the ABs, we have searched for genetic variants that alter antibacterial protein structure. We first looked at a number of Drosophila species closely related to Drosophila melanogaster and found that D. simulans has two major AB forms resolved by isoelectric focusing, while the sibling species D. mauritiana has four major forms, only one of which comigrates with an AB from D. simulans (Fig. 2). We mated D. mauritiana to D. simulans in reciprocal crosses and tested the offspring (Fig. 2). The F1 heterozygous females have five forms, a sum of the forms of both species. This co-dominant expression suggests that the genetic variants affect the structural genes, since if the variants involved genes for post-translational protein-modifying enzymes, we would have expected the pattern of one of the species to be dominant.

Having found genetic variation for AB structure in sibling species, we screened natural populations of D. melanogaster to identify isoelectric focusing variants within the species. So far we have identified four genotypes with variants of AB structure (Fig. 3). Stock At contains only AB8.7 and AB9.1, while stock Am has only AB7.1. The wild stock Oregon R has all three spots. Stock Cu lacks all three antibacterial proteins found in wild type but has a novel band at pI7.6. In addition, stock Cu is quite sensitive to infection. We are currently mapping these mutations to identify the chromosomal location of genes that affect AB structure.

Immune-specific mRNA. While genetic variants identify some antibacterial genes, specific mRNAs that become abundant only after a bacterial infection may identify others. To study the expression of genes whose activity is dependent upon bacterial infection and to isolate cloned copies of antibacterial genes, we designed a probe for Drosophila AB genes from a portion of sarcotoxins, antibacterial proteins from the flesh fly (19,20) that show homology to the cecropin family of ABs from Saturnid moths (21,22) (Table I).

We obtained two non-overlapping pools of oligonucleotides 17 bases long corresponding to the two sequences at the bottom of Table I to test as immune-specific probes. Pool #1 corresponded to the sarcotoxin sequence and Pool #2 corresponded to the cecropin sequence. Two day old adult male Drosophila were inoculated with Enterobacter cloacae and their RNA was extracted either 11 hours or 30 hours after inoculation. Total RNA was separated by electrophoresis and blotted to nitrocellulose. The blot was hybridized to end-labelled oligonucleotide probe. The results showed that oligonucleotide Pool #1 recognizes a major transcript band that is specific for inoculated animals (Fig. 4). In addition the probe recognizes one small and one large transcript that are present in lesser quantities in both control and immune animals. The ethidium bromide stained gel (Fig. 4) shows that approximately equal amounts of RNA were loaded in each lane. Parallel blots probed with Pool #2 did not show any immune-specific RNAs. We conclude that genes closely related to this part of the flesh fly sarcotoxin gene are

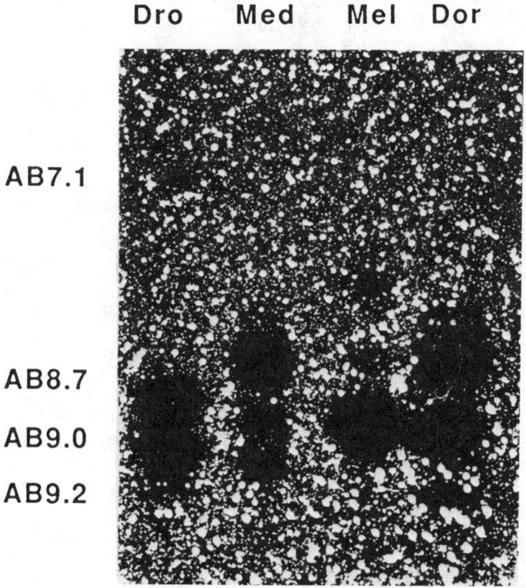

Figure 1. Different species of flies have different patterns of antibacterial activity after isoelectric focusing. Dro, *Drosophila melanogaster*; Med, *Ceratitis capitata*; Mel, *Dacus cucurbitae*; Dor, *Dacus dorsalis*. Antibacterial proteins (ABs) are labeled according to their isoelectric points.

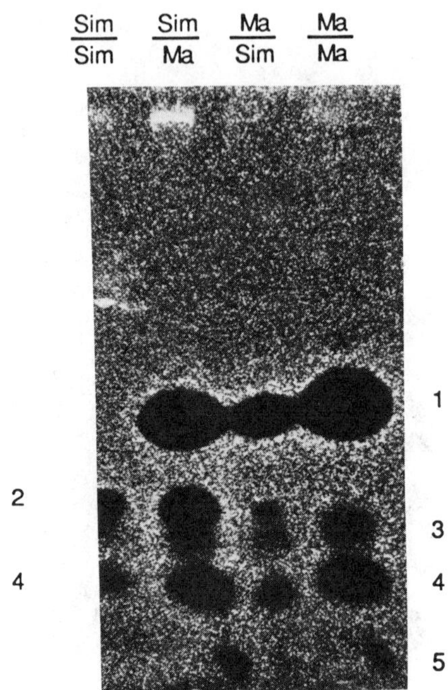

Figure 2. Genetic variants detected by isoelectric focusing are expressed codominantly. sim/sim, homozygous _Drosophila simulans_ females; sim/ma , female offspring of _D. simulans_ females mated to _D. mauritiana_ males; ma /sim, female offspring of _D. mauritiana_ females mated to _D. simulans_ males; ma /ma , homozygous _D. mauritiana_ females. Activity bands are numbered arbitrarily at the sides.

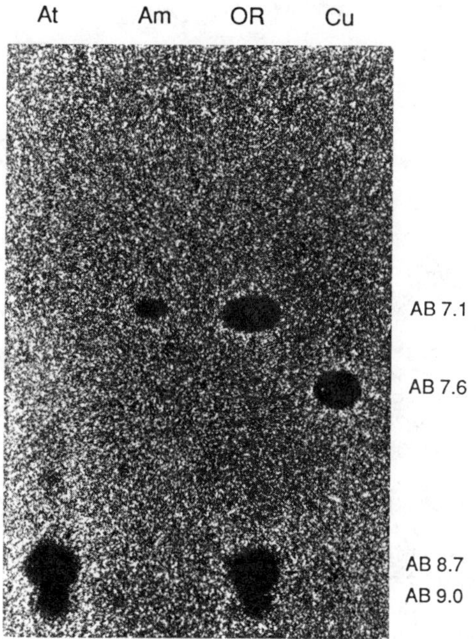

Figure 3. Genetic variants detected by isoelectric focusing in various stocks of Drosophila melanogaster.

Table I. The Oligonucleotide Pool

Polypeptide	Sequence	
Sarcotoxin IA	GWLKKIGKKIERVGQHTRDATIQGLGIAQQAANVAATAR	
Sarcotoxin IB	----------------------VI-V------------	(- = same as top row)
Sarcotoxin IC	---R-------------------V--------------	
Cecropin A	K-KLF ----K---NI--GIIKAGPAVAVVGGATQIAK	
Cecropin B	K-KVF ----KM-RNI-NGIVKAGPAIAVLGEAKAILS	
Cecropin C	-NPF -EL-K---RV---VVSAGPAVATVAQATALAK	
Consensus	KKIERV K	

Sequence chosen	K	K	I	E	R K	V
Codons	Lys AAG/A	Lys AAG/A	Ile AUU/C (AUA)	Glu GAA/G	Arg CGN AGA/G Lys AAG/A	Val GU

Pool #1	Pool #2
5' AC ACG CTC AAT CTT CTT 3' C G T T T	5' AC ATT CTC AAT CTT CTT 3' CC T G T T T

abundantly expressed in immune but not in control flies. We cannot yet correlate this message with a specific antibacterial protein.

To determine what cells make the immune-specific RNA we inoculated mid-third instar Drosophila larvae with bacteria and six hours later dissected them into fat bodies and fat body-free carcass. Total RNA was collected, separated by electrophoresis, transfered to nitrocellulose, and probed with labelled Pool #1 oligonucleotide. The results showed (Fig. 5) that intact inoculated larvae accumulate the immune-specific transcript while intact control larvae do not. The tissue dissection experiment showed that fat body cells of inoculated larvae contain transcripts homologous to the immune-specific probe, but the carcass does not. We conclude that fat body cells in Drosophila larvae accumulate a transcript that has homology to sarcotoxin when they have been inoculated with bacteria. We are currently cloning the responsible immune gene.

What Genes Are Necessary to Survive a Bacterial Infection? One strategy for learning how the immune system works is to identify genes that are necessary for proper functioning of the antibacterial immune system and to learn what these genes do to enable the fly to resist infection. Mutations in genes necessary for surviving a bacterial infection should result in animals that die when infected with bacteria at a dose that normal animals would survive. To find bacteria-sensitive lethal mutations we mutagenized males with gamma rays and set up lines from individual males bearing mutagenized X-chromosomes. Each line was tested for its ability to survive a bacterial injection by injecting bacteria into 5 males for each line and scoring for survivors. More than 62,000 males from over 12,000 lines have been individually injected with bacteria and tested for

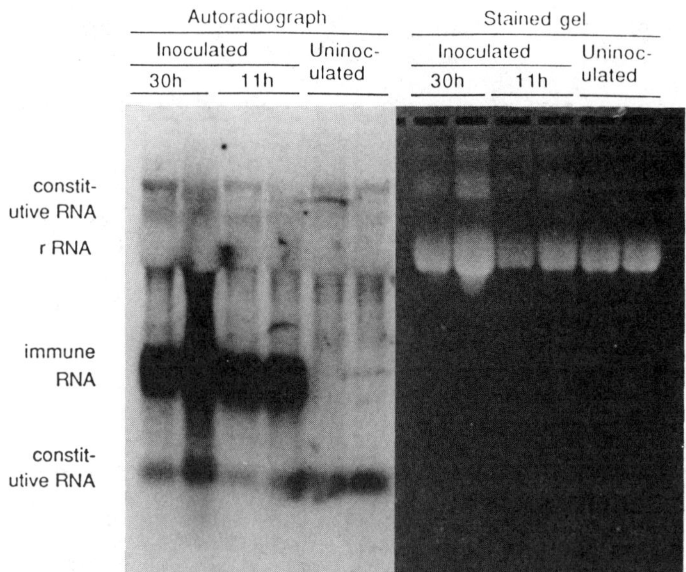

Figure 4. The oligonucleotide recognizes an immune-specific transcript. D. melanogaster were inoculated with bacteria or left as controls and incubated for 11 or 30 hours as indicated. RNA was collected and separated on the gel shown at the right stained with ethidium bromide and then hybridized to the oligonucleotide probe in a northern blot on the left.

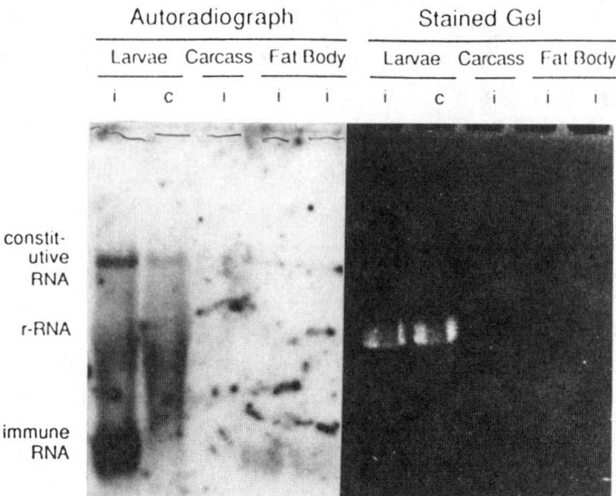

Figure 5. Drosophila fat body cells accumulate immune-specific RNA. Drosophila larvae were inoculated with bacteria and six hours later RNA was extracted from either whole larvae, from fat body, or from the carcasses minus fat body. These RNAs were separated on the gel shown on the right stained with ethidium bromide, then northern blotted and probed with the oligonucleotide on the left.

survival, and 10 X-linked bacteria-sensitive lethal (bsl) lines have been established. Table II presents data for three of the mutants.

Table II. Characteristics of Some Bacterial Sensitive Lethal Mutants

	Dead/inoculated					Isoelectric focusing bands			
Genotype	1st	2nd	3d	4th	Inhibition zone assay	7.1	7.6	8.7	9.0
bsl-4	4/5	4/5	4/5	5/5	+/-	+	+	+	+
bsl-6	5/5	4/5	5/5	5/5	++	+	-	+	+
bsl-9	5/5	4/5	5/5	5/5	+	+	+	+	+
Wild	0/5	0/5	0/5	0/5	+	+	-	+	+

Although we have just begun to study the bsl mutants, they have already given us an important and unexpected result. None of the mutants so far studied are totally lacking humoral antibacterial activity. In fact, bsl6 has greater than normal activity when inoculated with autoclaved bacteria. Another surprising finding was that some bsls have normal IEF patterns but others have an extra band at pI7.6 (Fig. 6). We do not yet know if this extra band is related to the bacteria-sensitive phenotype.

From our bacteria sensitive lethal mutations we can make the surprising conclusion that the presence of ABs detected after overlaying an isoelectric focusing gel with bacteria is insufficient for a fly to survive a bacterial attack. Therefore, at least some of our bsl mutations identify genes required for surviving a bacterial infection that are independent of the ABs. We hypothesize that the necessary function identified by the bsl mutants is a cellular immune function, perhaps phagocytosis of bacteria.

The result that bsl mutations are not missing ABs made us wonder if the ABs play any role in resistance to bacterial infection. To answer this question, we inoculated bsl-6 and bsl-4 flies with a sub-lethal dose of *E. coli* D31 bacteria to stimulate the production of ABs without causing a lethal infection, and 48 hours later challenged them by injecting a lethal dose of *E. coli* A585 bacteria. We found that immunized mutants survived, while unimmunized mutants died from the bacterial infection. Thus, we conclude that these bsl mutants lack a function that is necessary for naive animals to surmount a bacterial infection (perhaps a cellular function like phagocytosis), but which is unnecessary in immunized animals (presumably because immunized flies possess ABs which kill the bacteria).

Discussion

The results of these experiments 1) provide genetic variants for genes that control the structure of antibacterial proteins; 2) show that specific RNAs accumulate after a bacterial infection; and 3)

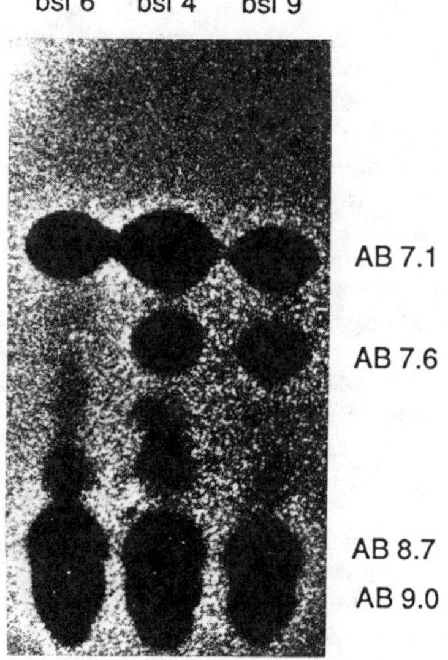

Figure 6. Bacteria-sensitive lethal mutants are inducible for antibacterial proteins.

provide mutations in genes whose activity is necessary to resist a bacterial infection.

Mapping the isoelectric focussing variants will determine the cytogenetic location of each AB. This will allow us to correlate individual AB proteins to cloned copies of AB genes which we are currently isolating and to design screens for mutations in each AB gene. If we are able to make mutations in individual AB genes, then we will be able to assess the phenotype of a fly that is genetically lacking each AB protein, and hence learn the physiological significance of each AB protein.

A potential practical outcome of research on insect immune systems would be the design of effective strategies to regulate populations of insect pest species by eliminating their antibacterial immune systems. To evaluate the likelihood that such a strategy would be effective, we must first determine if insects that lack an immune system will be susceptible to naturally occurring microbial pathogens (as are people who lack an immune system). Recombination between Drosophila stocks reported here should provide animals that are genetically lacking the ability to make any of the antibacterial proteins detectable by isoelectric focusing. We could then determine the susceptibility of these immune-crippled flies to bacterial pathogens, and hence test the potential effectiveness of an anti-immune strategy for insect control. Since the results reported here demonstrate similarities in the immune systems of Drosophila and Tephritid fruit fly pests, the results should be generalizable from the laboratory species to the agricultural pests.

The immune system of Drosophila seems to have two main subsystems, one involving the antibacterial proteins and the other identified by bacteria-sensitive lethal mutations. The humoral subsystem results in the production of diffusible antibacterial proteins of at least three species. Stock Cu, which lacks the ABs found in wild type Drosophila and which is correspondingly more sensitive to infection, blocks a part of this subsystem. That the ABs do help flies to survive bacterial infections was shown by inducing ABs to appear in bsl mutants by injection of dead bacteria and then finding that the inoculated mutants would survive an otherwise lethal injection of live bacteria. Evidently inoculation induces processes (perhaps secretion of ABs) that can overcome the deficiency in the bsl mutations. Similar experiments showed that inoculation protects locusts from a lethal dose of bacteria (19).

Some of the ABs may have structural homology to antibacterial proteins isolated from the flesh fly (20,21) since an oligonucleotide probe that can encode a portion of the sarcotoxin protein recognizes an immune-specific RNA in Drosophila fat body cells, the cellular origin of Drosophila's ABs (unpubl.). The induction mechanism must work relatively rapidly since we found immune-specific RNA in fat body cells within 6 hours after inoculation. Because of its homology to sarcotoxin, we assume that the immune-specific transcript detected by the oligonucleotide probe encodes an antibacterial protein.

Another subsystem of the immune response is blocked by our current bsl mutations. Since these mutants are unable to resist bacterial infection but nevertheless contain inducible ABs, this subsystem is at least partially independent of the ABs. We suspect from work with Drosophila (24-26) and other species (2,3) that this subsystem is the cellular immune system.

These data and those from other species suggest the following possible model for action of the immune system in Drosophila. When bacteria enter the animal, they are rendered nontoxix by a process we suspect is controlled by some of the bsl mutations; this process may be phagocytosis by hemocytes. Then in some way the blood cells signal the fat body that bacteria are present. In moths, this signal is bacterial cell wall components (26). The activated fat body begins to transcribe genes for the ABs, genes we suspect are identified both by mutations in stocks At and Am and by the immune-specific oligonucleotide probe. After a few hours, the concentration of ABs begins to rise in the hemolymph to provide further prophylaxis. The model suggests that phagocytosis provides the initial response, but one that might be overwhelmed by recurring bacterial infections, while the ABs provide a slower acting response that may destroy any surviving bacteria and prevent a second infection (10). Further analysis of the phenotypes of our current mutations and the induction of other bsl's on other chromosomes will provide valuable tests of various aspects of this model.

Finally, the current work provides bacteria-sensitive lethal mutations that point out pest control possibilities. Since bsl mutants die when injected with bacteria, they focus our attention directly on the immune steps at which the insect is particularly vulnerable. In future work we need to identify the cellular and molecular processes that are blocked by these mutations. If we could learn to artificially interfere with an immune step that is blocked by a bsl, then an effective insecticide might be developed. Because of the demonstrated similarities in the antibacterial immune systems of Drosophila melanogaster and fruit fly pests, it is hoped that such methods would lead to crop protection.

Literature Cited

1. Marchalonis, J.J. Immunity in Evolution; Harvard University Press: Cambridge, 1977.
2. Faye, I. J. Invert. Path. 1978, 31,19-26.
3. Ratcliffe, N.; Walters, J.B. J. Insect Physiol. 1983, 29,407-415.
4. Lackie, A.M. Develop. Compar. Immunol. 1981, 5,191-204.
5. Nappi, A.J. Parasitology 1981, 83,319-324.
6. Götz, P.; Vey, A. Parisitology 1974, 68,193-203.
7. Nappi, A.J. In Invertebrate Immunity, Mechanisms of Invertebrate Vector-parasite Relations Maramoresch, K. and Shope, R.E., Ed.;Academic: New York, 1975; p. 293-325.
8. Götz, P.; Boman, H.G. In Comprehensive Insect Physiology, Biochemistry, Pharmacology; Kerkut, G.A.; Gilbert, L.I. Ed.; Academic: New York, 1984 Vol. 3, p 454-485.
9. Boman, H.G.; Hultmark, D. Trends Biochemical Sci. 1981, 6,306-309.
10. Dunn, P.E. Ann. Rev. Entomol. 1986, 31,321-339.
11. Robertson, M.; Postlethwait, J.H. Develop. Compar. Immunol. 1986, 10,167-179.
12. Postlethwait, J.H.; Postlethwait, J.A.; Saul, S. J. Insect Physiol. 1987, In press.
13. Hultmark, D.; Steiner, H.; Rasmuson, T.; Boman, H.G.; Eur. J. Biochem. 1980, 106,7-16.

14. Richards, G.; Cassab, A.; Bourouis, M.; Jarry, B.; Dissous, C. EMBO J. 1983, 2,2137-2142.
15. Thomas, P. Proc. Natl. Acad. Sci. USA 1980, 77,5201-5205.
16. Maniatis, T.; Fritsch, E.F.; Sambrook, J. Molecular Cloning: A Laboratory Manual Cold Spring Harbor Laboratory, Cold Spring Harbor, New York, 1982.
17. Flyg, C.; Dalhammar, G.; Rasmussen, B.; Boman, H. Insect Biochem. 1987, 17,153-160.
18. Boman, H.G.; Nilsson, I.; Rasmuson, B. Nature 1972, 237,232-235.
19. Hoffmann, D. J. Insect Physiol. 1980, 26,539-549.
20. Okada, M.; Natori, S. Biochem. J. 1985, 229,453-458.
21. Okada, M.; Natori, S. J. Biol. Chem. 1985, 260,7174-7177.
22. van Hofsten, P.; Faye, I.; Kockum, K.; Lee, J.Y.; Xanthopoulos, K.G.; Boman, I.A.; Boman, H.G.; Engstrom, A.; Andreu, D.; Merrifield, R.B. Proc. Natl. Acad. Sci. USA 1985, 82,2240-2243.
23. Spies, A.; Karlinsey, J.E.; Spence, K. Comp. Biochem. Physiol. 1986, 83B,125-133.
24. Rizki, T. In Genetics and Biology of Drosophila; Ashburner, M.; Wright, T.R.F. Ed.; 1978, Vol 2b, p 397-452.
25. Carton, Y.; Bouletreau, M. Develop. Compar. Immunol. 1985, 9,211-219.
26. Dunn, P.; Dai, W.; Kanost, M.; Geng, C. Devel. Compar. Immuno. 1985, 9,559-568.

RECEIVED May 5, 1988

Chapter 14

Molecular Genetic Approach to the Study of Target-Site Resistance to Pyrethroids and DDT in Insects

Douglas C. Knipple, Jeffrey R. Bloomquist, and David M. Soderlund

Department of Entomology, New York State Agricultural Experiment Station, Cornell University, Geneva, NY 14456

The insect nervous system is the site of action of all major classes of insecticides currently in use, and as such is the site of insecticide resistance mechanisms that involve modified target sites. An altered voltage-sensitive sodium channel has been implicated as one of the mechanisms conferring resistance to pyrethroids and DDT. In this paper, the evidence providing the basis for our current understanding of this specific target site resistance mechanism will be summarized, and the need for molecular genetic approaches to its investigation will be discussed. Such approaches promise not only to enhance the implementation of rational resistance management strategies for compounds currently in use, but also to aid in the discovery and design of new compounds that act on other macromolecular targets.

Insect pests continue to compete effectively for human food and fiber supplies and to act as important vectors of human disease despite extensive efforts to control them. Because the use of chemical insecticides is at present our principal weapon against deleterious insects, the selection for significant levels of insecticide resistance in many populations of insect pest species threatens to compromise current crop protection and disease prevention strategies.

Our current understanding of insecticide resistance phenomena in insects and other arthropods is based upon genetic and biochemical evidence, which implicates a number of different mechanisms (1). These include metabolic mechanisms, which result in enhanced rates of insecticide detoxication or sequestration of insecticide, and also mechanisms involving reduced

0097–6156/88/0379–0199$06.00/0

insecticide penetration of the insect cuticle. Another class of resistance mechanism of particular concern, which is the focus of this paper, involves alterations of the neuronal targets of insecticide action. Unlike some forms of metabolic resistance, target site resistance can not be averted by the use of chemical synergists, and it often confers cross-resistance to entire classes of insecticides, effectively negating their usefulness in the control of resistant populations.

Concept of Target Site Resistance. The term "nerve insensitivity resistance" has been used to describe resistance phenomena that appear to be intrinsic to the insect nervous system. This term is useful, particularly as it relates alterations of normal neurophysiological processes to resistance phenomena, but we suggest that "target site resistance" is the most appropriate term for describing these resistance mechanisms. This term is also sufficiently general to be correctly applied to describe resistance to insecticidal compounds that have as their primary sites of action tissues other than the nervous system, such as the juvenile hormone analogue, methoprene (2). It relates directly to the mode of insecticide action, which involves a specific molecular interaction (binding) between a receptor (cell membrane macromolecule) and its ligand (insecticide molecule). Implicit in this concept is that structural changes in the macromolecular target site or changes in its membrane environment can result in a reduced binding affinity, which is manifested at the level of the intact organism as resistance to the adverse physiological effects of the insecticide.

Acetylcholinesterase (3), the GABA-receptor/chloride ionophore complex (4-6), and the voltage-sensitive sodium channel (7) have been shown to be the macromolecular target sites of the three major classes of chemical insecticides: respectively, organophosphorus and N-methylcarbamate compounds, cyclodienes, and pyrethroids and DDT. Reduced nerve sensitivity has been proposed as a mechanism of resistance for each of these classes of compounds in several insect species (1). In the remainder of this paper, we will restrict our attention to investigations of reduced nerve sensitivity to pyrethroids and DDT, and of the macromolecular complex implicated in this resistance, the voltage-sensitive sodium channel and its membrane environment. We will briefly review the evidence suggesting modifications of this important insecticide target site and discuss the development of strategies for elucidating the underlying molecular basis of the resistance to pyrethroids and DDT that these modifications are believed to confer.

Target Site Resistance to DDT and Pyrethroids in the Housefly

Reduced target site sensitivity has been described as a mechanism of pyrethroid resistance in several insect species. In the housefly, the resistance factor *kdr* (knockdown resistance) confers resistance to both the rapid paralytic (knockdown) and lethal actions of all pyrethroids tested to date, DDT, and DDT analogs (except carbinols) (8). Several alleles of *kdr* (including those designated *super-kdr*), which confer different levels of resistance to the above compounds, have been described and mapped genetically (9-10). The involvement of target site resistance in *kdr* houseflies was suggested by the failure of synergists to increase the toxicity of DDT and pyrethroids in whole animal bioassays (9). More rigorous experiments showed that penetration and metabolism of [^{14}C]-permethrin in two *kdr* strains was equivalent to a susceptible strain (11).

Electrophysiological studies have provided more definitive evidence that the *kdr* mechanism lies at the level of the nervous system and is therefore a target site phenomenon. Reduced sensitivity of the housefly nervous system to the neuroexcitatory effects of DDT and pyrethroids has been demonstrated in preparations of adult thoracic ganglia, adult motor neurons of the legs and indirect flight muscles, larval neuromuscular junction, and larval sensory nerves (12-16). In each case, the appearance of abnormal nerve function in preparations from *kdr* strains required longer periods of insecticide exposure or higher concentrations than in equivalent preparations from susceptible flies. These experiments established that the reduced sensitivity to pyrethroids and DDT in *kdr* houseflies is broadly distributed in the nervous system and is probably present throughout the course of development.

Investigations into the molecular mechanism of knockdown resistance have been aided by studies of the effects of pyrethroids on the nerve membrane action potential. Action potentials in nerve cells are generated by time and voltage-dependent ionic currents flowing through proteinaceous ion channels embedded in the nerve membrane (17). The voltage-sensitive sodium channel plays a central role in the initiation and maintenance of the depolarizing phase of the action potential by conducting a transient inward current of sodium ions. Studies utilizing a variety of vertebrate and invertebrate nerve preparations have shown that the principal actions of DDT and pyrethroids are to modify transient and resting sodium currents, resulting either in repetitive firing (Figure 1) or use-dependent nerve block depending on the compound and the nerve pathway investigated (18). More detailed experiments using voltage and patch clamp techniques demonstrated that both of these effects are due to prolongation of sodium currents, presumably through specific interaction of the insecticides with the gating apparatus of the sodium channel (18-19).

The sodium channel protein contains discrete but often interacting binding sites for several classes of potent neurotoxins, including DDT and pyrethroids (20). Ion flux and binding experiments have demonstrated that six classes of compounds bind to distinct sites on the sodium channel (20-21). Because of their specificity, these compounds are useful as probes of the *kdr* mechanism. Resistance to the lipophilic plant alkaloid, aconitine, in *kdr* houseflies has been shown in larval paralysis bioassays (Figure 2) (22) and in neurophysiological experiments (23). In contrast, there was no resistance to the venom of the scorpion, *Leiurus quinquestriatus*, which contains toxic polypeptides that prolong sodium current in a manner similar to pyrethroids, nor to the channel blocker tetrodotoxin or the local anesthetic procaine (22). These findings suggest the specific modification of two distinct receptor sites on the sodium channel molecule in *kdr* houseflies.

Three hypotheses have been put forth as explanations of the underlying mechanism of reduced neuronal sensitivity associated with the *kdr* phenotype. One of these is based on evidence that *kdr* strains possess neuronal membranes with fewer numbers of sodium channels (24-25). It has been proposed that this reduction in sodium channel number is sufficient to explain the *kdr* phenotype (26). However, Lund and Narahashi (27) determined that modification of less than 1% of the sodium channel population will achieve poisoning at the level of the nerve. Given this low effective level of receptor occupancy, we have calculated that even a 50% reduction in binding site number with no alteration in binding affinity can only confer about two-fold resistance (data not shown). Neurophysiological assays of nerves from *kdr* flies have shown that at least a tenfold higher pyrethroid concentration is required to produce disruption of nerve activity equivalent to that obtained in preparations from susceptible insects. A second hypothesis proposed by Chiang and Devonshire (28) is that reduced neuronal sensitivity to pyrethroids is due to alterations in the fluidity of the nerve membrane. This hypothesis is based on studies that clearly demonstrate higher transition temperatures in Arrhenius plots of a nerve membrane-associated enzyme activity from *kdr* strains compared to a susceptible strain. Moreover, the inheritance of altered membrane fluidity was correlated with the inheritance of the *kdr* trait. While these data provide a strong correlation between

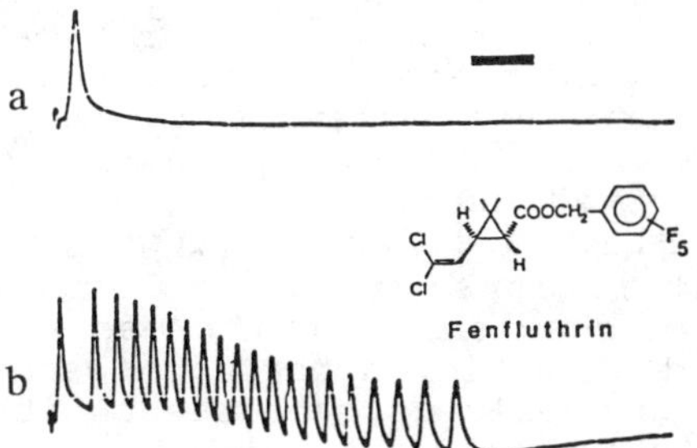

Figure 1. Responses in the house fly dorsolongitudinal flight muscles to motor nerve stimulation. In this preparation, action potentials in the muscle faithfully reflect activity present in the motor neuron. (a) Control preparations respond to stimulation by firing a single muscle action potential. (b) Treatment with fenfluthrin initiates high frequency burst discharge from a single stimulus.

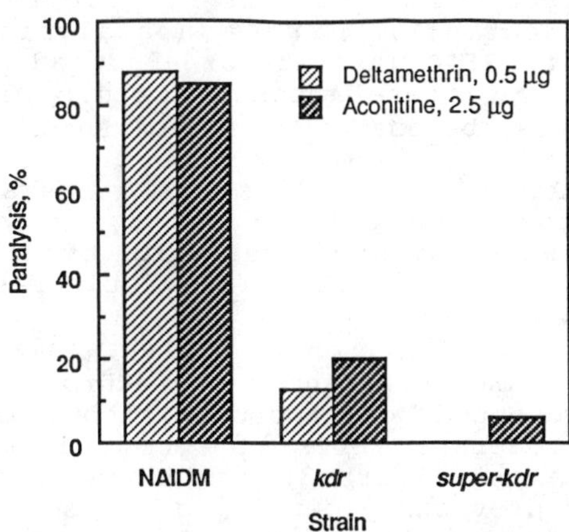

Figure 2. Responses of three strains of the housefly to single doses of deltamethrin and aconitine in a larval paralysis bioassay. Paralysis was assessed 10 min. after treatment and was defined as the loss of ability to perform a stereotypic curling movement. Resistance in the *kdr* and *super-kdr* strains is directed against both compounds. For each treatment group shown n = 60. The bar for deltamethrin in *super-kdr* is omitted because no paralysis was observed at dosages up to 20 µg/larva (n = 40). Data were taken from Ref. 22.

decreased membrane fluidity and resistance, they do not establish a causal relationship. A third hypothesis proposed by Salgado et al. (15) ascribes altered pyrethroid and aconitine sensitivity in *kdr* housefly strains to structural changes in the sodium channel molecule. In an effort to reconcile this hypothesis with the observed decreased membrane fluidity in *kdr* houseflies, Bloomquist and Miller (22) proposed that membrane alterations could reflect a compensatory homeostatic mechanism, which preserves normal nerve functions and is achieved through feedback regulation. While this hypothesis is consistent with the available data, there is at present no direct evidence for altered sodium channel structure in resistant insects.

Resistance to DDT and Pyrethroids in *Drosophila melanogaster*

Temperature-sensitive Paralytic Mutants. Behavioral mutants of *Drosophila melanogaster* that exhibit nerve conduction block at nonpermissive temperatures may also be relevant to the study of target site resistance to DDT and pyrethroids. Altered sodium channel properties have been implicated in mutations of four loci: *nap*^{ts} (no action potential); *para*^{ts} (paralysis); *sei*^{ts} (seizure); and *tip-E*^{ts} (temperature-induced paralysis, locus E) (29). Of these mutant strains, *nap*^{ts} is of particular interest because it has been shown to exhibit altered responses to some sodium channel-directed neurotoxins. Adult *nap*^{ts} flies are resistant to the alkaloid neurotoxin veratridine but hypersensitive to tetrodotoxin when compared to wild type flies in toxicity bioassays (30), and primary cultures of *nap*^{ts} neurons are resistant to the cytotoxic effects of veratridine *in vitro* (31). Hall (29) has also reported that *nap*^{ts} flies are resistant to pyrethroids, although no data were presented to support this conclusion. We have detected moderate resistance to deltamethrin in *nap*^{ts} adult females (Figure 3) using a surface contact exposure method previously developed for studies of DDT resistance in *Drosophila* (32). Further experiments documented a similar pattern of resistance to the pyrethroid fenfluthrin, and preliminary surveys completed to date show resistance to a variety of other pyrethroids and DDT (data not shown). Head membrane preparations from *nap*^{ts} flies have a reduced number of [^{3}H]-saxitoxin binding sites, which has been interpreted as reflecting a lower density of sodium channels in the nerve membrane (33). Hall and co-workers (29-30) have suggested that a reduction in sodium channel number is sufficient to explain the altered response to veratridine. This is analogous to the hypothesis advanced to explain the resistance of *kdr* houseflies to pyrethroids and DDT (26).

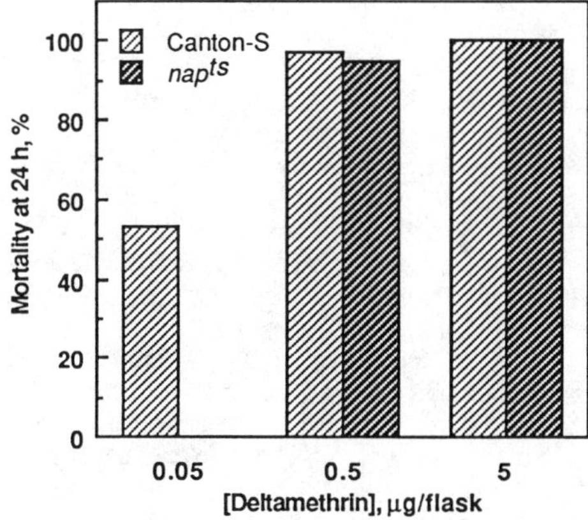

Figure 3. Toxicity of deltamethrin to the susceptible Canton-S and temperature-sensitive *nap*^{ts} strains of *Drosophila melanogaster*. Adult females were tested in a surface contact bioassay at three dosages of deltamethrin applied to 50 ml Ehrlenmeyer flasks. A dosage of 0.05 µg/flask was near the LD_{50} for Canton-S, but caused no mortality in the *nap*^{ts} strain. At 10x and 100x higher dosages, however, the toxicities were the same in both strains. For each treatment group shown n = 60.

The pharmacological properties of mutants of the three other loci implicated in sodium channel structure or processing are less well characterized than those of *nap*^{ts}. Although the responses of *para*^{ts} flies to veratridine intoxication are not known, *para*^{ts} neurons in culture exhibit moderate resistance to this compound at permissive temperatures and greater resistance at nonpermissive temperatures (31). The *sei*^{ts} and *tip-E*^{ts} mutations are also thought to involve alterations in sodium channel structure or function on the basis of their effects on the number and affinity of [^{3}H]-saxitoxin binding sites. In both *sei*$^{ts-1}$ and *tip-E*^{ts}, the number of [^{3}H]-saxitoxin binding sites was reduced relative to those in wild type flies, whereas *sei*$^{ts-2}$ showed a temperature-dependent increase in the K_D for [^{3}H]-saxitoxin binding (29). Further detailed characterization of the resistance of these strains to sodium channel-specific compounds is needed. In conducting these studies, it will be desirable to supplement surface contact bioassays with topical applications of neurotoxins to individual flies in order to rule out a behavioral component of any resistance observed.

Although it is tempting to ascribe the observed neurotoxin resistance in these strains to the mutations isolated on the basis of temperature-sensitive paralysis, there is no direct evidence that the observed resistance is conferred via reduced neuronal sensitivity. To address this issue, we have developed a physiological preparation to observe the alteration of motor nerve output to the indirect flight muscles of adult *Drosophila* in response to central nervous system stimulation during the course of insecticide intoxication. This preparation, which is based on similar studies of insecticide intoxicaton in houseflies (34) and on studies of the flight motor pathway in *Drosophila* (35), involves field stimulation of the brain or fused thoracic ganglion of immobilized adult flies with a pair of wire electrodes. Flight muscle motor nerve responses to these stimuli are measured as excitatory postsynaptic potentials in the flight muscle using a recording microelectrode inserted through the dorsal thoracic cuticle. The effects of the pyrethroid insecticide fenfluthrin on this pathway are indistinguishable from those observed in the housefly (Figure 1). In untreated flies, each stimulating pulse produced a single muscle action potential, whereas in fenfluthrin-treated flies each stimulus evoked a train of repetitive discharges in the motor nerve that were recorded as a series of muscle spikes. Current studies are directed at defining the latency and dose dependence of the fenfluthrin-induced repetitive discharges. Preliminary results indicate that burst discharges occur with a longer latency in the *nap*^{ts} strain than in wild type flies (our unpublished data). It will also be of interest to document the responses of *Drosophila* strains to aconitine to see whether the mechanism(s) conferring resistance in this species discriminate between alkaloid neurotoxins as the *kdr* mechanism does in *Musca*. These studies are essential to demonstrate a causal relationship between neurotoxin resistance in these strains and reduced neurotoxin sensitivity of the nervous system.

Insecticide Resistance in Wild *Drosophila* Populations. Besides the experimentally generated mutants implicated in altered sodium channel function, wild populations of *Drosophila melanogaster* provide another potentially valuable resource for the investigation of *kdr*-like resistance phenomena. It has been known for more than three decades that *Drosophila melanogaster* is able to develop resistance in response to insecticide pressure both in the laboratory and in the wild (36-39). Although *Drosophila* is not typically a target of commercial insecticide applications, its inadvertent exposure to them can be quite high in the contexts of some agricultural ecosystems. For example, in the apple and grape producing regions of western New York, where dicofol (containing as much as 10% DDT as a process contaminant) has been perennially applied at high rates in orchards, strains of *Drosophila melanogaster* that are extremely resistant (more than 1000-fold) to DDT have been selected in less than five generations from field-collected flies (32). The resistance phenotypes of these strains have been stably maintained in culture, in some cases for more than ten years without selection pressure (T. Glover, personal communication, and our unpublished observations).

Although the identification of metabolic detoxication mechanisms in this species has been reported by several laboratories (40-42), there are no published reports directly demonstrating the existence of a *kdr*-like mechanism in these or other wild *Drosophila* strains. Nevertheless, there is no *a priori* reason to expect that the occurrence of an altered target site of DDT and pyrethroids in this genetically highly polymorphic species is improbable. A precedent for the genetic isolation of an altered target site resistance mechanism in *Drosophila* exists in a strain of malathion resistant flies, which was shown to possess an altered acetylcholinesterase (42-43). Since it is possible to clone readily any locus that can be mapped in *Drosophila*, it would be of obvious interest to isolate genetically a *kdr*-like resistance mechanism from this species.

Molecular Properties of the Voltage-Sensitive Sodium Channel

Because altered sodium channels have been implicated in *kdr* and *kdr*-like resistance phenomena in insects, basic research on the biochemistry and molecular biology of this molecule, which plays a central role in normal processes of nervous excitation in animals, is of immediate relevance. The results of recent investigations of the voltage-sensitive sodium channels of vertebrate nerves and muscles have provided unprecedented insight into the structure of this large and complex membrane macromolecule. Sodium channel components from electric eel electroplax, mammalian brain, and mammalian skeletal muscle have been solubilized and purified (for a recent review, see Ref. 19). A large α subunit (ca. 260 kDa) is a common feature of all purified channels; in addition, there is evidence for two smaller subunits (β1 and β2; 37-39 kDa) associated with the mammalian brain sodium channel and for one or two smaller subunits of similar size associated with muscle sodium channels. Reconstitution experiments with rat brain channel components show that incorporation of the α and β1 subunits into phospholipid membranes in the presence of brain lipids or brain phosphatidylethanolamine is sufficient to produce all of the functional properties of sodium channels in native membranes (44). Similar results have been obtained with purified rabbit muscle (45) and eel electroplax (46) sodium channels.

Knowledge of the primary structure of the sodium channel protein has been advanced by the isolation and sequencing of the gene encoding the α subunit of sodium channels from electric eel (47) and of homologous genes from the rat (48). Hydropathic analysis of the inferred amino acid sequence of the eel protein revealed the existence of four major internally repeated homology units, each of which contains six segments with discrete properties, with the central one being notable for the occurrence of four evenly spaced negatively charged clusters of amino acids. This has led to the proposition of several detailed models, which attempt to correlate structural features with the known biochemical and biophysical properties of functional sodium channels (20,47,49-51). These models predict a pseudosymmetrical arrangement in the cell membrane of the four homologous domains, each comprising several membrane-traversing helices. The models also identify structural elements that are suggested to be involved in the voltage-sensitive gating of the channel, its inactivation, and cation selectivity. Comparative sequence analysis of the genes from rat (48) and more recently of a gene isolated from *Drosophila melanogaster* (52) has revealed a high degree of evolutionary conservation, particularly in those regions of the protein implicated in channel function. Interestingly, the *Drosophila* gene, which was isolated by homology to the eel sodium channel gene, maps to a cytogenetic locus (52) that is distinct from the four loci implicated in sodium channel function described above.

Comparative structural analysis of sodium channel genes has permitted the development of testable hypotheses concerning the neurotoxin recognition properties of the sodium channel protein. However, specific elements of the deduced structure have not yet been definitively correlated with the molecular recognition of sodium channel-directed neurotoxins by discrete binding domains. It is, thus, not possible at the present time to know which pharmacological properties are determined by the sodium channel protein *per se* and which are determined by interactions between it and crucial features of its membrane environment. Clearly, it will be necessary to analyze the effects of specific modifications of sodium channel structure in a defined membrane environment in order to address these and other questions relating to sodium channel function.

Several studies employing oocytes of the clawed frog, *Xenopus laevis*, for the *in vitro* translation of sodium channel encoding mRNAs (53-55) suggest that this experimental system may be particularly useful toward this end. The biophysical properties of sodium channels expressed in oocytes following injection of rat brain mRNA were similar to those of sodium channels in their native membrane environment, and were specifically inhibited by the sodium channel blockers tetrodotoxin and saxitoxin (55). Sodium channels encoded by mRNAs from rat skeletal muscle and eel electroplax have also been expressed in *Xenopus* oocytes (56-57). To date the expression of insect sodium channels in the *Xenopus* oocyte has not been reported, but the utility of this system for the translation and expression of insect acetylcholine receptor mRNA has recently been demonstrated (58). Successful application of this methodology to the expression of insect mRNAs encoding functional sodium channels offers a novel method to test some of the hypotheses for the molecular basis of the *kdr* mechanism.

Conclusions

Knowledge of the underlying molecular details of insecticide resistance mechanisms is, at present, at a very preliminary stage. To date there is only one case, i.e., that of elevated carboxylesterase levels in *Culex quinquefasciatus* resulting from gene amplification (59), in which a specific variation of a cloned gene has been associated with an insecticide resistance mechanism. Even in this instance, the functional role of the elevated carboxylesterase in insecticide detoxication has not been conclusively established. Elucidation of other resistance mechanisms at the molecular level will clearly require that the specific genes coding for gene products implicated in resistance mechanisms be identified and isolated. This task will be most easily accomplished in those systems where the level of genetic and biochemical knowledge for a given resistance phenomenon is well developed.

It is important to note that the general features of the biochemistry of target site resistance mechanisms are fundamentally different from those of metabolic resistance mechanisms. Generally, the protein products of genes involved in enhanced insecticide metabolism tend to be present in high levels in the insect and are biochemically relatively simple to assay by virtue of their enzymatic activities. In contrast, the target molecules of the three major classes of chemical insecticides are low abundance nerve membrane proteins, and their identification requires technically sophisticated assays of isolated subcellular membrane fractions, such as the binding of high specificity radiolabeled ligands or ion flux measurements. These features will require the adoption of radically different strategies for the molecular cloning of the genes implicated in the target site resistance category. With these considerations in mind, we expect that, in general, molecular approaches directed toward the elucidation of metabolic resistance mechanisms will be developed in the insect systems in which they occur and will be based primarily on straightforward biochemistry. Because of the low abundance and functional complexity of the molecules implicated in target site resistance mechanisms, experimental approaches developed for their investigation will be less direct and will necessarily draw more heavily upon the most technically favorable experimental systems.

Drosophila melanogaster provides a model insect system for such studies. It is the insect species most widely studied by geneticists and molecular biologists, owing to its ease of culture, a favorable genetic system with a legacy of decades of classical genetic and cytogenetic research, a small genome, and the availability of a procedure to introduce cloned sequences back into the germline (60). Among higher eukaryotes, its unparalleled integration of genetic and molecular approaches makes it the system of choice for the investigation of many fundamental biological processes. Despite these advantages and the clear demonstration of genetically conferred resistance to insecticides in both wild-type and laboratory stocks, it is, for the most part, a poorly recognized and underutilized resource for the investigation of insecticide resistance mechanisms. In addition, the availability of molecular probes derived from the *Drosophila* genome offers an obvious route to the isolation of homologous loci from economically important insect species in which target site mechanisms have been implicated in resistance phenomena.

There are two principal reasons for studying target site resistance mechanisms at the molecular level. The first is that the practical application of the knowledge of molecular mechanisms can be used to develop reliable diagnostic reagents for detecting the presence in individual insects of allelic variants conferring resistance. Specifically, characterization of the molecular differences between gene products of resistant and susceptible alleles could provide the basis for development of highly specific immunological or hybridization probes that permit the identification of the resistant genotype in individual insects. The success of continuing efforts to extend the useful life of existing insecticides by attempting to delay the onset of economically significant levels of resistance in insect populations would be greatly enhanced by the ability to determine accurately the frequency of resistance alleles and their distribution in large populations of wild insects.

Second, it is both necessary and desirable to develop new insecticidal compounds with improved efficacy and specificity. Decades of intensive research and development efforts to discover and develop new insecticides have resulted in a diminishing number of classes of insecticidal compounds with novel pharmacological properties (i.e., modes of action). This strongly suggests that the number of molecular targets suitable for insect chemical control is small, or, as a minimum, that other potential but as yet undiscovered target sites can not be identified readily by classical empirical approaches used for the discovery of new insecticidal compounds. It follows that research directed toward the discovery of new compounds, which act effectively on molecular target sites conferring resistance to existing compounds or which act on novel target sites, must be developed using mechanistic approaches involving molecular analyses at the level of individual genes and their corresponding primary gene products.

Literature Cited

1. Oppenoorth, F. J. In Comprehensive Insect Physiology Biochemistry and Pharmacology; Kerkut, G. A. and Gilbert, L. I., Eds.; Pergamon, Oxford, 1985; Vol. 12, Chapter 19.
2. Wilson, T. G. and Fabian, J. Dev. Biol. 1986, 118, 190.

3. Corbett, J. R.; Wright, K.; Baillie, A. C. In The Biochemical Mode of Action of Pesticides; 2nd Edition, Academic Press, London, 1984; Chapter 3.
4. Bloomquist, J. R.; Soderlund, D. M. Biochem. Biophys. Res. Commun. 1985, 133, 37.
5. Bloomquist, J. R.; Adams, P. M.; Soderlund, D. M. Neurotoxicology 1986, 7, 11.
6. Abalis, I. M.; Eldefrawi, M. E.; Eldefrawi, A. T. J. Toxicol. Environ. Health 1986, 18, 13.
7. Narahashi, T. In Cell. & Molec. Neurotox.; Narahashi, T., Ed.; Raven Press, New York, 1985; p 85.
8. Sawicki, R. M. In Insecticides (Progress in Pesticide Biochemistry and Toxicology, Vol. 5); Hutson, D. H., and Roberts, T. H., Eds.; Wiley, New York, 1985; Chapter 4.
9. Farnham, A. W. Pestic. Sci. 1977, 8, 631.
10. Farnham, A. W.; Murray, A. W. A.; Sawicki, R. M.; Denholm, I.; White, J. C. Pestic. Sci. 1987, 19, 209.
11. Nicholson, R. A.; Hart, R. J.; Osborne, M. P. In Insect Neurobiology and Pesticide Action (Neurotox '79); Society of Chemical Industry, London, 1980; p 465.
12. Miller, T. A.; Kennedy, J. M.; Collins, C. Pestic. Biochem. Physiol. 1979, 12, 224.
13. Ahn, Y.-J.; Shono, T.; Fukami, J. J. Pestic. Sci. 1986, 11, 591.
14. Scott, J. G.; Georghiou, G. P. Pestic. Sci. 1986, 17, 195.
15. Salgado, V. L.; Irving, S. N.; Miller, T. A. Pestic. Biochem. Physiol. 1983, 20, 100.
16. Osborne, M. P.; Hart, R. J. Pestic. Sci. 1979, 10, 407.
17. Hille, B. Ionic Channels of Excitable Membranes; Sinauer, Sunderland, MA, 426 pp.
18. Lund, A. E.; Narahashi, T. Pest. Biochem. & Physiol. 1983, 20, 203.
19. Chinn, K.; Narahashi, T. J. Physiol. 1986, 380, 191.
20. Catterall, W. A. Ann. Rev. Biochem. 1986, 55, 953.
21. Lombet, A.; Bidard, J.-N.; Lazdunski, M. Fed. Europ. Biochem. Soc. 1987, 219, 355.
22. Bloomquist, J. R.; Miller, T. A. Neurotoxicology 1986, 7, 217.
23. Salgado, V. L.; Irving, S. N.; Miller, T. A. Pestic. Biochem. Physiol. 1983, 20, 100.
24. Chang, C. P.; Plapp, F. W., Jr. Pestic. Biochem. Physiol. 1983, 20, 86.
25. Rossignol, D. P.; Pipenberg, P. K. Soc. Neurosci. Abst. 1986, 12, 1075.

26. Plapp, F. W. Jr. In Pesticide Resistance: Strategies and Tactics for Management; National Academy Press, Washington, DC, 1986; p 74.
27. Lund, A. E.; Narahashi, T. Neurotoxicology 1982, 3, 11.
28. Chiang, C.; Devonshire, A. L. Pestic. Sci. 1982, 13, 156.
29. Hall, L. M. In Tetrodotoxin, Saxitoxin, and the Molecular Biology of the Sodium Channel; Kao, C. Y. and Levinson, S. R., Eds.; New York Academy of Sciences, 1986; p 313.
30. Jackson, F. R.; Wilson, S. D.; Hall, L. M. J. Neurogenet. 1986, 3, 1.
31. Suzuki, N.; Wu, C. -F. J. Neurogenet. 1984, 1, 225.
32. Shepanski, M. C.; Glover, T. J.; Kuhr, R. J. J. Econ. Entomol. 1977, 70, 539.
33. Jackson, F. R.; Wilson, S. D.; Strichartz, G. R.; Hall, L. M. Nature 1984, 308, 189.
34. Adams, M. E.; Miller, T. A. Pest. Biochem. & Biophys. 1979, 11, 218.
35. Tanouye, M. A.; Wyman, R. J. J. Neurophysiol. 1980, 44, 405.
36. Bartlett, B. R. Can. Entomol. 1952, 84, 189.
37. Crow, J. F. J. Econ. Entomol. 1954, 47, 393.
38. King, J. C. J. Econ. Entomol. 1954, 47, 387.
39. Ogaki, M.; Tsukamoto, M. Botyu-Kagaku 1953, 18, 100.
40. Tsukamoto, M. Botyu-Kagaku 1960, 25, 156.
41. Waters, L. C.; Simms, S. I.; Nix, C. E. Biochem. Biophys. Res. Commun. 1984, 123, 907.
42. Morton, R. A.; Holwerda, B. C. Pest. Biochem. & Biophys. 1985, 24, 19.
43. Morton, R. A.; Singh, R. S. Biochem. Genet. 1982, 20, 179.
44. Messner, D. J.; Feller, D. J.; Scheuer, T.; Catterall, W. A. J. Biol. Chem. 1986, 261, 14882.
45. Kraner, S. D.; Tanaka, J. C.; Barchi, R. L. J. Biol. Chem. 1984, 260, 6341.
46. Rosenberg, R. L.; Tomiko, S. A.; Agnew, W. S. Proc. Natl. Acad. Sci. 1984, 81, 1239.
47. Noda, M.; Shimizu, S.; Tanabe, T.; Takai, T.; Kayano, T.; Ikeda, T.; Takahashi, H.; Nakayama, H.; Kanaoka, Y.; Minamino, N.; Kangawa, K.; Matsuo, H.; Raftery, M. A.; Hirose, T.; Inayama, S.; Hayashida, H.; Miyata, T.; Numa, S. Nature 1984, 312, 121.
48. Noda, M.; Ikeda, T.; Kayano, T.; Suzuki, H.; Takeshima, H.; Kurasaki, M.; Takahashi, H.; Numa,S. Nature 1986, 320, 188.
49. Kosower, E. M. Fed. Europ Bioch. Soc. 1984, 182, 234.

50. Greenblatt, R. E.; Blatt, Y.; Montal, M. Fed. Europ. Biochem. Soc. 1985, 193, 125.
51. Guy, H. R.; Seetharamulu, P. Proc. Natl. Acad. Sci. 1985, 83, 508.
52. Salkoff, L.; Butler, A.; Wei, A.; Scavarda, N.; Giffen, K.; Ifune, C.; Goodman, R.; Mandel, G. Science 1987, 237, 744.
53. Gundersen, C. B.; Miledi, R.; Parker, I. Nature 1984, 308, 421.
54. Sumikawa, K.; Parker, I.; Miledi, R. Proc. Natl. Acad. Sci. USA 1984, 81, 7994.
55. Noda, M.; Ikeda, T.; Suzuki, H.; Takeshima, H.; Takahashi, T.; Kuno, M.; Numa, S. Nature 1986, 322, 826.
56. Methfessel, C.; Witzemann, V.; Takahashi, T.; Mishina, M.; Numa, S.; Sakmann, B. Pflügers Archiv. 1986, 407, 577.
57. Thornhill, W. B.; Levinson, S. R. Biochem. 1987, 26, 4381.
58. Breer, H.; Benke, D. Molec. Brain Res. 1986, 1, 111.
59. Mouches, C.; Pasteur, N.; Berge, J.-B.; Hyrien, O.; Raymond, M.; Robert de St. Vincent, B.; de Silvestri, M.; Georghiou, G. P. Science 1986, 233, 778.
60. Rubin, G. M.; Spradling, A. C. Science 1982, 218, 348.

RECEIVED May 25, 1988

Chapter 15

Inhibition of Juvenile Hormone Esterase by Transition-State Analogs

A Tool for Enzyme Molecular Biology

András Székács[1,2], Bruce D. Hammock[1], Yehia A. I. Abdel-Aal[1], Matthew Philpott[1], and György Matolcsy[2]

[1]Departments of Entomology and Environmental Toxicology, University of California, Davis, CA 95616

[2]Plant Protection Institute of the Hungarian Academy of Sciences, Budapest, Hungary

A summary of Transition State Theory is presented, as it applies to the design of trifluoromethyl ketone esterase inhibitors. Possible mechanisms for the inhibition of the enzyme are discussed.

A new series of compounds, α,α'-alkanebis-thiotrifluoro-propanones was synthesized and showed excellent *in vitro* and moderate *in vivo* inhibition of the insect juvenile hormone esterase from the fifth instar larvae of *Trichoplusia ni* (cabbage looper). The potency of the above series was also screened for its ability to inhibit other esterases of toxicological and pharmacological significance. Trifluoroketones are discussed as an example of the importance of chemistry in biotechnology approaches.

It is generally recognized that the widespread use of insecticides has two serious side-effects: evolution of insecticide resistance in insects and danger to environmental and human safety. The continued use of insecticides requires stricter agricultural practices as well as the improvement of the chemicals themselves. Equally as important, however, new strategies for development of insecticidal agents are needed (1-3).

The elucidation of enzyme-substrate interactions has established new paradigms leading to the discovery of biologically active compounds. One such paradigm is the "Transition State Theory" as it applies to the mechanism of enzymatic reactions. Based on this theory, series of transition state analog inhibitors known as trifluoromethyl ketones have been synthesized in our laboratory. Our target has been an insect enzyme of developmental and reproductional importance, juvenile hormone esterase (4-12). The development of those extremely potent inhibitors served several aims: in addition to providing "traditional" inhibitors that can be used to block the enzyme and study its biochemical and physiological consequences, the new group of compounds led to powerful ligands for the high yield affinity chromatography

0097-6156/88/0379-0215$06.00/0

purification of this low abundance enzyme (13-15). The pure enzyme, juvenile hormone esterase (JHE) is a very suitable candidate for several biotechnological approaches from the use of the enzyme as a probe for endocrine regulation, to the more detailed research of the molecular biology and molecular genetics of the enzyme, itself. As discussed above the biological role of JHE in lepidopterous larvae was established using highly specific esterase inhibitors. Application of these compounds led to giant larvae which were locked in the feeding stage. The use of trifluoroketone affinity columns led to the purification of large amounts of the enzyme. This availability first led to a dose-response, linear with the quantity of the enzyme applied, observed for the direct injection of the enzyme into several insect species, showing that injected JHE has a clear anti-JH effect. Since such an effect is desirable in agriculture, it becomes important to clone the enzyme. Here again the affinity procedure from chemical approaches provided the protein necessary for development of molecular probes. The injection of purified enzyme in rabbits resulted in JHE specific antibodies needed for screening an expression library as well as for further immunoassay studies. The affinity purified protein also allowed classical aminoacid screening to be done which led to the synthesis of oligonucleotide probes for confirming positives from the expression library. Hopefully the cloned message can be inserted into baculovirus vectors which lead to precocious production of the enzyme. This, in turn, should result in anti-juvenile hormone effects such as cessation of feeding and developmental abnormalities.

These results and future applications of trifluoromethyl ketones show the importance of the "traditional" chemical optimization of compounds in various biotechnological approaches. In this presentation the authors wish to give a summary of the leading research paradigm, Transition State Theory resulting in trifluoromethyl ketones, a group of highly effective inhibitors of JHE. The same concepts can be applied to a variety of polarized carbonyls, carbamates, phosphates and phosphonates as inhibitors of esterases and proteases. Clear targets in the insecticide field will be enzymes involved in insecticide metabolism and neurohormone processing enzymes.

Juvenile Hormone Esterase

Our laboratory is concerned with targeting potential insecticides that disrupt normal development and metamorphosis in insects. Juvenile hormones (JHs), acting in concert with the steroid hormone ecdysone, are believed to control the timing of the larval-larval molts, larval-pupal and pupal-adult transformations of the insects. It has been demonstrated that the events leading to pupation are initiated by reduction of the JH titer in the hemolymph. In addition to a cessation of biosynthesis, this reduction in JH titer is controlled by degradative metabolism (16,17). Hydrolysis of the epoxide and ester functionalities present in active JH are two routes of degradation and subsequent inactivation of JH (18). The primary route of JH metabolism in the hemolymph of last stadium lepidopterous larvae is ester hydrolysis, and it is catalyzed by the enzyme juvenile hormone esterase (JHE). JHE has been shown to play a crucial role in initiating pupation in lepidopterous insects (19); selective inhibition of this enzyme prevents JH hydrolysis and causes a delay in the onset of pupation (20).

Transition State Theory and JH Ester Hydrolysis

In every chemical reaction the reactants are in equilibrium with an unstable activated complex, the transition state complex (TS), which decomposes to give the product. In his pioneering work, Linus Pauling pointed out that for promoting a reaction without influencing its equilibrium constant, the enzyme should have much higher affinity for the transition state of a reaction than for either the substrate or the product(s), thereby pushing a reaction in the desired direction by continuously removing its transition state (21). Based on this idea, extremely potent inhibitors can be developed for a given enzymatic reaction if one can synthesize "transition state mimics" (TSM): stable chemical compounds resembling the transition state (22).

Because transition states may have lifetimes of only several nanoseconds, in most cases, it is impossible to observe them directly. However, there are numerous lines of evidence for the existence of a tetrahedral-like transition state for non-enzymatic ester hydrolysis: a) substitution at a carbonyl group (as is the case of the hydrolysis of esters) most often proceeds by a tetrahedral mechanism, a second-order addition-elimination (for a review of this mechanism, see (23)); b) the kinetics are pseudo-first order either in the substrate or in the nucleophile, as predicted by the mechanism; c) for the ^{18}O labeled esters, the ^{18}O isotope is detectable in both products (in a "normal" S_N2 reaction all the ^{18}O isotopes should remain in the acid functionality)(24); d) in a few cases tetrahedral intermediates have been isolated or detected spectrally (25).

Ester hydrolysis of the JHs is shown in Equation 1. We have no exact thermochemical data measured for this reaction. In general, the hydrolysis of unsaturated long chain aliphatic acids is thermodynamically neutral under standard conditions (26,27) (i.e. the heat of hydrolysis for *cis*- and *trans*-oleic acid is -1.7 and +0.8 kJ/mol, respectively), which shows that these reactions are generally not favored based solely on thermodynamic considerations. In aqueous solution ester hydrolysis is driven by the high concentration of water. The juvenile hormones will be more resistant to acid or base catalysed hydrolysis because in these compounds conjugation with the α,β-unsaturation greatly stabilizes the carbonyl functionality (19). Therefore, the nonconjugated TS should be difficult to form, and this situation will be reflected by a high activation energy for the uncatalyzed hydrolysis of JH.

The ester hydrolysis catalyzed by hydrolase enzymes (E) proceeds according to Equation 2 and the free energy diagram is shown in Figure 1. After the preliminary noncovalent binding step ([ES]) the enzyme becomes complementary in structure to the substrate (or its TS) and forms a covalent adduct with it (ES). Release of the first product (MeOH) results in the acyl-enzyme (EX), which hydrolyses through a covalent enzyme-product adduct (EP) to the appropriate carboxylic acid and the free enzyme. It has been shown

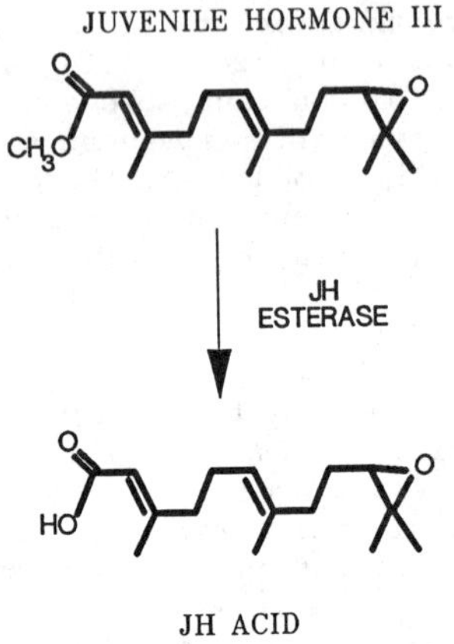

JUVENILE HORMONE III

JH ESTERASE (1)

JH ACID

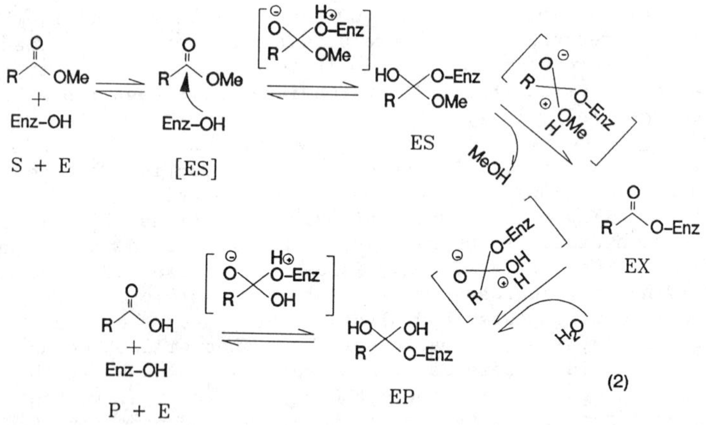

S + E [ES] ES MeOH EX H_2O EP P + E

(2)

that the rate limiting step in the reaction is the product release from the acyl-enzyme (28).

Any compound that is recognized by the enzyme as a TS will bind to it competitively with the substrate. It is important to note, however, that the mechanism of enzymatic reaction is not simply the reaction of the enzyme with the transition state of the uncatalyzed reaction; it can be seen by the cartoon in Figure 1, that the enzymatic hydrolysis of esters proceeds through several intermediates and transition states; a putative TSM compound might resemble any or several of these states. Thus, the concept of a "transition state mimic" is somewhat of a misnomer.

Trifluoromethyl Ketones as Transition State Mimics

An exemplary application of Transition State Theory in developing highly active inhibitors for esterase enzymes is the case of trifluoromethyl ketones (29). Replacing the alkoxy group of the carboxylic esters by a trifluoromethyl group results in highly polarized ketones which are sensitive to nucleophilic attack. Subsequently, in the presence of trace amounts of water, the keto-form will be hydrated and be in equilibrium with the corresponding geminal diol. The geminal diol is tetrahedral in geometry and in theory resembles the transition state of the uncatalyzed hydrolysis of esters (TSX). Thus, according to the original Pauling-theory, trifluoromethyl ketones should bind strongly to esterase enzymes. In this reaction they form hemiketals with the serine present at the active site of the enzyme; two possible reaction mechanisms are enzyme addition to the carbonyl (30) or condensation with the geminal diol (31) (enhanced by the hydrophobic aliphatic chain in the molecule) (Equation 3).

The most likely chemical reaction is addition to the carbonyl as shown in the top portion of Equation 3; however, in aqueous solution the majority of the compound exists as the geminal diol. If this situation predominates, either the equilibrium between the diol and carbonyl form must be fast on the time scale of our ability to measure enzyme reactions, or the I_{50}'s are in fact far lower than we have reported. Alternatively, the hydrated carbonyl could react directly with the enzyme. This could happen in two ways. In one case the TSM could be held near the catalytic site due to interaction with the R group. The relative abundance of the carbonyl might then be favored in a nonaqueous microenvironment. This process could even be accelerated catalytically. If there is enzyme involvement in production of the carbonyl from the geminal diol, then these TSMs could be considered "suicide" substrates. Another alternative explanation is a "normal" S_N2 type reaction between a serine anion and a protonated geminal diol with water as the leaving group. These alternative pathways are not mutually exclusive, but additional work on the kinetics and structural biochemistry of the interaction will be needed to indicate the predominant pathway.

The reaction results in an adduct with the enzyme (ETSM) with no ester C-OR bond present in the molecule to be cleaved. It is clear from Equation 3 that the hemiketal may only react by the release of the trifluoromethyl ketone by the enzyme through the transprotonation transition state, thereby acting as a reversible

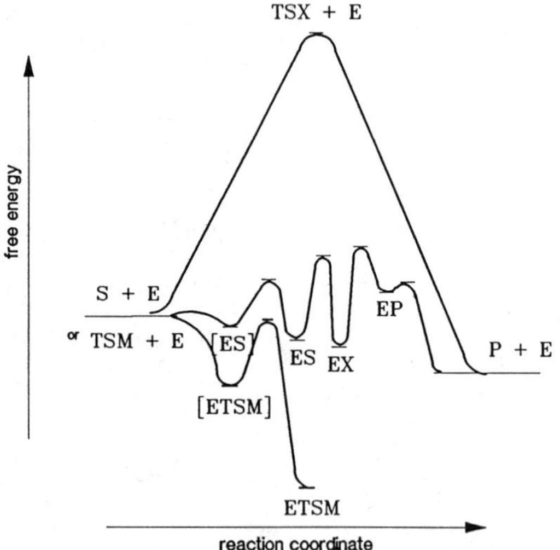

Figure 1. Schematic presentation of the free energy changes in non-enzymatic (ETX) and enzymatic reactions (EX) and in the reaction of a hypothetical transition state mimic (TSM) with the enzyme.

(3)

TSM + E [ETSM] ETSM

inhibitor. The enzyme-TSM complex (ETSM) is sterically and/or electronically similar to one of the enzyme-substrate Michaelis complexes (possibly to ES), but is favored energetically over the reaction between the enzyme and the substrate. A possible free energy diagram is shown in Figure 1.

Based on the above discussion, trifluoromethyl ketones should inhibit proteases such as chymotrypsin (32), and serine esterases, such as acetylcholinesterase (33,34), carboxylesterases (10), JHE and other esterases with varying selectivity. In a series of some juvenoid-like trifluoromethyl ketones and compounds of the structure A, 1,1,1-trifluoro-2-tetradecanone (TFT) was found to be highly active and selective against JHE (I_{50}: 1×10^{-7}M) as compared to α-naphthyl acetate esterase (α-NaE) or trypsin (4-6).

$CH_3(CH_2)_n$-C(O)CF_3 A

$CH_3(CH_2)_n$-SCH_2C(O)CF_3 B

Introducing a sulfur atom β to the carbonyl significantly increased the activity of the resulting compounds (B) on JHE and some but not all other esterases, possibly by bioisosterically mimicking (35) the α,β double bond of the natural JHs. Based on this finding, a series of aliphatic, aromatic (7,8) and terpenoid (9) derivatives were synthesized. The most active compound of these series was 3-octylthio-1,1,1-trifluoro-2-propanone (OTFP). In a recent report, by making a slight modification in the structure of the parent compound, two additional compounds with slightly higher activity have been synthesized (36). TFT and the aliphatic trifluoromethyl ketones appeared to be classical competitive inhibitors, while many trifluoropropanone sulfides were found to be reversible but slow and tight binding inhibitors of JHE (7,11,12).

The outstanding activity of some members of the thiotrifluoropropanones proved useful as ligands and eluting agents for the affinity purification of JHE (13-15). This new method made possible specific purification of JHE with unusually high yields, making possible new research in the areas of immunochemistry and molecular biology for this highly active but low abundant esterase.

In our laboratory, we feel that the scientific and practical use of these inhibitors, by no means, has reached a peak. An obvious need is to try different ligands for affinity elution. Therefore, a new class of transition state analogs was synthesized and tested for their inhibitory potential against JHE, cholinesterase and malathionase.

Synthesis

Compounds of the general structure C and D were synthesized and characterized (See Table I)

$$CF_3C(O)CH_2S\text{-}(CH_2)_n\text{-}SCH_2C(O)CF_3 \qquad C$$

n: 3 - 12 I - X

$$CF_3C(O)CH_2S\text{-}R\text{-}SCH_2C(O)CF_3 \qquad D$$

R: $-(CH_2)_2-O-(CH_2)_2-$ XI
$-(CH_2)_2-S-(CH_2)_2-$ XII
$-CH(CH_3)(CH_2)_2-$ XIII
$-CH(CH_3)(CH_2)_3-$ XIV

The compounds were synthesized according to a previously published procedure (7) with the modification that triethylamine was used to accelerate the reaction by neutralizing the HBr.

$$HS\text{-}R\text{-}SH + 2\ BrCH_2C(O)CF_3 \xrightarrow{Et_3N} CF_3C(O)CH_2S\text{-}R\text{-}SCH_2C(O)CF_3$$

The starting mercaptans were either purchased from Aldrich Chemicals or synthesized from the appropriate α,α'-dihaloalkane via the thiourea method (37). The yields were between 59% and 98% for compounds III and V, respectively, and in general higher than 85% depending on the speed of the addition of triethylamine. (Chemical details will be discussed elsewhere: Székács, A.; Hammock, B.D.; Abdel-Aal, Y.A.I.; Philpott, M.; Matolcsy, G.: Pest. Biochem. Physiol., submitted).

Enzyme Assays

The inhibition rate for all inhibitor assays was measured from the initial velocities within the linear time-activity relationships of the control and inhibited samples. The standard deviations were calculated for the I_{50} values of the compounds against the three enzymes used and the maximal values were used to compare the statistical variability within each assay test.

JHE: For the determination of JHE inhibition by the title compounds, the radiometric partition method (38) was used. Hemolymph JHE from Day 2 of the fifth instar larvae of *T. ni* was used (L_5D_2), diluted 1:500 with 0.08M phosphate buffer (pH=7.4 with 0.1% phenylthiourea to inhibit tyrosinases). The main reason for choosing this insect was that a great deal of effort has been put into the characterization of larval carboxylesterases and JHE in *T. ni* (39,40). In L_5D_2 larvae, the JHE titer is near its maximum (19). C10 ^3H labeled JH III (New England Nuclear) and unlabeled JH III (Calbiochem) were used as substrate solubilized in abs.ethanol.

Acetylcholinesterase (AChE): The lyophilized enzyme from electric eel (Sigma) was dissolved in 0.05M phosphate buffer (pH=7.4) at a concentration of 20 μg/mL. Acetylthiocholine iodide was used as substrate at a final concentration of $5x10^{-4}$M in buffer. 5,5′-Dithiobis-(2-nitrobenzoic acid) (DTNB) at a final concentration of $3.8x10^{-2}$M was used to monitor the released thiocholine according to a published procedure (41) with slight modification. Acetone was used as a solvent for the inhibitors.

Malathionase (ME): For measuring the inhibition of malathion esterase activity, general carboxylesterase from porcine liver (Sigma) was used at a final concentration of 16 μg protein/mL in 0.1M Tris HCl buffer (pH=7.5). The procedure involves an indirect determination of the malathionase activity by coupling the hydrolysis of malathion to the reduction of a tetrazolium dye (42). An acetone solution of malathion was used as substrate to a final concentration of $3x10^{-4}$M.

Table I. Inhibition of Esterases by α,α′-alkanebis-thiotrifluoropropanones

compound number	number of carbons	Molar I_{50} Value[a] [M]		
		JHE	AChE	ME
I	3	$9.87x10^{-7}$	$1.14x10^{-5}$	$3.63x10^{-6}$
II	4	$3.06x10^{-9}$	$8.44x10^{-6}$	$3.79x10^{-6}$
III	5	$7.00x10^{-9}$	$4.30x10^{-6}$	$2.38x10^{-6}$
IV	6	$1.50x10^{-8}$	$2.98x10^{-6}$	$9.02x10^{-7}$
V[b]	7	$1.68x10^{-8}$	$2.98x10^{-6}$	$4.89x10^{-6}$
VI[b]	8	$8.17x10^{-10}$	$7.78x10^{-6}$	$4.62x10^{-6}$
VII	9	$2.67x10^{-9}$	$1.34x10^{-5}$	$5.00x10^{-4}$
VIII[b]	10	$4.54x10^{-9}$	$4.35x10^{-6}$	$1.66x10^{-5}$
IX[b]	11	$1.33x10^{-8}$	$5.25x10^{-6}$	$1.03x10^{-3}$
X[b]	12	$8.15x10^{-9}$	$1.03x10^{-5}$	$3.02x10^{-3}$
XI	2-O-2	$2.16x10^{-8}$	$6.01x10^{-6}$	$8.58x10^{-6}$
XII	2-S-2	$4.94x10^{-9}$	$7.12x10^{-6}$	$3.53x10^{-6}$
XIIIb	3′	$4.68x10^{-7}$	$3.47x10^{-6}$	$4.56x10^{-6}$
XIV[b]	4′	$8.20x10^{-9}$	$8.12x10^{-7}$	$5.22x10^{-5}$

[a] JH III was used as a substrate for JHE from *T. ni*, acetylthiocholine iodide for electric eel AChE and malathion for carboxylesterase from porcine liver.
[b] The dithiol was prepared from alkyl-dibromide.

Summarized in Table I are the I_{50} values for α,α′-alkanebis-thiotrifluoroketones against JHE, AChE and malathionase. All the tested compounds showed much higher inhibitory potency against JHE than against the other two enzymes. The comparative potency in favor of JHE adds to the evidence that these trifluoromethyl ketones are transition state mimics (TSMs) of the enzyme natural substrates (JHs). However, the slopes of the inhibition curves for almost all the compounds were lower with JHE than with the other two enzymes. Possibly this is due to the fact that JHE from *T. ni* contains multiple JH catalytic sites (43-45), as it has been shown for the tobacco hornworm (*Manduca sexta*)(13). Except for compounds III and IV, the inhibition curves for malathionase inhibition had the highest slopes. The maximal values of the percentage standard error of pI_{50} were 0.77%, 0.18% and 0.66% of the mean for the JHE, AChE and malathionase assays, respectively.

The quantitative structure-activity relationships of the tested compounds are beyond the scope of this paper. However, the empirical structure-activity correlations based on the number of the carbon atoms in the alkyl chain is different for each of the esterases assayed.

It is worth noting that in the JHE assay the trend of the data showed similar structure-activity relationship to the 3-alkylthio-1,1,1-trifluoro-2-propanones (8), which adds some additional evidence for the rational design of these TSMs of JHs. In this assay, however, a definite peak occurs at carbon number eight, for the compound 1,1,1,16,16,16-hexafluoro-4,13-dithia-hexadecane-2,15-dione (I_{50}: $8.2x10^{-10}M$), which suggests that after reaching this particular size, the compound might be able to interact with the enzyme(s) in two catalytic sites (or highly hydrophilic parts) at the same time.

In contrast, cholinesterase did not seem to respond in a systematic fashion to the structural variation. Malathionase showed intermediate response with a generally decreasing potency as a function of the number of carbon atoms in the molecule.

In Vivo Assays

Using the irreversible inhibitor EPPAT, Sparks and Hammock (4) demonstrated that pupation could be delayed in *T. ni* presumably by inhibiting JHE and thereby maintaining an abnormally high JH and low prothoracicotropic hormone (PTTH) level. TFT (lacking the thioether moiety) failed to cause this effect while OTFP was effective in delaying pupation when repeated doses of 0.1-0.2 μmol were applied topically on day 1 and 2 of the fifth larval instar (L_5D_1 and L_5D_2)(7).

The two most effective *in vitro* inhibitors of JHE in the new series, VI and II, were screened *in vivo* for their ability to delay pupation in *Trichoplusia ni*, relative to the ability of OTFP, a response shown to be concurrent with the selective inhibition of JHE (7,9).

The compounds (2 μL of $1x10^{-1}M$ EtOH solution) were topically applied to *T. ni* on L_5D_1 and L_5D_2 at 4, 12 and 17 hrs ALO (after lights on) and the time of pupation was recorded. In the control group, larvae treated with 2 mL EtOH, 90-100% of larvae had pupated by L_5D_5 9 AM. At final doses of 0.2 μmol, VI was as effective as OTFP in delaying pupation relative to the ethanol controls, with only 45-75% of the larvae pupating by L_5D_5. At this concentration, treatment with II did not delay pupation significantly, 75-100% pupated by L_5D_5.

Conclusion

Although trifluoromethyl ketones acting on JHE will probably not result in effective insecticides (delaying pupation is generally not a desired biological effect in agriculture), continued research on them has led to several practical results as well as to new theoretical considerations. Knowledge gleaned by research with trifluoromethyl ketones will lead to a more complete understanding of the role of JHE in insect development. With this understanding of the mechanisms regulating normal development in insects, new avenues may be opened for targeting new and specific ways of insect control. This work with juvenile hormone esterase provides a nice example of chemistry, biochemistry and molecular biology as complementary technologies (19). These potent inhibitors were first used to demonstrate the essential role of JHE in insect development. A knowledge of the chemistry and biochemistry of the interaction of these transition state mimics with JHE led to the development of affinity purification systems for this low abundance enzyme (13). Large amounts of the pure protein allowed more detailed kinetic studies as well as the development of antibodies and nucleic acid probes used to clone the enzyme. It could well be that the resulting clones can be engineered into baculoviruses or other vectors for insect control.

In addition, studies from this and other laboratories illustrate that by mimicking hypothetical transition states one can develop exceptionally potent enzyme inhibitors. These compounds will advance our understanding of catalytic mechanisms, serve as probes for unraveling the roles of particular enzymes in catalytic processes, and allow their rapid purification. They are also promising commercially as pesticides, pesticide synergists, as well as pharmaceuticals. From such work it seems clear that both chemical and biotechnological approaches will be used alone and in combination in the development of biologically active materials.

Acknowledgments

The authors wish to express their sincere appreciation to Eric Dietze for the valuable consultations and to Kathleen Dooley for her careful work in typing the manuscript. This work was supported by grants ES02710-07, DCB-8518697, and 85-CRCR-1-1715 from NIEHS, NSF, and USDA, respectively. A.S. was supported by a fellowship from Fulbright Program #33917 (Institute of International Education). B.D.H. is a Burroughs Wellcome Scholar in Toxicology.

Literature Cited

1. Hollingworth, R.M. In Pesticides, Minimizing the Risks; Ragsdale, N.N.; Kuhr, R.J., Eds.; ACS Symposium Series No. 336; American Chemical Society: Washington, DC, 1986; pp. 54-76.
2. Metcalf, R.L. Ann. Rev. Entomol. 1980, 25, 219-56.
3. Hammock, B.D.; Soderlund, D.M. In Pesticide Resistance, Strategies and Tactics for Management; Glass, E., Ed.; (Natl. Acad. Press: Wash.D.C., 1986; pp.111-29.

4. Sparks, T.C.; Hammock, B.D. Pest. Biochem. Physiol. 1980, 14, 290-302.
5. Hammock, B.D.; Wing, K.D.; McLaughlin, J.; Lovell, V.M.; Sparks, T.C. Pest. Biochem. Physiol. 1982, 17, 76-88.
6. Sparks, T.C.; Hammock, B.D.; Riddiford, L.M. Insect Biochem. 1983, 13, 529-41.
7. Hammock, B.D.; Abdel-Aal, Y.A.I.; Mullin, C.A.; Hanzlik, T.N.; Roe, R.M. Pest. Biochem. Physiol. 1984, 22, 209-23.
8. Hammock, B.D.; Mullin, C.A. U.S. Patent 4 562 292, 1985.
9. Prestwich, G.D.; Eng, W-S.; Roe, R.M.; Hammock, B.D. Arch. Biochem. Biophys. 1984, 228, 639-45.
10. Ashour, M-B.A.; Hammock, B.D. Biochem. Pharm. 1987, 36, 1869-79.
11. Abdel-Aal, Y.A.I.; Roe, R.M.; Hammock, B.D. Pest. Biochem. Physiol. 1984, 21, 232-41.
12. Abdel-Aal, Y.A.I.; Hammock, B.D. Insect Biochem. 1985, 15, 111-22.
13. Abdel-Aal, Y.A.I.; Hammock, B.D. Arch. Biochem. Biophys. 1985, 243, 206-19.
14. Abdel-Aal, Y.A.I.; Hammock, B.D. Science 1986, 233, 1073-6.
15. Hammock, B.D.; Abdel-Aal, Y.A.I.; Hanzlik, T.N.; Croston, G.E.; Roe, R.M. In Molecular Entomology; Law, J.H., Ed.; Alan R. Liss, Inc.: New York, 1987 pp. 315-28.
16. Hammock, B.D.; Quistad, G.B. In Progress in Pesticide Biochemistry; Hutson, D.H.; Roberts, T.R., Eds.; Wiley: New York, 1981; Vol.1, pp. 1-83.
17. de Kort, C.A.D.; Granger, N.A. Ann. Rev. Entomol. 1981, 26, 1-28.
18. Slade, M.; Zibitt, C.H. In Insect Juvenile Hormones: Chemistry and Action; Menn, J.J.; Beroza, M., Eds.; Academic Press: New York; 1972; pp. 155-76.
19. Hammock, B,D.; Abdel-Aal, Y.A.I.; Ashour, M.; Buhler, A.; Hanzlik, T.N.; Newitt, R.; Sparks, T.C. In Human Welfare and Pest Control Chemicals-Approaches to Safe and Effective Control of Medical and Agricultural Pests; Sasa, M.; Matsunaka, S.; Yamamoto, I.;Ohsawa, K., Eds.; XVIII. Pesticide Science Symposium and Public Lectures; Pesticide Science Society of Japan: Tokyo, 1986; pp.53-72.
20. Hammock, B.D. In Insect Physiology Biochemistry and Pharmacology; Kerkut, G.A.; Gilbert, L., Eds.; Pergamon Press: New York, 1985; pp. 431-72.
21. Pauling, L. Amer. Sci. 1948, 36, 51-8.
22. Wolfenden, R. Ann. Rev. Biophys. 1976, 45, 271-306.
23. Jencks, W.P. Catalysis in Chemistry and Enzymology; McGraw-Hill: New York, 1969; pp. 463-554.
24. Bender, M.L.; Thomas, R.J. J. Am. Chem. Soc. 1961, 83, 4183-9.
25. Bender, M.L. Chem. Rev. 1960, 60, 53-113.
26. Cox, J.D.: Pilcher, S. Thermochemistry of Organic and Organometallic Compounds; Academic Press: London, New York, 1970; pp. 140-493.
27. Lange's Handbook of Organic Chemistry, 13th Ed.; Dean, J.A., Ed.; McGraw-Hill: New York, 1984; pp. 9-70 - 9-99.
28. Walsh, C. Enzymatic Reaction Mechanisms; Freeman: San Francisco, 1979; pp.53-107.

29. Abdel-Aal, Y.A.I.; Hammock, B.D. In Bioregulators for Pest Control; Hedin, P.A., Ed.; ACS Symp. Ser. No. 276; American Chemical Society: Washington, DC, 1985; pp. 135-60.
30. Stein, R.L.; Strimpler, A.M.; Edwards, P.D.; Lewis, J.J.; Manger, R.C.; Schwartz, J.A.; Stein, M.M.; Trainor, D.A.; Wildonger, R.A.; Zottola, M.A. Biochemistry 1987, 26, 2682-9.
31. Abeles, R.H. Drug Dev. Res. 1987, 10, 221-34.
32. Imperiali, B.; Abeles, R.H. Biochemistry 1986, 25, 3760-7.
33. Brodbeck, U.; Schweikert, K.; Gentinetta, R.; Rottenberg, M. Biochim. Biophys. Acta 1979, 567, 357-69.
34. Gelb, M.H.; Svaren, J.P.; Abeles, R.H. Biochemistry 1985, 24, 1813-7.
35. Thornber, C.W. Chem. Soc. Rev. 1979, 8, 563-80.
36. Linderman, R.J.; Leazer, J.; Venkatesh, K.; Roe, R.M.: Pestic. Biochem. Physiol. 1987, 29, 266-77.
37. Speziale, J.A. In Organic Syntheses; Rabjohn, N., Ed.; Wiley: New York, 1963; Coll. Vol. IV, pp. 401-3.
38. Hammock, B.D.; Roe, R.M. Methods Enzymol. 1985, 111B, 487-94.
39. Wing, K.D.; Sparks, T.C.; Lovell, V.M.; Levinson, S.O.; Hammock, B.D. Insect Biochem. 1981, 11, 473-85.
40. Yuhas, D.A.; Roe, R.M.; Sparks, T.C.; Hammond, A.M.,Jr. Insect Biochem. 1983, 13, 129-36.
41. Ellman, G.L.; Courtney, K.D.; Andres, V., Jr.; Featherstone, R.M. Biochem. Pharmacol. 1961, 7, 88-95.
42. Talcott, R.E. Tox. Appl. Pharmacol. 1979 47, 147-50.
43. Jones, D.; Jones, G.; Rudnicka, M.; Click, A.; Streekrishna, S. Experientia 1986, 42, 45-7.
44. Jones, D.; Jones, G.; Click, A.; Rudnicka, M.; Streekrishna, S. Comp. Biochem. Physiol. 1986, 85B, 773-81.
45. Hanzlik, T.N.; Hammock, B.D. J. Biol. Chem., 1987, 262, 13584-91.

RECEIVED May 4, 1988

PLANT GENE EXPRESSION

Chapter 16

Research in Biotechnology at the Agricultural Research Service, U.S. Department of Agriculture

Jerome P. Miksche[1] and G. Ram Chandra[2]

[1]Agricultural Research Service, U.S. Department of Agriculture, Beltsville, MD 20705
[2]Plant Molecular and Genetics Laboratory, Plant Genetics & Germplasm Institute, Agricultural Research Service, U.S. Department of Agriculture, Beltsville, MD 20705

Within the framework of the Agricultural Research Service (ARS) mission, the term Biotechnology is defined simply as those biological means used to develop processes and products employing organisms or their components. Those biological means include bioreactors, bioreactions, immunolocalization of biologically active cells and enzymes, cell tissue and organ cultures, genetic engineering, gene transfer, recombinant DNA technology and hybridoma techniques (1). The new biotechnology methodologies present approaches and precision never before available to agricultural researchers to address the many difficult challenges.

What are the biotechnology efforts in the ARS and where are they located; the profiles of our scientists as related to industry and university colleagues; and an overview of research efforts they are pursuing will be covered in this chapter.

The Agency expends 30 million dollars annually on biotechnology related research at 130 locations across the United States. This effort is represented by approximately 200 projects. The research is directed accordingly:

$300,000	Soil and Water
11.4M	Plant Productivity and Protection
9.9M	Animal Production and Protection
8.4M	Product Quality

The research is conducted by ARS scientists located on many of the Land Grant universities and ARS Research Centers located in several geographic areas. A strong concerted thrust was developed to dissect the complexities of plant gene expression at the Albany, CA, location. The research conducted at the Beltsville, MD, location is a good example of the diversity of the ARS-biotechnology related research projects that are focused to solve several complex problems in agriculture (1). Much of this research is interdisciplinary involving several fields of sciences and commodities with one major objective, i.e., to develop new technologies to produce and market value-added U.S. agricultural products (2).

ARS doesn't have a formally structured biotechnology program, but it uses the tools of molecular biological sciences including recombinant DNA technology to direct research efforts to find solutions to high priority problems. These priorities (Table I) are partly formulated by agribusiness industrial groups, commodity organizations, user organizations, and congressional mandates to the Agency and ARS scientists. The National Program Staff finalizes the priorities and directs research programs to address these priorities in a timely and efficient manner.

Table I. Biotechnology Research Directed to Solve High Priority Problems

A.	Quality of Natural Resources
B.	Reduce High Cost of Farm Production
C.	Improving Crop Protection
D.	Enhancing Market Value and Quality of Farm Products
E.	Improve Marketing and Export of Agricultural Products
F.	Promote Human Health Through Nutrition

There are about 1,000 worldwide biotechnology firms, with 400 located in the United States. Many "buy outs," failures, and shifts have taken place since 1980, but generally there is a gradual increase in the number of firms. This gradual increase in biotechnology indicates industrial interest and permanence. The Agency is cognizant of this pattern and it desires to deliver research answers to the agribusiness biotechnology marketplace and ultimately to the grower, as this is the most efficient way to transfer knowledge to help the consumer.

The increased emphasis in biotechnology and subsequent knowledge transfer to industry require a match-up of talent between ARS and industry. Tables II and III demonstrate dovetailing the percent of ARS scientists in general biotechnology associated industrial areas and a listing of specific commercial areas, respectively. It is evident from the tables that the major short and long term biotechnology research areas are being addressed (3).

Table II. Distribution of ARS Scientists by Industries and Associated Areas

Industries and Areas	Percent
Pharmaceutical	19
Chemical	10
Fermentation Genetic Engineering	4
Food Processing	3
Applications to Plants in Agriculture	44
Animal Improvement/Disease	13
Use of Genetically Engineered Organisms in the Environment	7

The agricultural problems before us will seek resolution through cooperation with university investigators. Table IV lists the percent of ARS projects dealing with biotechnology on the basis of scientific fields. The scientific disciplines related to biotechnology are given in Table V. It is evident that the biological and agricultural associated disciplines are represented more strongly than are engineering and medical science disciplines.

Table III. Industry and Associated ARS Research Areas

Pharmaceutical	Chemical
Antibiotics	Fermentation
Antibodies	Fertilizers
Antigens (Vaccines)	Biological Processing
Enzymes	Pesticides
Growth Hormones	Polymers
Other Hormones	Industrial Chemicals by Biological Technologies
Interferons	
Lymphokines/Cytokines	
Protein Pharmaceuticals	
Non Protein	
Genetic Engineering and Fermentation	**Food Processing**
Living Whole Cell Fermentation	Single Cell Protein Production
Isolated Enzymes	Baking, Brewing, and Winemaking
Energy Production from Biomass	Microbial Polysaccharides
	Enzymes in Food Processing
	Sweetener, Flavors, and Fragrances

Table IV. Distribution of Scientists by Fields

Field	Percent
Agricultural Sciences	23
Biological Sciences	62
Engineering	7
Biomedical Sciences	8

Table V. Academic Disciplines of ARS Scientists*

Agricultural Sciences		Biological Sciences		Engineering		Medical Sciences	
Respondents Total 346		Respondents Total 909		Respondents Total 98		Respondents Total 109	
Discipline	Percent	Discipline	Percent	Discipline	Percent	Discipline	Percent
Agronomy	11	Bacteriology	46	Agriculture	3	Environmental	
Animal Husbandry	3	Biochemistry	13	Biomedical	7	Health	4
Animal Science	9	Botany	5	Chemical	2	Medicine	6
Animal Nutrition	2	Cell Biology	8	Fermentation	10	Oncology	5
Crop Science	18	Ecology	1	Genetic	78	Parasitology	14
Environmental		Embryology	1	Sanitary or		Pathology	21
Science	2	Entomology	5	Environmental	1	Pharmacology	4
Food Science	5	Genetics	11			Toxicology	10
Forestry	1	Immunology	5			Veterinary	
Horticulture	9	Microbiology	7			Medicine	45
Plant Pathology	17	Molecular Biology					
Poultry Science	4	Plants	7				
Soil Science	3	Animals	5				
Veterinary		Microorganisms	6				
Science	16	Parasitology	2				
		Physiology					
		Plants	7				
		Animals	4				
		Microorganisms	3				
		Virology	4				
		Zoology	1				

* The scientists had an opportunity to select more than one academic discipline.

Research Overview

Water Quality

Safe and economic disposal of pesticide wastes to assure quality groundwater is a great need in this country. The yearly usage of pesticides in the United States is approximately 370,000 tons. Genetically engineered and native microbes that degrade pesticides will safely meet this need to clean out pesticide application equipment and handle on farm excessive residual pesticides. Currently, only a few genes encoding degradative enzymes have been isolated and characterized, and more require development to address the critical water quality challenge.

Plant Production and Protection

Plant Production. The first research priority of ARS as well as other agriculturally related organizations, public and private, is production efficiency. This implies a low dollar grower input with high quality and yield output. Many of the genes related to growth, form, and yield are multiple and operate in a complex developmentally regulated fashion.

The significance of a complex integrated gene system is difficulty in gene transfer as compared with single gene constructs and delivery. Therefore, certain projects are addressing multiple gene transfer methods.

Elucidation of the biochemistry and metabolism of several key compounds and pathways in higher plants is mandatory to implement effective transfer of complex genes. This is particularly relevant to directing the regulatory enzyme systems that are related to partitioning and distribution of desirable carbohydrate, lipid, and protein molecules to specific plant organs. The emphasis of some studies is directed towards the movement of lipids and protein to produce more nutritious soybean seeds, sugar molecules to sugarbeet roots, and rubber to leaves or stems of guayule or hevea.

Indispensable to gene transfer of desirable single and complex gene systems are plant cell and tissue culture techniques. Predictable regeneration of plants with wanted and known genotype is of prime importance to successful application of crop improvement through gene transfer. The cardinal factor to regeneration of agronomically important crops from cells will lie in the technology to control the mutation and destruction rate of genes in culture stressed cells. The general result of the stress is somaclonal variation which is not desirable when engineered crop plants with uniform genotype and phenotype are the wanted product. Scientists are addressing the molecular basis of formation of somaclonal variation to improve plant tissue culture techniques.

Plant Protection. Protection of plants from weeds, fungi, insects and bacteria with pesticides has been a costly grower input and the challenge is to change that to reduce costs in farming.

Pesticide use has been the most effective component of disease and pest control. Overall, the use of pesticides to control disease infestations is rapidly changing because of the following reasons: (1) pesticide cost; (2) pesticide impact on the environment, i.e., humans, animals and plants; (3) registration of pesticides and associate regulatory problems; and (4) increased resistance of pests and pathogens to pesticides.

Genetics and breeding are the desirable approaches to solving pest problems, many of which have been successful, but the results are generally slow to attain. It is evident that a lack of basic information about pathogens exists as well as host-disease interaction mechanisms. A lack of communication between pathologists and breeders has generally existed over the years because of temporal and spatial factors; i.e., the coordination of and pathogen effect studies between the various disciplines present serious scientific man-year continuity problems. Part of the answer to the challenges before us will be met with tissue culture and molecular gene transfer technology coupled with continued breeding and genetic studies. Techniques involving *in vitro* methods that address disease problems are successful and tissue culture systems to rigorously select for resistance of fungal and bacterial diseases of plants are being addressed.

Biological control may provide an opportunity to replace some pesticides for control of specific pests. In addition to the classical biocontrol techniques of collecting and release, the challenge is to develop new technologies to develop biocontrol agents. Specifically, the use of genetic engineering and other manipulative techniques may be useful in transferring biocontrol molecules. Basic research on host pathogens and host insect interactions will also be needed to make this approach successful.

The basic genomic structure and function of pathogens is another aspect of disease issues. Fungal, bacterial, viral and viroidal genomes are small and correspondingly simpler in structure than those found in higher eucaryotes. The smallness and simplicity of pathogenic organisms suggest feasibility of molecular studies that could elucidate factors that control virulence and toxin production. Knowledge of virulence characters will aid in the specification of plant resistance mechanisms to pathogenic organisms. The complexities of biochemical host-pathogen interaction initiated upon pathogenic invasion of susceptible and resistant hosts cannot be analyzed without genetic information of pathogens.

Plants that display herbicide tolerance, stress tolerance, and disease resistance are being researched today and some systems are ready for field testing. A disease resistant plant has not yet been developed and this research is the area being pursued.

Animal Production and Protection

Animal Production. Production of livestock that are leaner, faster growing, more disease free and more genetically sound than current breeds at less cost to the producer meets production efficiency goals to be attained. The animal scientists are researching metabolic and hormonal mechanisms to manipulate the rates nutrients are distributed into muscle, fat, and bone.

The major research problems with animals are to develop transgenic animals that display increased feed efficiency, a lean carcass, improved reproductive capacity, increased disease tolerance and an enhanced growth rate. All of these would lead to lower dollar input with a greater profit for the farm operator. This is the production efficiency for which to strive and when implemented will yield greater profit for the entire agribusiness sector with enhanced enivonmental results.

Bovine somatotropin is already available in the marketplace and second generation somatotropins are being developed to increase milk and production growth rates. Swine growth hormones and how they regulate or control partitioning of nutrients for the production of leaner meat is researched with the challenge to clearly understand the hormonal regulation of partitioning nutrients to the right constituent.

Animal Protection. The development of disease free animals is the ultimate goal for researchers. The tools of biotechnology are available which will help set the course to attempt to reach that objective. Most of the current research activities are pursuing disease detection and protection avenues. These areas of research are important to begin to untangle the many animal diseases as well as developing products for the veterinary pharmaceutical market.

Monoclonal antibodies are being developed to identify, isolate, and characterize proteins and epitopes of biological importance, improve ELISA tests, detect latent infections and offer passive therapeutic protection.

Recombinant DNA technology is being employed to develop hybridization probes for *in vitro* disease diagnosis and *in situ* detection of disease agents in animal tissues; detection of markers for disease susceptibility and resistance; develop supportive efforts for protein sequencing, peptide mapping, nucleotide sequencing, and gene mapping.

Vaccine production is pursued with the objectives to genetically engineer vaccines with specific gene deletions reducing virulence; produce viral and bacterial vectored vaccines; and develop subunit vaccines using purified antigens obtained from *in vitro* expression vectors such as pseudorabies purified protein vaccine and foot-and-mouth disease vaccine.

The continued development of efficient and reduced disease free animals is significant and has major consequences to the farm animal operator.

Meeting the researchable problems requires solutions through the improvement of innate immune systems of farm animals to reduce, if not eliminate, many animal diseases and associated costs.

Determination of the biochemical pathways associated with nutrient partitioning to muscle, bone, fat, and other animal tissues and organs through research programs will facilitate the subsequent engineering of nutrient distribution and regulation of animal growth.

Researchers are developing methods and procedures that will increase the reproductive capacities and potential of animals.

Insect pests cause considerable damage to animal livestock and crop plants. Molecular entomology is a rapidly growing field and the challenges to reduce insect caused problems are being addressed by ARS entomologists. The same biotechnology tools used in plant and higher animal research are employed.

Construction of sex-linked genetic markers in major economic pests will facilitate expansion of genetic control methods such as the sterile insect technique and backcross sterility in *Heliothis virescens*.

Research in molecular genetics of insect neurohormones, specifically neuropeptides, will facilitate rapid advances especially in identifying the pheromones and cloning neuropeptide genes into microbes for production purposes and into plants for novel control strategies.

The development of transformation systems will yield major breakthroughs in establishing advanced insect control strategies.

Product Quality

Industrial Products from Agricultural Commodities

Agribusiness, like the pharmaceuticals industry, is big business. Apart from edible food products, U.S. agriculture will be keyed more and more to the manufacture of industrial nonfood products. These comprise commodity/bulk chemicals (generally of petroleum origin) and specialty high-value products. Examples of the latter are health-care and cosmetic products.

Already in the U.S. today, about half a billion dollars worth of industrial oils from plants are being used. Research work is in progress to augment the use of oils, starch, and dairy products from surplus agricultural commodities. It is expected that the results of this research will be a key element in a strategic restructuring, namely, a vertical integration of agriculture with businesses based on chemicals and high-value proprietary products from this vast renewable resource.

Starch Conversion

Starch from surplus corn has been converted into a superabsorbent material known as SUPER SLURPER. This is being marketed under license by several companies for use in removing water from refined petroleum products and as an absorbent in bandages and diapers. Biodegradable plastic films of starch mixed with other polymers find application as mulches in agriculture and as a packaging material for food. Heavy metals can be absorbed in industrial waste treatment.

Biotechnological methods are being used for converting starch to materials having properties similar to those of imported gums and so reduce imports. Starch is being reacted chemically in an extruder reactor system similar to that used for making plastics. It is anticipated that the reaction products will be suitable as coatings, films, surfactants, and foams.

Other applications being developed are composites with synthetic polymers for use as semipermeable membranes. Pesticides will be made safer to handle and their release controlled by encapsulation with starch. The latter application is important in the light of various concerns about the contamination of groundwater by pesticides.

Vegetable Oils and Animal Fats

Research on soybean oil and tallow is directed toward the manufacture of materials which will find application in lubricating oils, plasticizers, surface coatings, and cosmetics. Genetic engineering techniques are being used for tailored cleavage of oils and fats. In conjunction with various chemical catalytic systems, it is planned to produce novel materials of higher market value.

Dairy Material Conversion

Whey from the dairy industry is a waste product. Research is being conducted on biotechnological systems for upgrading lactose in whey to human health-care and cosmetic products.

Wool and Leather

For the manufacture of top quality U.S. wool, it is crucial to develop techniques for removing contaminants such as grease, vegetable matter, stains, and dark fibers.

For the leather industry, valuable insights are being gained into novel, improved methods and materials for converting hides to leather. For example, the interaction between tanning and salts used in leather manufacture is being elucidated. Minor constituents such as collagens and various other components can have a profound influence on leather quality.

Novel Crops

Some consideration is being given to the development of novel crops as alternatives or adjuncts to conventional agronomic species. Often, such plants produce oils and other chemicals which would find a market in the plastics, detergents, and lubricating oil industries.

Certain plants are potential sources for pharmaceutical materials. Grown in cell culture, a biotechnological reactor process can be developed for making these high-value materials.

Cotton

There is a wide range of research work directed toward the improvement of cotton product industries. For example, there is work at the molecular level on modifying the internal crystalline structure of cotton fibers to improve their processing and utilization characteristics. At the pilot scale, spinning quality is being analyzed to find out how factors such as culturing of the cotton plant, harvesting, ginning, and processing affect quality.

Work is in progress to produce wrinkle resistant fabrics that can be dyed. Desirable thermal storage properties are being built into cotton fibers by using special polymers, a feature which is attracting keen industrial interest.

The on-going ARS biotechnology research discussed in this chapter is consistent with the mission of our Agency and with the recommendations made by the National Research Council, Board of Agriculture, on the subject of "New Directions for Biosciences in Agriculture."

Literature Cited

1. Augustine, Patricia C.; Danforth, Harry D.; Bakst, Murray R. BARC Symposium X, "Biotechnology for Solving Agricultural Problems," Agricultural Research Service, U.S. Department of Agriculture, Beltsville, MD, 1985; pp 1-6.
2. Agricultural Biotechnologies, Strong Acceleration of Research Programs at Beltsville, U.S. Department of Agriculture, Agricultural Research Service, Beltsville, MD, 1984.
3. New Directions for Biosciences Research in Agriculture, U.S. Department of Agriculture, Agricultural Research Service, and National Academy of Sciences, National Academy Press, Washington, DC, 1986.

RECEIVED May 11, 1988

Chapter 17

Regulation of Gene Expression in the Sunflower Embryo

C. A. Adams, K. E. Koprivnikar, R. D. Allen, E. A. Cohen, C. L. Nessler, and T. L. Thomas

Department of Biology, Texas A&M University, College Station, TX 77843

The sunflower embryo is a convenient system for the comparative analysis of gene expression in development. Here we demonstrate the utility of this system in the analysis of several genes with different expression patterns and functions. These include the seed storage protein genes of sunflower that encode super-abundant mRNAs and genes that encode less abundant actin, tubulin, and β-lipoxygenase mRNAs. Relatively small changes (~3-fold) in steady state levels of actin, tubulin, and β-lipoxygenase transcripts are observed during embryogenesis. In contrast, very large changes (~100-fold) are observed for seed storage protein mRNAs over the same developmental interval. Comparisons of steady state mRNA levels and relative nuclear RNA synthesis rates of these genes indicate that for the most part mRNA steady state levels are controlled at the level of transcription, although in some cases post-transcriptional control is important. These results are significant in understanding the role of gene regulatory hierarchies in development.

Regulation of gene expression in development is achieved by a variety of mechanisms in different tissues and organisms. Control at the level of transcription may be accomplished by the presence or absence of tissue-specific trans acting factors, by holding the chromatin containing the gene in a "silent" configuration, or in some cases by methylation of the gene (reviewed in 1). Post-transcriptional controls may involve differences in the rate of pre-mRNA processing or in the stability of an mRNA within the cell (2), and control of the rate at which mRNA is assembled into polysomes (3) (translational control). Studies of these controls are plentiful in animal systems (1-3) but there are relatively few plant systems that have been studied as extensively. Sunflower is an excellent system for the study of gene regulation during embryogenesis. A single inflorescence provides embryos of different

0097-6156/88/0379-0240$06.00/0

stages spanning early and late developmental events; furthermore the size of the sunflower inflorescence facilitates the isolation of large numbers of developmentally staged embryos. Thus biochemical amounts of RNA and protein can be isolated from nuclei for the study of transcription and isolation of *trans* acting factors. In addition, a number of genes have been characterized in sunflower. Two sunflower 11S (helianthinin) seed storage protein (SSP) genes (HaG3 and HaG10) have been cloned and changes in expression during embryogenesis have been described (*4-6*). Also the structure of a 2S (albumin) SSP gene (HaG5) from sunflower has been determined and shown to have a slightly different expression pattern from that of the helianthinins (*5-7*). In this paper we describe the control of gene expression in sunflower embryogenesis with respect to HaG10, HaG5, actin, tubulin, and β-lipoxygenase (β-LOX) genes. Comparisons of steady state levels of mRNAs for these genes with relative rates of transcription at different stages of development indicate that mRNA steady state levels are controlled mainly at the level of transcription although in some cases post-transcriptional control may be important.

Materials and Methods

Growth of Plants. Sunflower seeds (*Helianthus annuus* L. cv. Giant grey stripe, Northrup King Seed Co. Minneapolis, MN) were germinated in sterile soil and grown in the greenhouse for one week before transplanting to the field. Plants were grown in the field during the summer and in growth chambers with 16 h day length at 23°C at other times of the year. Embryos were harvested in liquid N_2 and stored at -80°C until needed.

RNA Isolation. RNA was isolated as previously described (*4*) except that tissue was homogenized using a Polytron (Brinkman Instruments, Westbury, NY). Poly A^+ RNAs were purified by oligo(dT) cellulose chromatography (*8*).

Northern Blot Hybridizations. Total or polyA^+RNAs were denatured with dimethyl sulfoxide and glyoxal, size fractionated on 1% agarose gels, and transferred to nitrocellulose filters. Labeled DNA probes were denatured by boiling and hybridized to filters for 16 hr at the appropriate temperature in 50% formamide, 25 mM phosphate buffer pH 6.8, 5X SET (1X SET = 150 mM NaCl, 20 mM Tris pH 7.8, 1 mM EDTA), 0.1% sodium dodecylsulfate (SDS), 10% dextran sulfate, 5X Denhardt's solution (1X Denhardt's = 0.02% ficoll, 0.02% BSA, 0.02% polyvinylpyrrolidone, 0.02% SDS), 100 μg/ml denatured herring testis DNA, 50 μg/ml polyadenylic acid and 10 μg/ml polycytidylic acid. After hybridization the filters were washed for 1h in 4X SET wash (4X SET + 0.025M phosphate buffer, 0.2% SDS), 1 h in 2X SET wash, and 1 h in 1X SET wash all at 50°C. Tubulin probed blots were washed in 50% formamide/ 5X SSC at 55°C for 2h. Filters were air-dried and exposed for autoradiography at -80°C with an intensifying screen. Radioactivity in each band on the filters was measured by excising the band and counting by liquid scintillation or densitometry of autoradiographs using a Bio-Rad Model 620 video densitometer.

Probes for HaG5, HaG10, actin, and β-lipoxygenase consisted of DNA fragments labeled with ^{32}P by nick translation. HaG5 and HaG10 probes consisted of 0.7kb and 1.2kb Eco RI cDNA fragments respectively. The sequences of each of these probes lies entirely within the coding sequences of the genes (5,6). Actin probe was a three kilobase (kb) soybean genomic clone fragment isolated from pSAc3 (9) which was a gift of Dr. Richard Meagher (University of Georgia, Athens, GA). The probe for β-lipoxygenase was a cDNA clone from pea and was kindly provided by Dr. Rod Casey (John Innes Institute, Norwich, Great Britain). Tubulin probe consisted of a 1.6 kb transcript produced from a soybean tubulin genomic clone inserted into pSP65 (Promega Biotec, Madison, WI) using Sp6 RNA polymerase according to the instructions of the manufacturer. This tubulin construct was kindly provided by Dr. Mark Guiltinan (10).

Nuclei Isolation and Runoff Assays. Nuclei were isolated and purified by the procedure of Luthe and Quatrano (11) except that all buffers were adjusted to pH 8.6. Nuclear runoff reactions were carried out in a volume of 50 µl containing 100mM $(NH_4)_2SO_4$, 30mM Tris HCl pH 8.6, 7mM $MgCl_2$, 2.5mM in each of GTP, CTP, ATP, 3mM creatine phosphate, 25 ng/ml creatine phosphokinase, 3mM β-mercaptoethanol, 1 unit/µl RNasin, and 50 µCi ^{32}P-UTP (New England Nuclear, Boston, MA, USA). Reactions were started by mixing all components, bringing to room temperature and then adding 10µl of nuclei (10^5 nuclei). Samples were incubated at 25°C for up to one hour. Reactions were stopped by addition of ice cold 150mM sodium pyrophosphate/3mM UTP with vigorous mixing and allowed to sit on ice for 10 minutes. One ml of ice cold 5% (w/v) TCA was then added to each sample, vortexed and incubated on ice for 10 minutes. TCA precipitable counts were collected on Whatman GF/C glass filters, washed extensively with 5% TCA/150mM sodium pyrophosphate, then with ethanol, and finally with 2 ml of diethyl ether. After air drying, radioactivity on the filters was determined by liquid scintillation counting.

Isolation of Heterogeneous Nuclear RNA from ^{32}P-UTP Labeled Nuclei. Heterogeneous nuclear RNA (hnRNA) was isolated by the method of Lee et al. (12) with the following modifications. Polyvinyl sulfate was eliminated from the lysis buffer, 0.75g of CsCl was added to the homogenate after the proteinase K step, and the samples centrifuged over the CsCl pad for 22 h at 50,000 rpm at 15°C in a TLA 100.3 rotor using a Beckman tabletop ultracentrifuge. The resulting RNA pellet was dissolved in 6M guanidine hydrochloride/50mM sodium citrate pH 7.0 and precipitated by the addition of 2.5 volumes of ethanol/-80°C overnight. After pelleting in an Eppendorf centrifuge the RNA was dissolved in 50µl of water and 75µl of 5M potassium acetate pH 6.0 was added. The samples were then held at -20°C overnight and RNA was pelleted in an Eppendorf centrifuge for 30 min. The pellets were washed with 70% ethanol, dried *in vacuo* and redissolved in hybridization buffer.

Dot and Slot Blot Hybridizations. Slot blots of plasmid DNAs containing clones for HaG5, HaG10, actin, tubulin, and β-lipoxygenase were made by denaturing the DNAs in 0.5M NaOH at 100°C for 10 min followed by addition of an equal volume of ice-cold neutralization solution (1M NaCl, 0.3M sodium citrate, 0.5M Tris HCl pH 8.0, 1M HCl) (13) and filtering through nitrocellulose using a slot blot manifold from Schleicher and Schuell (Keene, NH). Dot blots containing total RNA from staged embryos were made in the same manner except that the heating in alkali step was eliminated and a Minifold (Schleicher and Schuell) was used. Blots were air dried and then baked at 80°C for 4 h. Before use the blots were pre-hybridized in the same manner as Northern blots except that for slot blots pre-hybridization was done for 48 h, 500,000 cpm of hnRNA from 8-, 13- or 25-days post-fertilization (DPF) embryo nuclei were then added to each filter and hybridization carried out for 72 h at 40°C. Slot blots were then washed in the same manner as northern blots and placed on film for 96 h with an intensifying screen.

Results

SSP Genes Exhibit Very Similar Patterns of Expression. Figures 1 and 2 demonstrate that HaG5 and HaG10 have identical expression patterns both qualitatively (Figure 1) and quantitatively (Figure 2) from 8- through 25-DPF except that HaG5 expression is detectable by 5 DPF while HaG10 expression is not detectable until 7 DPF. Both genes reach a peak in expression at around 13 DPF and then decline to very low levels by 25 DPF with Ha10 mRNA being somewhat more prevalent at 25 DPF than that of Ha5. Both are undetectable in dry seed by RNA gel blot analysis (data not shown). Since the gel blot method gave more accurate quantitative data we continued to use it for the analysis of the expression pattern of other genes of interest.

Actin, Tubulin, and β-Lipoxygenase have Similar Expression Patterns that are Different from those of SSP Genes. Table I summarizes the results of Northern gel blot analysis of polyA$^+$RNA from embryos at 8, 13 and 25 DPF with respect to the expression of actin, tubulin, and β-LOX. The degree of expression of respective transcripts are reported as percent of maximum cpm of probe hybridizing to total RNA. Each of these genes shows the same general expression pattern although actin radioactivity declines more rapidly than that of tubulin, which in turn declines more rapidly than β-LOX.

Table I. Expression patterns for actin, tubulin, and β-lipoxygenase genes during sunflower embryogenesis

		Expression level at indicated DPF[a]		
Gene	Transcript size	8	13	25
Actin	1.7 kb	100	43	30
Tubulin	1.6 kb	100	60	25
β-LOX	4.0 kb	100	65	50

[a]These numbers are the average of two separate measurements.

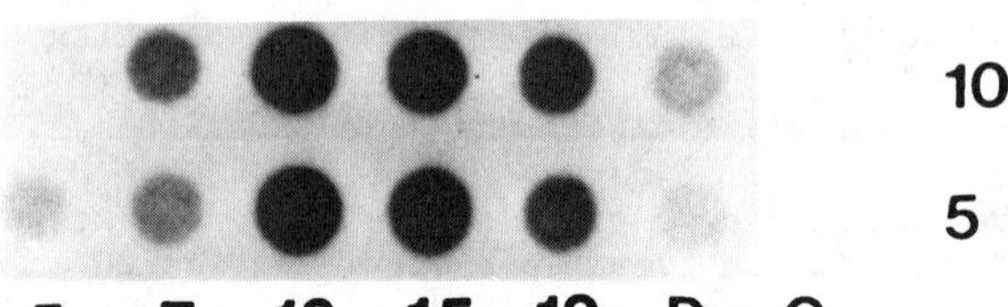

Figure 1. Dot blot analysis of HaG5 and HaG10 expression. One µg of total RNA from embryos at the indicated day post-fertilization (DPF) was bound to each dot on nitrocellulose and the dot blots were then hybridized to the appropriate probe, washed and exposed to film overnight. "Dry" = mature dry seed, Germ. = germinating dry seed.

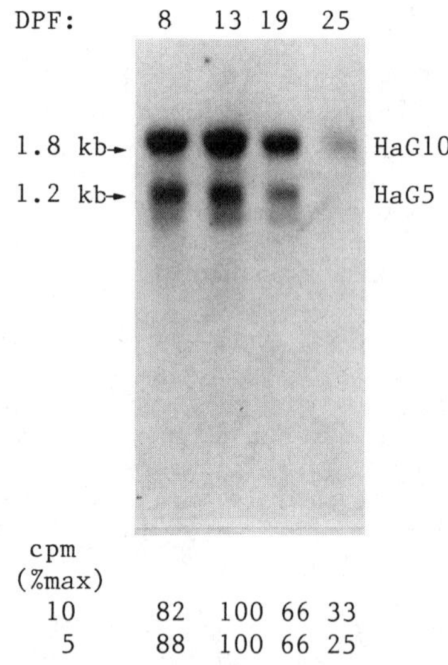

Figure 2. Northern gel blot of total RNA from staged embryos. Numbers indicate the relative cpm of probe bound (expressed as percent of maximum) at each stage of development for HaG5 and HaG10.

Prevalence of Specific Transcripts in Heterogeneous Nuclear (hn) RNA Isolated from ^{32}P-UTP-Labeled Nuclei. Table II shows the results of measurement of transcript levels in hnRNA from developmentally staged nuclei labeled with ^{32}UTP. Both HaG5 and HaG10 nuclear RNA reach a peak at 13 DPF and decline to very low levels by 25 DPF. β-LOX transcript levels are highest in 8 DPF nuclei and diminish in a manner similar to the steady state RNA levels (Table I). Neither actin nor tubulin transcripts could be detected above background in these experiments. Values shown are percent of maximum cpm of hnRNA from nuclei of different DPF embryos that hybridized to slot blots containing at least a 100-fold excess of clone DNA specific to the gene indicated.

Table II. Relative rates of transcription for different genes during embryogenesis

	Relative transcription rate at indicated DPF[a]		
Gene	8	13	25
HaG5	10	100	5
HaG10	10	100	5
β-LOX	100	60	10

[a]Values at different DPF for each gene may be compared, comparisons between genes at specific DPF are not valid. Numbers are the average of two different experiments.

Discussion

The genes in this study may be divided into two groups according to overall expression patterns determined by measuring steady state mRNA levels. The SSP genes show a definite peak in expression at around 13 DPF while the actin, tubulin, and β-LOX mRNAs appear to peak early in embryo development (at or before 8DPF) and then gradually decrease through 25 DPF. These two groups also differ in the absolute abundance of their mRNAs. SSP gene mRNAs were easily detectable in total RNA whereas the use of polyA+RNA is required for the detection of actin, tubulin, and β-LOX transcripts. Differences in the structure of these genes and their correlation to expression pattern and level is currently under investigation.

The most informative comparison is that of steady state mRNA levels to the relative transcription rate for each gene. In the case of both SSP genes, the difference in the relative transcription rate at 8 and 13 DPF is qualitatively similar to the relative steady state levels of mRNA. However the difference in the level of transcription between 8 DPF and 13 DPF is large compared to the difference in steady state mRNA levels between these points. This suggests that both transcriptional and post-transcriptional controls are operative. The latter possibly involves mRNA stability within the cell. In the case of the β-LOX gene the transcriptional levels follow the same pattern as the steady state mRNA levels indicating almost exclusively transcriptional control of this gene. For both actin and tubulin the levels of transcription were undetectable in similar experiments but these transcripts were easily detectable in steady state mRNA. Thus post-transcriptional control of mRNA levels may be important for these genes.

The SSP genes studied here belong to small gene families, various members which will cross-hybridize with the probes used (5,6). These results thus reflect contributions from more than one gene in each family. This is probably true for the actin (9), tubulin (10), and β-LOX (14) genes as well. The relative importance of transcriptional or post-transcriptional regulation of expression may vary for individual genes but the methods used here do not allow us to distinguish between different gene family members. Gene specific probes are being constructed for some of these genes for use in studying differential regulation of members of the same family.

Our results suggest that a hierarchy of gene regulation is operative during higher plant growth and development and furthermore that the relative importance of each hierarchical level depends on the particular genes involved. It may be that different levels of control are utilized for different genes because of specific physiological consequences. Present studies are focusing on the role of *trans*-acting factors and *cis*-acting elements in differential gene expression in sunflower. Our ultimate goal is to understand, in detail, the mechanisms that control gene expression and their physiological consequences in higher plants.

Literature Cited

1. Davidson, E. H. "Gene Activity in Early Development", 3rd ed.; Academic Press: New York, 1986, pp. 1-44.
2. Cabrera, C. V., Ellison, J. W., Moore, J. G., Britten, R. J., Davidson, E. H. *J. Mol. Biol.* 1982 *179*, 75-111.
3. Infante, A. A., and Heilmann, L. J. *Biochemistry* 1981 *20*, 1-8.
4. Allen, R. D., Nessler, C. L., Thomas, T. L. *Plant Molec. Biol.* 1985 *5*, 165-173.
5. Cohen, E. A. Masters thesis, Texas A&M University, 1986.
6. Allen, R. D., Cohen, E. A., Vonder Haar, R. A., Orth, K. A., Ma, D. P., Nessler, C. L., Thomas, T. L. In: *Molecular Approaches to Developmental Biology*; Firtel, R. A.; Davidson, E. H., Eds.; Alan R. Liss Inc.: New York, 1987; pp. 145-424.
7. Allen, R. D., Vonder Haar, R. A., Cohen, E. A., Adams, C. A., Ma, D.-P., Nessler, C. L., Thomas, T. L. *Mol. Gen. Genet.* 1987 *210*: 211-218.
8. Aviv, H., Leder, P. *Proc. Natl. Acad. Sci.* USA 1977 *69*: 1408-1412.
9. Shah, D. P., Hightower, R. C., Meagher, R. B. *Proc. Natl. Acad. Sci. USA* 1982 *79*, 1022-1026.
10. Guiltinan, M. J., Velten, J., Bustos, M. M., Cyr, R. J., Schell, J., Fosket, D. E. *Mol. Gen. Genet.* 1987 *207*, 328-334.
11. Luthe, D. S., and Quatrano, R. S. *Plant Physiol.* 1980 *65*, 305-308.
12. Lee, J. J., Calzone, F. J., Britten, R. J., Angerer, R. C., Davidson, E. H. *J. Mol. Biol.* 1986 *188*, 173-183.

13. Maniatis, T., Fritsch, E. F., Sambrook, J. Molecular Cloning. A Laboratory Manual; Cold Spring Harbor, New York, 1982, p. 331.
14. Start, W. G., Ma, Y., Polacco, J. C., Hildebrand, D. F., Freyer, G. A., Altschuler, M. Plant Molec. Biol. 1986 7, 11-23.

RECEIVED May 25, 1988

Chapter 18

Molecular Dynamics of the 32,000-Dalton Photosystem II Herbicide-Binding Protein

Autar K. Mattoo[1], Franklin E. Callahan[1], Bruce M. Greenberg[2,3], Pierre Goloubinoff[2,4], and Marvin Edelman[1,2]

[1]Plant Hormone Laboratory, Agricultural Research Service, U.S. Department of Agriculture, Beltsville, MD 20705
[2]Plant Genetics Department, The Weizmann Institute of Science, Rehovot, 76–100, Israel

The ubiquitous 32kDa thylakoid protein, the psbA gene product, is a major product of the plastid translational machinery and an integral component of the photosystem II reaction center. The protein is also the target site for triazine and urea-type herbicides. It is first synthesized as a precursor on stromal lamellae where it is processed to the mature 32kDa form. Following processing, the 32kDa protein is translocated to the topologically distinct granal lamellae. Sometime during its translocation and/or integration within the granal photosystem II, the protein undergoes covalent modification with palmitic acid. This modification occurs in the amino-terminal half of the protein and is inhibited by atrazine and diuron herbicides. Light-intensity dependent degradation of the protein occurs after its assembly in the granal lamellae, and results in a membrane-associated 23.5kDa product derived from the amino terminal two-thirds of the parent protein. The cleavage site is localized to a functionally active and phylogenetically conserved region between amino acid residues 238 and 248. Contiguous with the proposed cleavage domain is an α-helix-destabilizing region, bordered by arginine residues 225 and 238. This region, characteristic of rapidly catabolized proteins, seems to have evolved along with oxygenic photosystem II, since it is absent from the analogous protein in the non-oxygenic photosynthetic bacteria.

[3]Current address: Department of Biology, University of Waterloo, Ontario N2L 3G1, Canada
[4]Current address: Central R&D Department, E. I. du Pont de Nemours and Company, Wilmington, DE 19898

0097–6156/88/0379–0248$06.00/0

Chloroplasts encode one of the most studied of plant gene products, a unique protein with an apparent molecular weight of 32,000 on SDS-polyacrylamide gels. The 32kDa protein has a conserved primary structure (1) in the algae and higher plants tested (2). The recognition of this plastid protein as a monogenic target for the most commonly used photosystem II herbicides, atrazine (3) and diuron (4), and as one of the most rapidly turning over plant proteins (5) has generated considerable interest among molecular biologists and biotechnologists because of the potential for genetic engineering. In addition, the recent demonstration of the 32kDa protein as an essential component of the higher plant photosystem II reaction center (6,7) has added a new dimension to this interest with still broader implications for crop improvement in agriculture.

Biosynthesis, Translocation and Posttranslational Modifications

Synthesis. The 32kDa protein is coded for by the psbA gene located in the plastid genome (8). The psbA gene has been sequenced from several higher plant chloroplasts, algae and cyanobacteria and, like its protein product (2), is highly conserved (Fig. 1). The coding of the higher plant psbA gene sequence predicts a hydrophobic protein with 353 amino acid residues and a molecular weight of about 39,000 (8). The protein is, in fact, synthesized on membrane-bound ribosomes (9). It is first synthesized as a precursor and has been identified as such *in vitro*, *in organello* and *in vivo*. Processing of the precursor to the mature 32kDa form was shown *in vivo* and *in organello* by pulse-chase experiments (12).

The apparent size of the precursor is about 33.5-34.5kDa on SDS-polyacrylamide gels which is somewhat different from the molecular weight predicted from the deduced open reading frame. This discrepancy could result if the protein behaved in an anomalous manner on SDS-PAGE. Alternatively, it was suggested that translation of 32kDa mRNA initiates at the second methionine in the open reading frame at position 37 (met37) instead of the one at position 1 (met1). This would result in a protein with a molecular weight of 34.5kDa (15-17). The latter possibility has been experimentally discounted by a recent study that used full length, truncated and mutated constructs of psbA gene from *Solanum nigrum* as templates in an *in vitro* transcription/translation system (18). It was found that when the met37 codon, ATG, was mutagenized to AAG (which codes for lysine) or to AGG (which codes for arginine), while still retaining the first ATG codon (met1), the translation product had the same mobility on SDS-PAGE as the 33.5kDa precursor protein. On the other hand, deletion at the 5'-end of the gene to remove the codon for met1 without affecting the met37 codon produced a truncated, 29kDa protein. Thus, translation of mRNA for the 32kDa protein initiates at met1. Recent immunological evidence showing met1 and met37 as part of the mature protein (19), and localization of a post-translational phosphorylation at position 2 of the open reading frame with the generation of acetyl-O-phosphothreonine as the N-terminal residue in the mature 32kDa protein (20), support the contention that initiation of translation occurs at met1. It would seem then that, as is the case with other hydrophobic photosynthetic membrane proteins (30), the 32kDa protein behaves anomalously on SDS-PAGE.

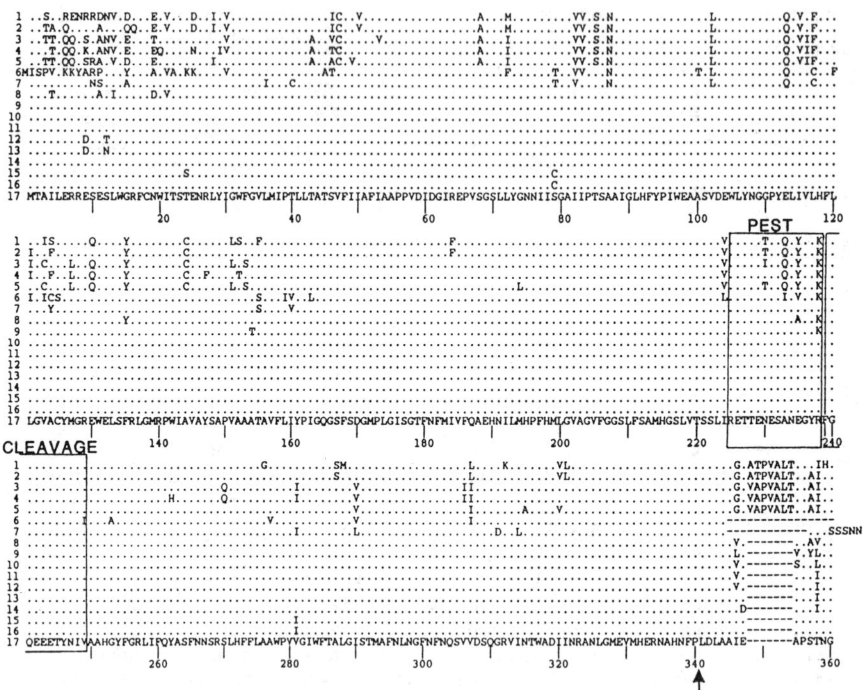

Figure 1

Comparison of deduced amino acid sequences from psbA genes. Note the conservation of the PEST-like region contiguous with the α-helix-destabilizing, primary cleavage domain in the 32kDa protein. The arrow indicates the probable site where the precursor piece on the carboxy-terminus is cleaved (22). psbA sequences were compiled as follows: (1) Anacystis nidulans copy I, and (2) copy II/III (41). (3) Anabaena 7120 copy I, and (4) Anabaena 7120 copy II (42). (5) Fremyella diplosiphon (43). (6) Euglena gracilis (44). (7) Chlamydomonas reinhardtii (45). (8) Marchantia polymorpha (46). (9) Zea mays (47). (10) Sinapis alba (48). (11) Brassica napus (49). (12) Medico sativa (50). (13) Pisum sativum (51). (14) Glycine max (52). (15) Petunia hybrida (53). (16) Amaranthus hybridus (15). (17) Solanum nigrum is the reference sequence (54) and is identical to the Spinacia oleracea, Nicotiana debneyi (8) and Nicotiana tabacum (55) psbA sequences. (Adapted from ref. 56).

Precursor Processing. The precursor protein is synthesized on stromal lamellae and is processed there to the 32kDa mature form (21). Processing of the precursor is a posttranslational event (12) and takes place at the carboxy terminus (22). Following processing, the mature protein translocates to spatially-distinct chloroplast membranes, the grana, where functional photosystem II reaction centers are mainly located (21, 23).

The mechanism of translocation is not understood. Any mechanism put forward to explain the lateral diffusion of the protein has to reconcile the fact that the 32kDa protein on stromal lamellae is integrated in an orientation that seems similar to the protein integrated within the granal lamellae. This is shown by experiments involving trypsinization of the two membrane types following their isolation from pulse-labeled Spirodela plants. Previously it was established that the membrane-bound 32kDa protein yields, upon trypsin digestion, two membrane-associated fragments of 22kDa and 20kDa on SDS-PAGE (24). Using this technique, we have found identical *in situ* trypsinization patterns of the stromal lamellae versus granal-associated 32kDa protein (Fig. 2). Moreover, even in the precursor state, the stromal-lamellae associated protein shows similar trypsin-sensitive domains (P. Goloubinoff, unpublished). Thus, these data raise the possibility that membrane integration and orientation are independent of processing and translocation of the 32kDa protein.

Membrane Integration. What is the driving force behind the integration process? An analysis of several decoded amino acid sequences of the protein shows considerable variability in the carboxy-terminal region. Notwithstanding this variability, a negative charge of at least 1 is maintained in the precursor piece of the species examined. Asp342 is conserved in all these cases while a high frequency of glu or asp at position 347 is observed (25). The integration event could be driven by the negative charge on the precursor piece of the 33.5kDa protein. Such a model for integration of the mature 32kDa protein within stromal lamellae is consistent with the membrane trigger hypothesis (26), in the present case, being applied to a protein that undergoes processing at the carboxy terminus. The interaction of the protein with stromal membrane lipids could possibly help in this integration process. In this context, it is worth noting that stromal membranes have a different complement of glycolipid molecular species as compared to that found in the granal membranes (27). Whether this difference is favorable for integration of the 32kDa protein in stromal membranes is not known.

Possible Signals for Translocation. The question arises: What signals enable stromal 32kDa protein to move to the granal lamellae? At 30 μmole.m^{-2}.s^{-1} of white light, the half life of the newly-synthesized 32kDa protein on stromal membranes is between 9-18 min while the newly translocated granal 32kDa protein has a half life of 6-12 hours (21, 28). This difference in the life times of its presence in stromal versus granal membranes is partially reflected in western blot analysis of steady state levels of protein at these two distinct locations, viz., the amount of stromal 32kDa protein is one-fifth to one-tenth of that found in the granal fraction. In addition to its modification by phosphorylation (discussed above), the 32kDa protein also undergoes another post-translational modification, viz., with palmitic acid (21, 28). The acyl linkage is resistant to acidic hydroxylamine and to conditions of denaturation, electrophoresis and fixation of the protein on SDS-polyacrylamide gels, suggesting an ether or amide linkage. The palmitoylation of the protein is light stimulated and rapid, being observed within 1 min of radiolabeling of Spirodela plants with ^{3}H-palmitic acid. Both atrazine and diuron, that bind the 32kDa protein, inhibit acylation of the protein (29). These data suggest that such herbicides might compete for sites on the protein that are palmitoylated, or cause a conformational change in the protein such that acylation sites are no more accessible to the enzyme that ligates palmitic acid to the protein. Light, diuron or heat are known to cause conformational changes in the protein particularly in the region bounded by T22 and T20 (4, 30). An acylation site appears to be associated with the characteristic, trypsin-derived membrane-associated protein fragments, T22 and T20 as well (21). Furthermore, acylation is restricted to the 32kDa protein associated with the granal lamellae. Speculations on the role of palmitoylation in the dynamics of the 32kDa protein include: 1. Promotes proper functional integration within the granal photosystem II reaction center; and 2. acts as a rapid translocation signal that destabilizes the protein on stromal lamellae allowing its lateral diffusion to grana.

Degradation

Once the 32kDa protein is translocated to the granal lamellae it undergoes a light-dependent degradation. Its half life in the grana is dependent on the light intensity at which the plant is cultivated, while the protein is stable in the dark (31). Degradation occurs in visible light, where both photosystem I and II are functional as well as in far red light, where photosystem I is predominantly active (32). Both atrazine and urea-type herbicides bind to the protein and characteristically inhibit its degradation (31, 32). The light-intensity dependent degradation of the 32kDa protein results in the production of a membrane-associated fragment of 23.5kDa (33). The primary cleavage occurs about 2kDa carboxy-terminal to arg225, somewhere between arg238 and ile248 (Fig. 1). This domain of the 32kDa protein is localized in the functionally-active hydrophilic loop between helices IV and V (33), is rich in glutamate residues, and has been conserved phylogenetically (Fig. 1). It is the region of the protein where quinone and herbicides are postulated to bind and where single amino acid changes that are altered in herbicide mutants of higher plants and algae exist. Interestingly, adjacent to this proposed site of primary cleavage is a 14 amino acid sequence bordered by arg225 and arg238 and rich in glutamate, serine and threonine residues (Figs. 1 and 3). This region resembles similar-type regions called 'PEST' (showing preponderance of proline-P, glutamate-E, serine-S and threonine-T residues and bounded by positively charged amino acids) that are commonly found in rapidly degraded proteins (34). PEST regions are suggested to

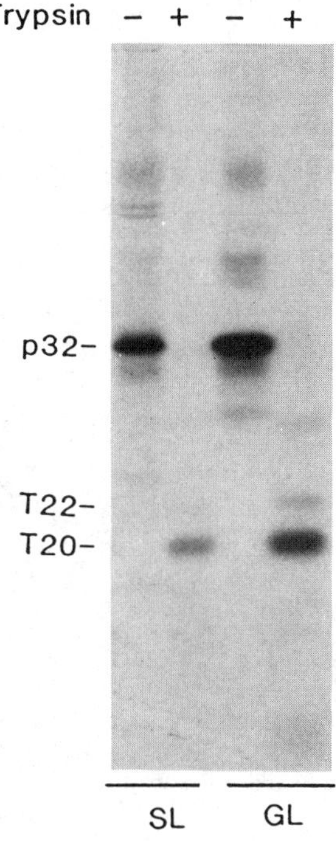

Figure 2

In situs trypsin digestion patterns of ^{35}S-methionine-labeled stromal and granal membranes. Spirodela plants were pulse labeled in the light with [^{35}S]methionine for 3 min, washed with non-radioactive growth medium and incubated further for 20 min with non-radioactive methionine (1mM). The plants were harvested, homogenized and whole thylakoids isolated. Fractionation of thylakoids into stromal (SL) and granal (GL) lamellae was achieved by the method of Leto *et al.* (37). Each membrane sample was digested with(+) or without(–) trypsin (24) and then prepared for SDS-PAGE. The positions of undigested 32kDa protein [p32] and trypsin fragments [T22 and T20] are indicated.

Lc	PFHMLG I SLF F TTAWAL AMHG ALVLS------AANPVKGKTMRTPDHEDT
32kDa	PFHMLG VAGVFGGSLFS AMHG SLVTSSLIRETTENESANEGYRFGQEEET

PEST region Cleavage domain

Figure 3

Comparison of aligned sequences in the PEST-like region of the 32kDa photosystem II reaction center protein with those of L subunit of *Rhodopseudomonas capsulata* (Lc). Adapted from ref. 38.

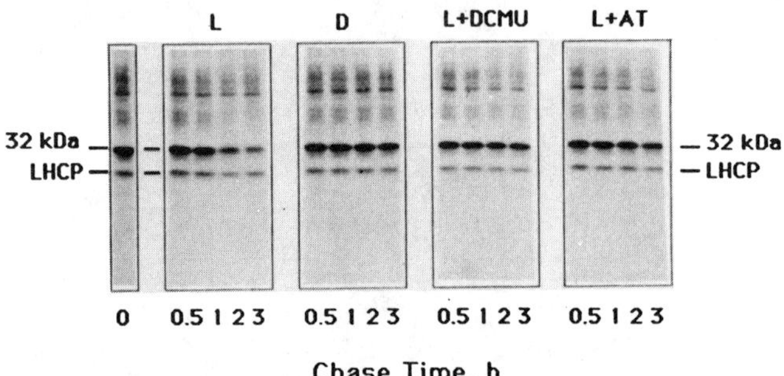

Figure 4

Degradation of the 32kDa protein in isolated Spirodela membranes. Spirodela plants were pulse labeled with [^{35}S]-methionine for 2 h and homogenized with STN/10mM Mg buffer (4). Membranes were centrifuged, washed with 10mM tris-glycine, pH 8.5, containing 10mM $MgCl_2$ and suspended in STN/10mM Mg buffer. Incubations were carried out in the dark (D) or at 50μM DCMU or 50μM atrazine (AT) for the times indicated. Samples were then subjected to SDS-PAGE. The positions of 32kDa protein and the light harvesting chlorophyll a/b apoprotein (LHCP) are indicated.

function as primary determinants for degradation of the proteins harboring them. In the chloroplast genome, which has been fully sequenced (40), a PEST-like region is found only in the 32kDa protein (33). The close proximity of the proposed cleavage site with the PEST-like region in the 32kDa protein suggests that the latter may be involved in regulating the degradation of the 32kDa protein. The PEST-like region is absent from the analogous protein from the non-oxygenic bacterial photosynthetic reaction center (35) viz., the L subunit (Fig. 3). In contrast to the 32kDa protein which is rapidly degraded, the L subunit is relatively stable. This raises the possibility of a linkage between evolution of oxygenic photosystem II and the introduction of the PEST-like region in the 32kDa protein. Genetic or engineered alteration of the site of cleavage and the PEST-like region in the 32kDa protein should prove valuable in studying the structure-function relationships of this reaction center component.

Mimicking In Vivo Degradation in Isolated Thylakoids

An *in vitro* system is being developed to study degradation of the 32kDa protein in isolated thylakoids. The requirements for a successful *in vitro* system include: 1. the protein should be degraded without affecting other membrane components; 2. protein degradation should be light dependent; and 3. herbicides (atrazine and diuron) that bind to the 32kDa protein should inhibit the degradation. Results in Fig. 4 demonstrate degradation of the 32kDa protein in isolated thylakoids. The results meet the three criteria mentioned above. The detection of the 23.5kDa product in this system is still under investigation. From the steady state ratio of 23.5kDa product to 32kDa protein found *in vivo*, the rate of 23.5kDa product formation was estimated to be 4-fold higher than the parent protein (33). However, since protein synthesis and reutilization of radioactivity do not occur *in vitro* during the chase (as they do *in vivo*), the absence of the 23.5kDa product in this system is not really remarkable. The data in Fig. 4 further suggest that the activity responsible for the degradation of the 32kDa protein is located in the thylakoids (36). Thus, this model system should now be useful to test inhibitors, activators, etc. on their effects on the 32kDa protein degradation and obtain a handle on the kind of protease or autocatalytic mechanism that leads to degradation.

Acknowledgments

We thank Sudhir Sopory for comments on the manuscript. Results of experiments presented here were supported in part by a United States-Israel Binational Agricultural Research and Development (BARD) grant to AKM and ME.

Literature Cited

1. Whitfeld, P.R.; Bottomley, W. *Ann. Rev. Plant Physiol.* 1983, *34*, 279-310.
2. Hoffman-Falk, H.; Mattoo, A.K.; Marder, J.B.; Edelman, M.; Ellis, R.J. *J. Biol. Chem.* 1982, *257*, 4583-7.

3. Steinback, K.E.; McIntosh, L.; Bogord, L.; Arntzen, C.J. Proc. Natl. Acad. Sci. USA 1981, 78, 7463-7.
4. Mattoo, A.K.; Pick, U.; Hoffman-Falk, H.; Edelman, M. Proc. Natl. Acad. Sci. USA 1981, 78, 1572-6.
5. Mattoo, A.K.; Edelman, M. In Frontiers of Membrane Research in Agriculture; St. John, J.B.; Berlin, E.; Jackson, P.C., Eds.; Rowman and Allanheld: New Jersey, 1985; pp 23-34.
6. Nanba, O.; Satoh, K. Proc. Natl. Acad. Sci. USA 1987, 84, 109-12.
7. Marder, J.B.; Chapman, D. J.; Telfer, A.; Nixon, P.J.; Barber, J. Plant Mol. Biol. 1987, 9, 325-33.
8. Zurawski, G.; Bohnert, H.-J.; Whitfeld, P.R.; Bottomley, W. Proc. Natl. Acad. Sci. USA 1982, 79, 7699-703.
9. Ellis, R.J. Biochim. Biophys. Acta. 1977, 463, 185-215.
10. Minami, E. -I.; Watanabe, A. Arch. Biochem. Biophys. 1984, 235, 562-70.
11. Herrin, D.; Michaelis, A. FEBS Lett. 1985, 184, 90-95.
12. Reisfeld, A.; Mattoo, A.K.; Edelman, M. Eur. J. Biochem. 1982, 124, 125-129.
13. Ellis, R. J. Ann. Rev. Plant Physiol. 1981, 32, 117-37.
14. Grebanier, A.E.; Coen, D.M.; Rich, A.; Bogorad, L. J. Cell Biol. 1978, 78, 734-46.
15. Hirschberg, J.; McIntosh, J. Science 1983, 222, 1346-9.
16. Hirschberg, J.; Bleecker, A.; Kyle, D.J.; McIntosh, L.; Arntzen, C.J. Z. Naturforsch. 1984, 39C, 412-20.
17. Cohen, B.N.; Coleman, T.A.; Schmitt, J.J.; Weissbach, H. Nucl. Acids Res 1984, 12, 6221-30.
18. Eyal, Y.; Goloubinoff, P.; Edelman, M. Plant Mol. Biol. 1987, 8, 337-43.
19. Sayre, R.T.; Andersson, B.; Bogorad, L. Cell 1986, 47, 601-8.
20. Michel, H.; Hunt, D.F.; Shabanowitz, J.; Bennett, J. J. Biol. Chem. 1988, 263, 1123-30.
21. Mattoo, A.K.; Edelman, M. Proc. Natl. Acad. Sci. USA 1987, 84, 1497-1501.
22. Marder, J.B.; Goloubinoff, P.; Edelman, M. J. Biol. Chem. 1984, 259, 3900-8.
23. Wettern, M. Plant Sci. Lett. 1986, 43, 173-7.
24. Marder, J.B.; Mattoo, A.K., Edelman,M. Methods Enzymol. 1986, 118, 384-96.
25. Edelman, M.; Goloubinoff, P.; Marder, J.B.; Fromm, H.; Devic, M.; Fluhr, R.; Mattoo, A.K. In Molecular Form and Function of the Plant Genome; van Vloten-Doting, L.; Groot, G.S.P.; Hall, T.C. Eds.; Plenum: New York, 1985; pp. 291-300.
26. Wickner, W. Ann. Rev. Biochem. 1979, 48, 23-45.
27. Norman, H.; St. John, J.B.; Callahan, F.E.; Mattoo, A.K.; Wergin, W.P. In The Metabolism, Structure and Function of Plant Lipids; Stumpf, P.K.; Mudd, J.B.; Nes, W.D., Eds.; Plenum: New York, 1987, pp 193-5.
28. Callahan, F.E.; Edelman, M.; Mattoo, A.K. In Progress in Photosynthesis Research; Biggens, J. Ed.; Nijhoff: The Hague, 1987; Vol. 3; pp. 799-802.
29. Callahan, F.E.; Mattoo, A.K. Plant Physiol. Suppl. 1987, 83, p. 113.
30. Wiest, S.C. Physiol. Plant. 1986, 66, 527-35.

31. Mattoo,A.K.; Hoffman-Falk, H.; Marder, J.B.; Edelman, M. Proc. Natl. Acad. Sci. USA 1984, 81, 1380-4.
32. Gaba, V.; Marder, J.B.,; Greenberg, B.M.; Mattoo, A.K.; Edelman, M. Plant Physiol. 1987, 84, 348-52.
33. Greenberg, B.M.; Gaba, V.; Mattoo, A.K.; Edelman, M. EMBO J. 1987, 6, 2865-9.
34. Rogers, S.; Wells, R.; Rechsteiner, M. Science 1986, 234, 364-8.
35. Deisenhofer, J.; Epp, O.; Miki, K.; Huber, R.; Michel, H. Nature 1985, 318, 618-24.
36. Reisman, S.; Ohad, I. Biochim. Biophys. Acta. 1986, 849, 51-61.
37. Leto, K.J.; Bell, E.; McIntosh, L. EMBO J. 1985, 4, 1645-53.
38. Youvan, D.C.; Bylina, E.J.; Alberti, M.; Begusch, H.; Hearst, J.E. Cell 1984, 37, 949-57.
39. Kirsch, W.; Seyer, P.; Herrmann, R.G. Curr. Genet. 1986, 10, 843-55.
40. Shinozaki, K. et al. EMBO J. 1986. 5, 2043-9.
41. Golden, S.S.; Brusslan, J.; Haselkorn, R. EMBO J. 1986, 11, 2789-98.
42. Curtis, S.; Haselkorn, R. Plant Mol. Biol. 1984, 3, 249-58.
43. Mulligan, B.; Schulte, N.; Chen, L.; Bogorad, L. Proc. Natl. Acad. Sci. USA 1984, 81, 2693-7.
44. Karabin, G.D.; Farley, M.; Hallick, R.B. Nucl. Acids Res. 1984, 12, 5801-12.
45. Erickson, J.M.; Rahire, M.; Rochaix, J.D. EMBO J. 1984, 3, 2753-62.
46. Ohyama, K. et al. Nature 1986, 322, 572-4.
47. Larrinua, I.M.; McLaughlin, W.E. In Progress in Photosynthesis Research; Biggins, J. Ed.; Nijhoff: The Hague, 1987; Vol. 4; pp. 649-52.
48. Link, G.; Langridge, U. Nucl. Acids Res. 1984, 12, 945-58.
49. Reith, M.; Straus, N.A. Theor. Appl. Genet. 1987, 73, 357-63.
50. Aldrich, J.; Cherney, B.; Merlin, E. Nucl. Acids Res. 1986, 23, 9537.
51. Oishi, K.K.; Shapiro, D.R.; Tewari, K.K. Mol. Cell Biol. 1984, 4, 2556-63.
52. Spielmann, A.; Stutz, E. Nucl. Acids Res. 1983, 11, 7157-67.
53. Aldrich, J.; Cherney, B.; Merlin, E.; Christopherson, L.A.; Williams, C. Nucl. Acids Res. 1986, 23, 9536.
54. Goloubinoff, P.; Edelman, M.; Hallick, R.B. Nucl. Acids Res. 1984, 24, 9489-96.
55. Sugita, M.; Suguira, M. Mol. Gen. Genet. 1984, 195, 308-13.
56. Goloubinoff, P. Ph.D. Thesis, The Weizmann Institute of Science, Rehovot, Israel, 1988.

RECEIVED August 5, 1988

GENETICALLY ENGINEERED MICROBIAL PESTICIDES

Chapter 19

Development of Genetically Improved Strains of *Bacillus thuringiensis*

A Biological Insecticide

Bruce C. Carlton

Ecogen, Inc., 2005 Cabot Boulevard West, Langhorne, PA 19047

The last decade has seen a growing public awareness of the potential hazards of synthetic chemicals used to control pests of crop plants, and a renewed interest in developing biological products as alternatives to chemical pesticides. Biological pesticides offer a number of advantages to synthetic chemicals, including lack of polluting residues, high levels of safety to non-target organisms, lower development and registration costs, and a reduced likelihood of pest resistance. Despite these advantages, the development of new biologicals has been slowed by production problems, inconsistent performance, and a lack of improved strains. New methodologies of genetics and molecular biology offer approaches for the development of genetically improved strains having superior efficacy properties. This presentation focuses on the applications of these new genetic technologies for generating improved versions of the bacterial insect pathogen, *Bacillus thuringiensis* (BT). The history of BT as a biological insecticide is reviewed, including the nature of the insecticidal determinants and its production and mode of action. The complexity of BT strains is discussed, both the insecticidal activities and the properties of the crystalline inclusions that constitute the active components. The genetics of these complex spectra has been shown to be reflected in complex plasmid-based inheritance patterns of the genes involved. The plasmid localizations of individual insecticidal genes have been demonstrated by alterations affecting individual plasmids (curing, conjugation) as well as by direct molecular probing of the plasmid arrays with segments of known insecticidal genes. Development of new strains by a combination of plasmid curing and conjugal transfer is described, and the regulatory issues involving the environmental introduction of genetically altered strains are discussed.

0097–6156/88/0379–0260$06.00/0

The control of pests of food and fiber crops is essential to attaining acceptable yields and quality. In 1987, nearly $20 billion was spent worldwide on products for combatting plant diseases, weeds, and insects. Despite this enormous input, nearly 40% of the world's food and fiber was lost to these pests.

Adding to the problems of pest control has been the increasing concern expressed by a wide array of environmental groups. These organizations have focused on the inherent problems of certain pesticide products, which cause either acute or chronic toxicity to non-target organisms, including man, and their tendency to accumulate in the food chain and in sources of potable water. In the past, these potential problems were not generally recognized, allowing the widespread application of hazardous chemicals into the environment.

In 1978, Rachel Carson's book, Silent Spring, signalled the beginning of an era of intensive scrutiny of both the benefits and risks of chemical pesticides. Her graphic portrayal of the hazards of these materials capitalized on a public mood that was beginning to grasp the significance of the aftermath of the dropping of nuclear bombs on Japan, and caused public opinion to become centered on the detrimental effects of inherently dangerous chemicals to non-target organisms. At this same time, a more broadly-based grass-roots "anti-pollution" movement was developing. The practice of the chemical and pesticides industries in disposing of toxic by-products of their businesses in what came to be known as "toxic waste dumps" helped to focus unfavorable publicity on industry. If all of this weren't enough, the Bhopal tragedy of 1984 ultimately served to underscore the potential dangers of chemical pesticide products to even the most casual observer.

In recent years, the realizations of Silent Spring have impacted significantly upon the pesticide industry. The registration process has undergone extensive revamping and a number of chemicals have been banned or restricted in many parts of the world. The industry has not yet accepted a mandate to develop alternative products, however. There are several reasons for this slow response. One is the very strong lobby of the pesticide industry in Congress. The companies that sell chemical pesticides have billions of dollars committed to production facilities, inventories, marketing programs, and research programs that typically take 7 to 10 years to bring a new product from discovery to commercialization. A second, and more fundamental reason, for industry's slow response has been the conceptual inability to adopt new approaches to pest management.

Are there feasible alternatives to chemical pesticides? And if there are, can the alternatives compete with current chemical pesticide products? Can they be applied at the same costs and in the same ways that farmers understand? A fairly superficial perusal of the literature reveals a lengthy list of biological agents that have been shown to have potential for controlling various pests. These include microbial pathogens such as viruses, bacteria, fungi, and protozoans (Table I). Within some of these groups (e.g., bacteria and fungi) there have been discovered different species or strains that can control different pests, such as plant pathogens and insects. Many of these organisms have been known for years. *Bacillus thuringiensis*, for example, was recognized as a

Table I
Biological Agents with Pest Control Activities

Agent	Target
Bacteria	
Bacillus popillae	Japanese beetle grubs
Bacillus thuringiensis- var. kurstaki	Lepidopteran insect larvae
Bacillus thuringiensis- var. israelensis	Dipteran insect larvae
Pseudomonas fluorescens	Plant fungal pathogens
Bacillus sphaericus	Mosquito larvae
Pasteuria penetrans	Plant parasitic nematodes
Viruses	
Insect baculoviruses	
Heliothis NPV's	Budworms and bollworms
Lymantria dispar NPV	Gypsy moth larvae
Neodiprion sertifer NPV	Pine sawfly larvae
Laspeyresia pomonella GV	Codling moth larvae on fruit trees
Fungi	
Colletotrichum gloeosporioides	Northern jointvetch
Alternaria casei	Sicklepod
Phytophthora palmivora	Milkweed vine
Metarhizium anisopliae	Crickets, spittlebugs, stalk borers, etc.
Beauvaria bassiana	Wide variety of insects
Trichoderma spp.	Rot fungi, soil-borne fungal pathogens tree diseases
Neozygites spp.	Spider mites
Protozoans	
Nosema spp.	Grasshoppers and locusts
Nematodes	
Neoplectana spp.	Variety of insects
Heterohabditis spp.	

pathogen of certain caterpillars as early as 1901. Another type of biological control agent that has received significant attention is that of predators and parasites. Examples such as ladybird beetles, parasitic wasps, and the use of sterilized males have been suggested as means for controlling insect pests. Still a third group of products is the attractants and deterrents, such as pheromones and baits.

How have these biologicals impacted the pesticide industry to date? The answer is, very little. Of a total worldwide pesticide market of about $20 billion, biologicals account for only about $50-$60 million - less than half of a percent. Why has there been so little market acceptance of these products? To understand this we need to compare both the advantages and the disadvantages of biological pesticides as compared to synthetic chemicals (Table II).

Biologicals, because of their natural derivations, have the distinct advantages of being biodegradable, so that they do not leave toxic residues or by-products to contaminate the environment. They have a very high level of safety for humans, animals, fish, and other non-target organisms, principally because they act by very different modes of action than most organic chemical pesticides that attack metabolic systems shared by both pest and non-pest organisms. This safety advantage is very important, because it impacts to a large extent the cost of development and registration of a new pesticide product. Most chemical pesticides have a 7-10 year development time frame, with registration costs ranging from $20-$30 million. This expense is, in large part, due to the concern over possible high animal toxicities of such materials that necessitates long-term toxicological testing on experimental animals. Biologicals, because of their high target specificity, typically require only short-term (Tier I) toxicological tests, with the potential for registration within a year of submission and a total registration cost of $150-$250 thousand.

Another, and increasingly important, advantage of biologicals is a lower likelihood of inducing resistance in populations of the target pest. Over the past 50 years, the accelerated and prolonged use of organic synthetic chemical pesticides has resulted in an increasing incidence of high levels of resistance to these materials on the part of their targets. A recent summary has reported that at least 447 species of insects and mites, 100 species of plant pathogens, and 48 species of weeds are now resistant to control by chemical pesticides (1). In many instances, resistance induced to one class of chemicals leads to cross-resistance to other chemicals, due to alterations in enzyme systems of the host organisms that can inactivate a range of chemical compounds. Although it may be that biologicals will also induce resistance if used in greater quantities over prolonged periods of time, available data from laboratory experiments and a recognition of the inherent diversity of biological agents would predict that such resistance, if it does occur, would be slower and more restricted than chemical resistance.

With all these advantages of biological agents why, then, has their acceptance as alternative products been so slow? There are several reasons. First is that the growth requirements of many of them (e.g., fungi, protozoans) are so complex that they could not in the past be produced in quantities and at a cost that would allow them to effectively compete in the pesticide marketplace. Second, the fragile biological nature and biodegradability of biopesticides has led to an

Table II

Comparison of Chemical and Biological Pesticides

	Chemicals	Biologicals
Environmental persistence	Often high	Little or none
Field efficacy	High	Variable
Activity spectrum	Can be very broad	Moderate-highly specific
Mammalian toxicity and carcinogenicity	Moderate-High	None
Other non-targets	May be high	None
User cost	Acceptable	Have been high in the past - becoming competitive
Development times and costs	High (7-10 yrs; \$10-\$30 million)	Low (3-5 yrs; \$1-\$3 million)
Reentry and tolerance limitations	Yes	No
Probability for resistance	Moderate-high	Low

inherent inconsistency in their performance that for many of them has not yet been solved. Third, all of the biological products developed commercially to date have been native, unmodified strains and isolates. Little attempt was made to genetically alter the native strains to generate variants that either are more effective or that have otherwise improved properties. In large part, the major reason for inactivity at this level has been a lack of methodologies to improve these materials in specific and targeted ways. This situation is now changed with the development of new genetic technologies such as gene splicing, protoplast fusion, somaclonal and gametoclonal variation, and whole-organism regeneration from cultured cells. These technologies have made possible the genetic manipulation of biological systems to a level of sophistication that has never before been possible.

The remainder of this presentation will focus on the applications of some of these technologies toward the development of improved versions of one type of biological pesticide, the bacterial insect pathogen *Bacillus thuringiensis*.

BACILLUS THURINGIENSIS AS A BIOLOGICAL INSECTICIDE

Of all the biological agents that have been evaluated as pest control products, the most successful by far has been the bacterial insect pathogen *Bacillus thuringiensis* (or BT as it is commonly referred to). Approximately 90-95% of the sales of all biological pest control products worldwide are of this bacterial agent, totalling approximately $50-$55 million in annual sales. BT products are of two types. The most predominant is BTk (k for variety *kurstaki*), and is sold under several trade names (e.g., Dipel, Thuricide, Bactospeine, Biobit, etc.) for control of a number of lepidopteran (caterpillar) pests of importance on forest trees, vegetable crops, cotton, and ornamentals. These products have been sold since the early 1960's and account for the bulk of BT applications. In the late 1970's, another BT, called BTi (variety *israelensis*, for the location of its discovery), was developed as a product for the control of a number of dipteran (fly) vectors of human diseases, including mosquitos that disseminate malaria, yellow fever, and Dengue fever, and blackflies that transmit onchoceriasis (African river blindness). This product, sold under tradenames such as Teknar, Vectobac, etc., has achieved North American sales of $3-$5 million, although its most extensive use has been promoted by agencies such as AID and the World Health Organization in underdeveloped countries of Africa, Central and South America, and Asia.

Despite the fairly long history of BT, its expansion as a pesticide product has been very slow. In fact, usage of the BTk products has actually decreased over the last 10 years, due primarily to the capture of the vegetable insecticide market by chemical pesticides. The major reasons for this failure to expand have already been alluded to, namely inconsistency of performance (especially when compared to the chemicals) and past inability to compete in price. Another reason is that almost all of the BTk products are produced from a single strain, HD-1, isolated by a U.S. Dept. of Agriculture laboratory in Texas in 1969 (2). Although its activity on some pest insects (e.g., gypsy moth larvae, cabbage loopers) is good, on others such as the cotton bollworm its effectiveness is minimal. Only one new product (Javelin) has appeared on the market; a product based on a different strain isolated by researchers at the U. S. Forest Service laboratory in Hamden, CT.

Interestingly, the level of effort to develop new biopesticide products by the major agrichemical companies has been minimal. Most of the effort directed at BT research in both large and small companies has been focused on the molecular biology of the genes for the insecticidal determinants (proteins termed delta-endotoxins), and on vectoring the genes for these proteins into either plants or alternative bacterial hosts. What has been lacking is a sustained and intensive effort to identify and further develop novel strains for specific insect applications. This is the direction that we have taken at Ecogen Inc.

Before discussing improvement strategies for BT, a brief description of the nature of the product and its mode of action are appropriate. BT is a common gram-positive soil bacterium that shares many properties with its close relative *B. cereus.* BT has been isolated in soil samples from many different parts of the world, and has also been isolated from dusts of crop grains and from carcasses of a variety of insects, including some for which the BT apparently has little or no pathogenic activity. BT, and most bacteria in the genus *Bacillus*, are somewhat unique in the bacterial kingdom in their ability to form endospores; typically ovoid, highly refractile and heat-resistant structures that encapsulate the DNA of the bacterium. These spores, once released from the dead mother cell, can lie dormant for indefinite periods until they encounter favorable conditions that allow them to germinate and regenerate the rapidly-dividing vegetative cells. What makes BT truly unique is the production, in sporulating cells, of an inclusion body referred to as a parasporal crystal. This crystal is typically diamond-shaped (bipyramidal) in appearance, although different strains produce a wide array of crystal shapes including cuboidal, spherical, and other types. The composition of the crystal is a highly condensed, high molecular weight protein (typically 75-150 Kd in mass).

For commercial production, the BT cells are grown in large fermentors in complex media that support high cell densities and ultimate sporulation of the cells. Cell lysis releases the spores and crystals into the growth medium, and they are recovered by either centrifugation or other techniques that concentrate the particulates. Depending on the desired formulation, the concentrate is either spray-dried and formulated into wettable powder or oil flowable, or formulated directly from concentrate into an aqueous flowable. Application is with standard spray equipment, either ground or aerial.

In its mode of action BT acts as a stomach poison (i.e., it must be ingested by a sensitive insect to bring about the insecticidal response). As the insect feeds on the foliage (or in the case of mosquito larvae, filters the particulates from the water) the crystals, which are insoluble in water at neutral pH, are solubilized in the typically alkaline conditions of the insect midgut. Solubilization dissociates the high molecular weight protoxin molecules, and there follows an activation process, believed to be a proteolytic cleavage of the protoxin protein, to generate the active toxin. Although the mode of action is not fully understood, it is hypothesized that the activated toxin binds to receptor sites on the gut epithelial cells, disrupting the normal gradient of ions and pH from the hemolymph across the epithelial cell layer to the gut lumen and leading to osmotic effects. Cytologically the manifestations of the toxin's effect are swelling of the epithelial cells, followed by bursting and eventual sloughing off of the entire midgut lining, causing the insect to literally starve to death. There are other symptoms of the toxin that occur very soon after ingestion, such as paralysis of the mouth parts and the gut, which cause the insect to cease feeding. There is also evidence, some of it conflicting, that the spores may play a role in the insecticidal response, possibly by initiating a septicemia in the insect hemolymph that may augment the effects of an otherwise sublethal dose of the toxin. It is worth noting also that, since the insect must eat the toxin crystals for them to be effective, it may be important in formulating the BT to consider additives that either improve the palatability or that could mask the effect of possible anti-feeding substances in the formulation to maximize efficacy.

COMPLEXITY OF BT IN NATURE

One of the first points to emphasize in considering the opportunities for developing improved strains of BT is to recognize that this bacterial group is much more diverse than just a BTk strain for caterpillar control and a BTi strain for fly larvae control. The most widely-recognized criterion for BT classification is based on immunological differences in surface flagellar antigens, which de Barjac (3) has used to identify more than 20 different subspecies (varieties). Other criteria, such as toxin crystal antigens (4), biochemical properties (5), and plasmid arrays (6) have shown that there is considerable diversity, even between strains of the same flagellar type (serovar). Of particular importance in this strain diversity is the wide spectrum of insecticidal activities between strains. Most BT's have activities that vary over a range of at least 100X on different target insects. By far the most prominent group of susceptible insects are larvae of lepidopteran species (the caterpillar forms of moths and butterflies). A second group showing susceptibility (to different BT strains) are the dipterans, particularly several species of mosquitos and blackflies. More recently (7, 8), discoveries have been made of BT strains having insecticidal activity against coleopteran (beetle) larvae, including the Colorado potato beetle, elm leaf beetle, boll weevil, and black vine weevil. These observations suggest that there may be, in nature, BT strains having activity against a broad array of insect species, if one looks hard enough to find them.

In addition to the insecticidal diversity between strains of BT, there is often considerable diversity of insecticidal activities within a single strain. As an example, consider the strain designated HD-1, which is the U.S. Dept. of Agriculture isolate that serves as the basic strain used in most of the current products sold for caterpillar insect control. This strain, in addition to its caterpillar activity, has a low but detectable level of activity against mosquitos. Examination of the crystalline inclusions produced by HD-1 reveals the typical, rather large bipyramidal structure and, in addition, a smaller cuboidal body, often located at one apex of the bipyramidal crystal (9). If the crystal proteins are solubilized by incubation at high pH (e.g., 10-11) in the presence of a reducing agent and run on a suitable matrix for separation of the proteins, three major fractions are observed (Figure 1). Two of these are high molecular weight proteins of 135 and 140 Kd (designated P1 proteins). Strain HD-1 produces about twice as much of the somewhat larger protein than the 135 Kd material, as judged by band intensities. The third major band (designated P2) is smaller, about 65-68 Kd. Depending on the conditions of solubilization and fractionation, a variety of less prominent protein bands are observed. These are believed to be due to the variability in solubilization and to the effects of bacterial proteases that are carried along with the crystals during the recovery process.

If the major protein fractions are recovered and bioassayed on various test insects, the following observations are made. The two high molecular weight P1 protein fractions have activity against only lepidopteran (caterpillar) insect larvae. The smaller (P2) fraction has activity against both lepidopteran and dipteran insects (10). This protein, which can be demonstrated to be a single component, is bifunctional in its activity and clearly accounts for the low-level dipteran activity in the HD-1 strain. This strain thus is complex in the sense that it produces two types of crystals (bipyramidal and cuboidal), containing a complex of at least three separable proteins - two high molecular weight P1 proteins produced in different amounts and a smaller P2 protein.

How is this biochemical diversity reflected in the genetics of insecticidal protein determination? As has been shown by many groups (see (11) for review) the genes encoding these insecticidal proteins are not localized, for the most part, on the essential "chromosome" of the BT organism. Rather they have been shown to be carried on a variety of non-essential, extrachromosomal plasmids. These toxin-encoding plasmids, typically moderate (~40-60 Mdal) to large (>100 Mdal) in size, often have the capability of self-transmissibility from one BT strain to another by a conjugation-like mechanism (12, 13). The localization of genes encoding insecticidal toxins to specific plasmids can be accomplished by several methods. Historically, the loss of insecticidal crystal product associated with the "curing" (loss) of a specific plasmid was utilized as circumstantial evidence for the association of toxin genes with plasmids (14). The conjugal transfer of toxin-synthesizing ability from a crystal-producing strain to a non-producing strain, associated with transfer of a specific plasmid, has also been used to localize toxin-encoding genes (15). More recently, the direct molecular probing of plasmid arrays with segments of cloned toxin genes has provided a more definitive method for identifying plasmids that carry such genes (16, 17).

Using a combination of these methodologies, the HD-1 strain has been shown to harbor two plasmids that encode insecticidal toxin genes (Figure 1) (6, 17, 18). One, a moderate-size plasmid of 44 Mdal, contains a gene that is referred to as a 5.3-type gene, meaning that molecular probing with a known toxin gene segment identifies at least a portion of the gene on a particular restriction enzyme digestion fragment (Hind III) of 5.3 Kb. A comparison of HD-1 and a derivative that has lost this plasmid reveals the concomitant loss of the 5.3 Kb hybridizing band as well as the smaller of the two high molecular weight P1 toxin proteins (i.e., the 135 Kd protein). A second plasmid involved in toxin protein synthesis in HD-1 is much larger (~115 Mdal). Loss of this plasmid from HD-1 leads to a loss of expression of the more intense P1 toxin protein band of about 140 Kd, as well as the P2 protein of 65-70 Kd. In addition, two restriction enzyme fragments of 6.6 and 4.5 Kb are lost, as well as the DNA segments corresponding to the P2 gene. Thus, this large plasmid contains two different P1 genes and a P2 gene.

In summary, the complexity of the HD-1 strain can be stated as follows: it contains at least four insecticidal toxin genes, three of the P1 type (that specify proteins active on lepidopteran insects and are of high molecular weight) and a P2 gene (whose protein is much smaller and has bifunctional activity on both lepidopteran _and_ dipteran insects). These

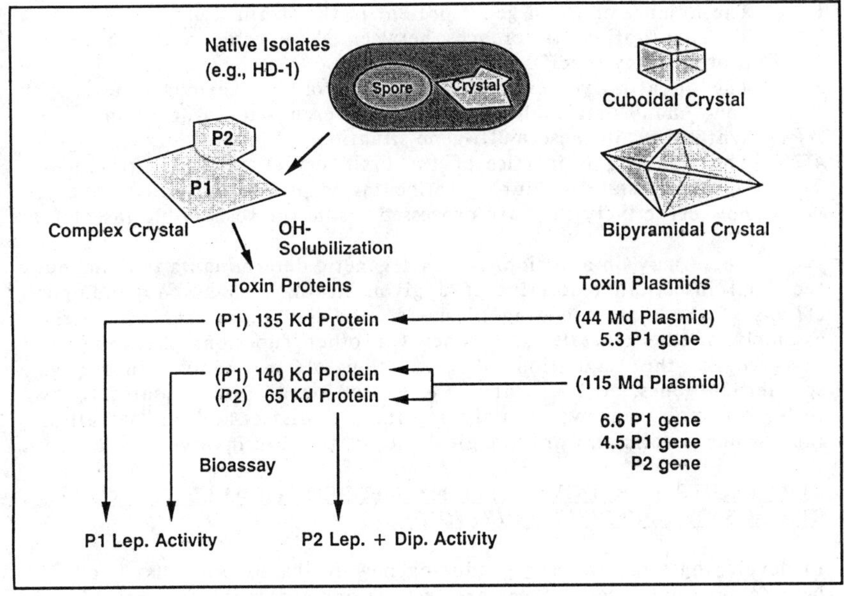

Figure 1. Evidence for Complex Genes in BT var. *kurstaki* Strains.

four genes are distributed on two plasmids; one P1 gene on a moderate-sized 44 Mdal plasmid and the other two P1 genes plus P2 on a large ~115 Mdal plasmid. Conjugation experiments with HD-1 have shown that the 44 Mdal plasmid is readily capable of self-transmission to other strains of BT, conferring upon them the ability to produce the 5.3-type P1 protein. The larger plasmid transfers very poorly, if at all. As will be described later, this transmissibility of certain toxin-encoding plasmids can be used as a non-recombinant means of constructing new toxin gene combinations having either higher levels of activity against selected target insects, or a different spectrum of activity.

To summarize the complexity of BT in nature, one can state that there are several factors that influence the insecticidal activity of a given strain. These are:

1. The number of toxin genes present in the strain.
2. The qualitative differences between these genes and the toxin proteins they specify.
3. The quantitative levels of expression of the various genes (i.e., how many molecules of the different toxin proteins are synthesized in these multi-gene situations).
4. The resulting properties of the toxin crystals that are produced (i.e., their stability during fermentation production and recovery; how effectively they are processed inside the susceptible insects).

There may, in addition, be other genetic determinants that impact the final insecticidal activity of a given strain, such as copy number effects of the toxin-encoding plasmids, modifier genes that affect the assembly of the crystals, and genes for other functions that may be involved in the regulation of expression of the toxin genes (e.g., sporulation genes, etc.). Clearly, the expression of insecticidal activity in BT is complex and we are only beginning to dissect and unravel all of the various genetic and physiological factors that are involved.

STRATEGIES FOR DEVELOPMENT OF GENETICALLY IMPROVED STRAINS OF *BACILLUS THURINGIENSIS*

In developing strategies for producing new BT-based products that will be a commercial success, there are two major hurdles that emerge. First are the technical problems, i.e., what is required to generate strains that can compete with chemical insecticides? The second, and no less formidable, hurdle is the ability to move such strains through the regulatory process. Technical success in developing a "super strain" will be of little value if the product cannot be registered within a reasonable time frame and at an acceptable cost. Conversely, an easy regulatory path does not make a commercial success if the product cannot compete on its merits.

At Ecogen Inc., we have developed a series of strategies that we believe will allow us to develop a first generation of strains of BT that will be commercial successes in several important insect-control markets (e.g., forestry, vegetables, cotton, potatoes) within the very near future. In addition, we have a parallel series of strategies that we believe will continue to assure a flow of second generation products into the insecticide pipeline over the longer term. Both of these strategies capitalize on the exploitation of genetic diversity of BT insecticidal genes and their activities; the divergence in strategies revolves around the genetic methodologies that are utilized to manipulate these genes into new product combinations.

The strategies for first generation product development are as follows:

- Expand sources of insecticidal genes;
- Identify plasmids harboring active genes;
- Characterize insecticidal activities;
- Construct new strains by non-recombinant methodologies;
- Optimize fermentation production and recovery processes;
- Develop formulations that maximize field efficacy.

A few comments will amplify on how each facet of the strategy fits into the overall product development process. Expanding the sources of insecticidal genes is being undertaken by an extensive screening program to isolate new strains of BT from a highly diverse array of ecological niches (e.g., soils, grain dusts, dead insects, etc.). To date we have accumulated more than 4000 novel isolates from this activity. These strains are initially bioassayed on an array of insects representing important economic targets (e.g., cotton, corn, and forestry pests). Once a strain is identified as having an interesting level of activity against a particular target, it is then further characterized to identify the genes responsible for the activity of interest. This characterization involves the identification of the relevant toxin-encoding plasmid(s), transfer of the plasmid(s) (where possible) into a standard background strain of BT to assess the insecticidal activities of individual genes, and molecular probing of both intact plasmids and restriction enzyme digests of total cellular DNA to assess the number, localization, and type of insecticidal genes present.

Once these data are available, new strain constructs are made by the methodologies of selective plasmid curing and conjugal transfer. Both of these non-recombinant methodologies are clearly implicated in the basis of diversity of BT strains in nature. This realization served to distinguish these genetic manipulations from that of *in vitro* gene splicing (recombinant DNA), and has aided greatly in effectively moving these strains through the regulatory processes of the Environmental Protection Agency and into field trials. Ecogen was, in fact, the first company to receive permission from the EPA to field test live, genetically modified bacteria without having to file an Experimental Use Permit.

A few comments are in order regarding the relative merits of the plasmid curing and conjugal transfer methods for strain construction. The rationale with the plasmid curing approach is to generate a superior derivative of a strain that has a generally good activity against a particular insect or group of insects. The approach is to isolate variants that have lost one or more plasmids carrying toxin genes of low activity, and to obtain a higher level expression of the remaining, more active toxin genes (Figure 2). That this can be accomplished is probably due to the existence of gene dosage compensation mechanisms in BT that allow for a maximal level of toxin protein synthesis per cell. If one or two genes are removed from a strain initially possessing several such genes, then the remaining ones can be expressed at higher levels. Regardless of the actual mechanisms involved, we have found that many such partially-cured variants do, in fact, have significantly higher insecticidal activities than their parental counterparts. The conjugal transfer

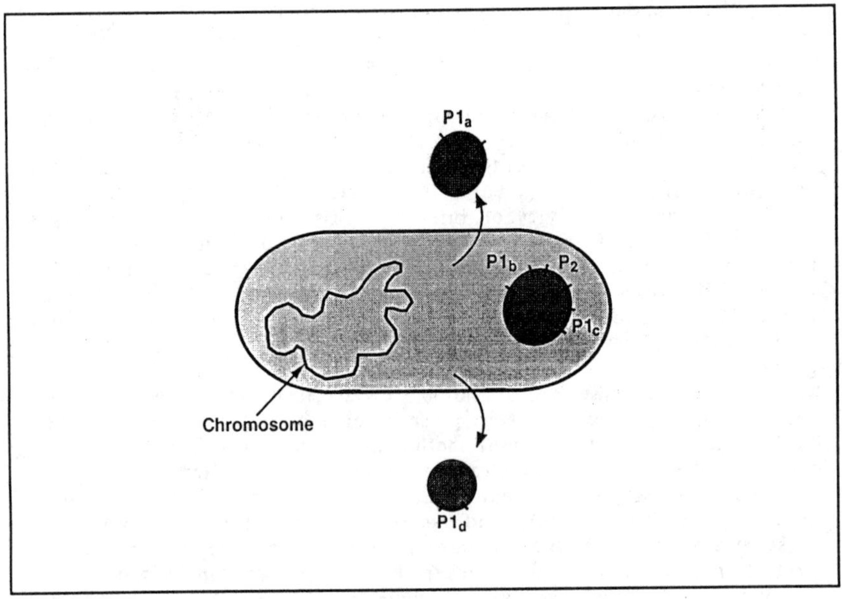

Figure 2. Strain Improvement by Plasmid Curing.

approach, in contrast, is more versatile in that it allows the combination, in a single strain, of insecticidal genes from totally different origins (Figure 3). With this approach one can start with a strain having, for example, a single toxin plasmid harboring a high activity gene. Mating this strain with one having a different high activity toxin plasmid can generate a transconjugant which now has both plasmids. This strain can be mated with still a third strain harboring a high activity plasmid, and so on. We have been able to use this process to generate strains having up to four toxin-encoding plasmids, all derived from different BT origins. The combination of plasmids appears to generate synergistic gene/toxin interactions leading to much greater toxicity levels.

An additional example of the utility of the conjugal transfer approach to new strain construction for developing a broader spectrum product is illustrated by the example of a strain we have constructed as a potato insecticide product (Figure 4). Although the most destructive insect pest of potatoes in North America is the Colorado potato beetle, there are several lepidopteran insects such as European corn borer and potato tuber moth that can be significant pests on this crop. An ideal product would thus have a broad activity against several such target insects. With this goal in mind, we isolated a strain of BT that had acceptable levels of activity against the Colorado potato beetle. Unfortunately, this strain had no detectable lepidopteran insect activity, although its Colorado potato beetle activity was localized to a transmissible plasmid. We mated this strain with a recipient which had been shown to have good activity against the European corn borer, and derived a transconjugant that now had activity directed against both types of insects. This new strain has been field-tested and shows exceptional promise as a general purpose potato insecticide.

Using the combined techniques of selective plasmid curing and conjugal plasmid transfer we have been able to generate new strains having insecticidal potencies in laboratory assays many times higher than the HD-1 activity against certain target insects. We believe that these potency increases, in conjunction with improvements in production and formulation methodologies, will result in products that can effectively compete with synthetic insecticides within a very short time-frame. Field evaluations of such strains over a two-year period have confirmed that they perform better than current commercial BT products and in some cases are equal to the chemical pesticides currently used for those applications.

STRATEGIES FOR DEVELOPING SECOND GENERATION PRODUCTS

In looking to the future availability of much improved BT-based bioinsecticide products we have also developed a molecular biology effort that will assure a continuing array of more effective products. Recognizing that the availability of natural genetic diversity may be limiting, and that the BT insecticidal gene combinations that can be generated by plasmid curing and/or conjugational methodologies are restricted by the localizations of insecticidal genes to either the same or to non-transmissible plasmids, we have conducted a parallel program employing the techniques of molecular biology. Selected insecticidal genes of interest are isolated by molecular cloning techniques and then subjected to various *in vitro* manipulations to further improve their properties. The major objectives in these approaches are: 1, to isolate

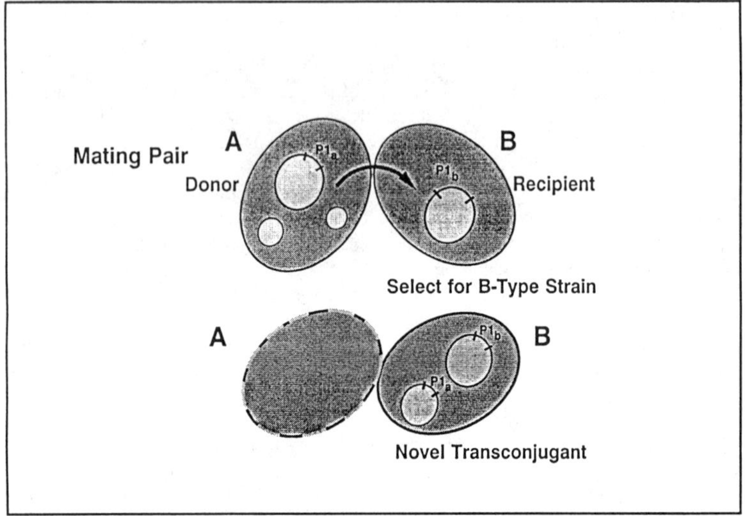

Figure 3. Strain Improvement by Conjugal Transfer.

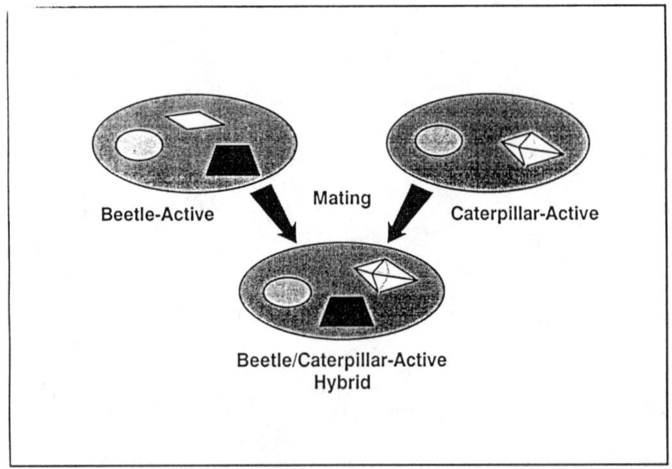

Figure 4. Construction of a Broader Spectrum BT Strain.

genes having inherently superior activities against specific target insects; and 2, to devise ways to improve their activities by further genetic modifications. Previous work on P1 types of BT insecticidal genes has revealed that these genes exist as families of related sequences, in which there is extensive homology (i.e., near-identity) for the 5' and 3' regions (19). The genes show major divergence within the central regions, which presumably are important for both activity and specificity, although the exact sequences have not yet been localized. Among the novel genes we have isolated and characterized is that for a bifunctional P2 toxin (20). This gene specifies a protein found in the cuboidal crystal referred to previously, with activity against both caterpillars and mosquitos. Comparison of the nucleotide sequence of the P2 gene with those for P1 genes shows no detectable homology, although at the protein level there is a region encompassing about 100 amino acids showing 38% homology to a P1 gene. Studies of this type will allow the delineation of those gene sequences important for insecticidal activity and specificity, and the alteration of these sequences by techniques such as site-directed mutagenesis in order to generate variants having improved properties. These altered genes can then be expressed either individually or in combinations in a variety of host-vector systems for a variety of pest control applications. Among those that have been suggested are:

- Expression in BT or its close relatives (*B. cereus, B. megaterium*).
- Insertion into plant-inhabiting epiphytes or endophytes.
- Insertion into systems capable of proliferating in aquatic environments (e.g., blue-green algae, for delivery of mosquito control activities).
- Insertion into the genomes of plants, making them directly insecticidal.

Although these applications have considerable appeal, there are a number of current and potential problems in their utilization. First is the continuing debate regarding the dissemination of recombinant organisms into the environment. Despite extensive laboratory evidence accumulated over the past decade that the use of recombinant DNA methodologies has not generated any organisms exhibiting unpredicted pathogenicity or uncontrolled growth, there is still the lingering concern of the "unknown" that leads regulatory agencies to be extremely cautious in approving environmental release of recombinant organisms. In addition, these concerns have been reflected in various legal maneuverings (injunctions, etc.) by environmental awareness groups to slow down the process of field testing of recombinant organisms. These actions have led to extended delays in the approval process for field testing of products such as the ice-minus strains of *Pseudomonas syringae* by the University of California and Advanced Genetics Sciences, and the *Pseudomonas fluorescens* strain containing a BT insecticidal gene constructed by scientists at Monsanto. At this point in time, it is difficult to predict whether the current levels of caution on field testing of recombinant organisms will be relaxed in the foreseeable future. Recent studies, such as the one released in 1987 by a committee appointed by the National Academy of Sciences (21), have evaluated the risks associated with the use of recombinant DNA. This panel concludes that "assessment of the risks of introducing R-DNA organisms into the environment should be based on the nature of the organism and the environment into which it will be introduced, not on the method by which it was modified." This report also concludes that "there is no evidence that unique hazards exist either in the use of R-DNA techniques or in the movement of genes between unrelated organisms." Taking this rationale one step further, and based upon the guidelines for containment dictated by the Recombinant DNA Advisory Committee (RAC), experiments in which organisms are constructed by recombinant DNA and in which the entire source of DNA is from the same organism are exempted from the guidelines, and should be considered very low risk for purposes of environmental dissemination. Thus, just as natural evolution has produced strains of BT harboring several insecticidal toxin genes on two or more plasmids, *in vitro*-constructed BT strains containing novel combinations of such genes on BT-derived cloning vectors should not be considered to be inherently different. History will judge whether the acceptance of recombinant products will proceed quickly or whether the process will be long and arduous.

A second potential limitation of the delivery of BT insecticidal activities via either other microbial hosts or directly into plants is the ability of these alternative delivery systems to express acceptable levels of the insecticidal delta-endotoxin proteins. It is quite clear that the expression of these genes in BT is very high, leading to 25-35% of the total sporulating cell protein. It has also been shown that non-BT hosts, such as *E. coli*, utilize different and much weaker transcriptional promoters of the BT toxin genes, so that expression levels are very much lower (22). Thus, in organisms such as pseudomonads and other non-BT bacterial hosts, it will undoubtedly be necessary to fuse the BT genes to high-expression promoters to obtain acceptable levels of expression. In the case of direct expression in plant genomes, the problem of expression may be further complicated by the properties of the protoxin, which in BT is aggregated into the typical crystalline inclusion bodies. Expression of a full-length BT toxin gene may produce inclusions that could literally "choke" a plant's vascular system. In fact, the reports of successful expression of BT toxin genes in plants have utilized truncated genes lacking the 3' end (23, 24) which may help to insure that the insecticidal product remains in a soluble state, more akin to the activated toxin. A further consideration of the effectiveness of delivering the BT insecticidal activity in the plant is that of developmental regulation, i.e., expression of the activity in those cells and tissues where it will have the greatest effect. It would be of little value, for example, to have high levels of expression of a Heliothis-active gene in cotton roots and little or no expression in the leaves and fruiting structures where the insects cause their damage. While progress is being made in identifying tissue specific promoters in plant systems, the proper coupling of the promoters to the insecticidal genes may not be straightforward, nor is there any guarantee that tissue-regulated expression of such genes will not lead to undesirable effects on crop yield or other parameters.

CONCLUSIONS

What are the most likely possibilities for the delivery of BT insecticidal activities in the foreseeable future? In this author's opinion the greatest opportunity is in the development of new, genetically-manipulated strains of BT applied to crops, forests, and for dipteran insect control by conventional application methods. Our experience at Ecogen has shown that new strains can be constructed that have substantially increased potencies compared to existing products. The current registration standards for BT insecticides (Table III) are straightforward and the toxicological testing, as long as no unexpected problem arises, is confined to Tier I levels. Each test can be completed in 30 days or less, and the total cost is modest. Given the potentially significant limitations, both technical and regulatory, in the application of BT insecticidal activities via alternative delivery systems, I see the greatest opportunities for product expansion over the next several years to be in the form of BT itself. The major question that I perceive in this regard is whether recombinant strains will constitute a significant fraction of the new products. Regardless, future prospects for the expansion of BT and other biological pesticide products to a significant position in the pest control market are very optimistic.

Table III

Current Registration Standard for BT Insecticides

- Data Requirements
  - Tier I mammalian toxicology limits
  - Use pattern requirements
  - Non-target organisms
    - Avian acute oral
    - Freshwater fish
    - Freshwater aquatic invertebrates
    - Plants
    - Beneficial insects
    - Estuarine and marine animals (for dipteran applications by var. israelensis)
  - Product Composition
    - Active ingredients
    - Inerts
- Freedom from beta-exotoxin
- No ground water concerns
- Exemption from tolerance requirements
- Isolated restrictions to comply with endangered species requirements.

Literature Cited

1. Georghiou, G. P. In Pesticide Resistance: Strategies and Tactics for Management; National Academy Press; Washington, D.C., 1986, pp. 14-43.
2. Dulmage, H. T. J. Inv. Path. 1970, 15, 232-239.
3. de Barjac, H. In Microbial Control of Pests and Plant Diseases 1970-1980; Academic Press: New York, 1981, Chapter 3, pp. 35-43.
4. Krywienczyk, J.; Angus, T. A. J. Inv. Path. 1967, 9, 126-128.
5. Baumann, L.; Okamoto, K.; Unterman, B. M.; Lynch, M. J.; Baumann, P. J. Inv. Path. 1984, 44, 329-341.
6. Carlton, B. C.; Gonzalez, J. M., Jr. In Molecular Biology of Microbial Differentiation; Hoch, J. A., Setlow, P., Eds.; American Society of Microbiology: Washington, D. C., 1985, pp. 246-252.
7. Krieg, V. A.; Huger, A. M.; Langenbruch, G. A.; Schnetter, W. Z. ang. Ent. 1983, 96, 500-508.
8. Herrnstadt, C.; Soares, G.; Wilcox, E.; Edwards, D. Bio/Technology 1986, 4, 305-308.
9. Yamamoto, T.; Iizuka, T. Arch. Biochem. Biophys. 1983, 227, 233-241.
10. Yamamoto, T.; McLaughlin, R. E. Biochem. Biophys. Res. Comm. 1981, 103, 414-421.
11. Carlton, B. C.; Gonzalez, J. M., Jr. In The Molecular Biology of the Bacilli, Vol. II; Dubnau, D. A., Ed.; Academic Press: New York, 1985; Chapter 8, pp. 211-249.
12. Gonzalez, J. M., Jr.; Carlton, B. C. In Genetic Exchange: A Celebration and a New Generation; Streips, U. N., et al., Eds.; Marcel Dekker: New York, 1982; pp. 85-95.
13. Chapman, J. S.; Carlton, B. C. In Plasmids in Bacteria; Helinski, D. R., et al., Eds.; Plenum Publishing Corp.: New York, 1985; pp. 453-467.
14. Gonzalez, J. M., Jr.; Dulmage, H. T.; Carlton, B. C. Plasmid 1981, 5, 351-365.
15. Gonzalez, J. M., Jr.; Brown, B. J.; Carlton, B. C. Proc. Natl. Acad. Sci. USA 1982, 79, 6951-6955.
16. Lereclus, D.; Lecadet, M.-M.; Ribier, J.; Dedonder, R. Mol. Gen. Genet. 1982, 186, 391-398.
17. Kronstad, J. W.; Schnepf, H. E.; Whiteley, H. R. J. Bact. 1983, 154, 419-428.
18. Whiteley, H.R.; Schnepf, H. E.; Kronstad, J. W.; Wong, H. C. In Genetics and Biotechnology of Bacillus; Ganesan, A.T. and Hoch, J. A., Eds.; Academic Press: New York, 1984; pp. 375-386.
19. Whiteley, H. P.; Schnepf, H. E. In Annual Review of Microbiology; Ornston, L. N., et al., Eds.; Annual Reviews: Palo Alto, 1986; pp. 549-576.
20. Donovan, W. P.; Dankocsik, C. C.; Gilbert, M. P.; Gawron-Burke, M. C.; Groat, R. G.; Carlton, B. C. J. Biol. Chem. 1988, 263, 561-567.
21. Committee on the Introduction of Genetically Engineered Organisms into the Environment. Introduction of Recombinant DNA-Engineered Organisms into the Environment: Key Issues. National Academy Press, Washington, D.C., 1987.

22. Wong, C. W.; Schnepf, H. E.; Whiteley, H. R. J. Biol. Chem. 1983, 256, 1960-1967.
23. Fischhoff, D. A.; Bowdish, K. S.; Perlak, F. J.; Marrone, P. G.; McCormick, S. M.; Niedermeyer, J. G.; Dean, D. A.; Kusano-Kretzmer, K.; Mayer, E. J.; Rochester, D. E.; Rogers, S. G.; Fraley, R. T. Bio/Technology 1987, 5, 807-813.
24. Vaeck, M.; Reynaerts, A.; Hofte, H.; Jansens, S.; De Beuckeleer, M.; Dean, C.; Zabeau, M.; Van Montagu, M.; Leemans, J. Nature 1987, 237, 33-37.

RECEIVED June 1, 1988

Chapter 20

Transgenic Crop Varieties Resistant to Insects

Mark Vaeck, Arlette Reynaerts, Herman Höfte, and Herman Van Mellaert

Plant Genetic Systems N.V., J. Plateaustraat 22, B–900 Ghent, Belgium

The bt2 gene from Bacillus thuringiensis berliner 1715 encodes a 130 KDa protein that is highly toxic to lepidopteran larvae. The active toxin is a 60 KDa polypeptide, derived from the NH2-terminal half of the molecule. Truncated genes encoding the active toxin and gene fusions containing 5'-fragments of bt2 fused to the neomycin phosphotransferase gene (neo) of Tn5 were constructed. The latter encode stable fusion proteins which are toxic and which exhibit specific NPTII enzyme activity. Chimeric genes consisting of these modified toxin genes, flanked by regulatory sequences, were transferred to tomato, potato and tobacco plants, using Agrobacterium vectors. Transgenic plants expressed sufficiently high levels of active toxin to be protected from feeding damage caused by insect larvae. The insect resistance trait was stably inherited in subsequent generations.

Commercial preparations of B.t., obtained through conventional fermentation techniques, have been used for more than two decades as biological insecticides (1). They exhibit desirable properties such as high insect toxicity and environmental safety. B.t. does not affect non-target insects and is completely nontoxic to vertebrates. Nevertheless its use has been limited due to high production costs, limited stability in field conditions and a too narrow insecticidal spectrum. The insecticidal activity of B.t. relies in the crystalline inclusions which are produced upon sporulation. The crystals contain insecticidal proteins, delta-endotoxins, which affect the midgut epithelium of sensitive insects. The exact mechanism of their toxic activity is still unknown.

0097–6156/88/0379–0280$06.00/0

Most B.t. strains are selectively toxic towards a variety of lepidopteran larvae. One B.t. subspecies, called israelensis, is highly toxic to dipteran larvae (2) and is used for the control of mosquitoes and blackflies. Recently, a B.t. subspecies tenebrionis, expressing toxicity towards Coleoptera, has been isolated (3).

With the advent of biotechnology, interest in B.t. has greatly increased. Using new plant vector systems, foreign genes can be transferred and expressed into plants. In this paper we present some data on the succesful introduction of B.t. genes in different crop species. The B.t. genes are expressed at insect controlling levels and provide the transgenic plants with a defense mechanism against devastating insects.

Results and Discussion

Expression of B.t. Genes in Transgenic Plants. We have used Agrobacterium mediated T-DNA transfer to express several modified genes, derived from the Lepidoptera specific Bt2 gene in tomato, tobacco and potato. Particularly successful was the expression of genes that contained the NH2-terminal half of bt2 fused to the neo gene. These encode fusion proteins that exhibit both insect toxicity and neomycin phosphotransferase activity. The latter was used to select, through selection for high kanamycin resistance, cells that expressed substantial amounts of insecticidal protein (4).

High toxicity resulting in 80-100% mortality of M. sexta larvae, was observed in tobacco plants expressing fusion proteins or a truncated bt2 gene (Table I). None of the plants transformed with the full length bt2 gene produced insect killing activity. Also tomato and potato plants transformed with a fusion gene or a truncated bt2 gene exhibited significant insecticidal activity. Greenhouse experiments revealed that the obtained transgenic plants were protected against insect feeding damage. They showed very limited damage, restricted to feeding areas of a few mm^2, whereas nontransformed control plants were entirely consumed within 10 days. B.t. protein levels in the insect resistant plants ranged from 7 to 40 ng/mg total protein content. The data are summarized in Table I.

Inheritance of the Insect Resistance Trait. Copy numbers or the T-DNA inserts in transgenic plants were determined using Southern blotting. Plants expressing highest levels of B.t. protein contained around 5 copies of T-DNA. From one such plant, 15 F1 progeny were assayed and they all exhibited 100% insect killing.

We also analysed inheritance of the new trait in plants with one or two T-DNA inserts. Km resistance and insecticidal activity were linked and the inheritance patterns were fully compatible with simple Mendelian rules for a dominant trait.

Selection of Insecticidal Genes. Although in the presently generated transgenic plants, B.t. protein levels are sufficiently high to control M. sexta, other insect species,

Table I. Insect killing activity in transgenic plants expressing B.t. genes

Bt gene	Ti plasmid	Percentage of plants causing a mortality of 0-40%	40-80%	80-100%
TOBACCO				
Bt:NPTII fusion	pGS1152	24	20	56
Bt2 intact	pGS1161	100	0	0
Bt2 truncated	pGS1163	37	3	60
POTATO				
Bt:NPTII fusion	pGS1152	18	38	44
Bt2 truncated	pGS1163	54	30	16

Mortality in first instar _M. sexta_ larvae is determined in a six day feeding assay

The recombinant Ti plasmids are described in Vaeck et al., 1987.

which are less sensitive to Bt2 toxin, may need higher levels of expression to be effectively controlled. Or alternatively, one may consider the use of other B.t. toxins more active against such insects.

Even within the lepidopteran pathotype, large differences in the insecticidal spectra of _B.t._ toxins exist. For example, screening of a large collection of _B.t._ strains has lead to the identification of strains highly toxic to _Spodoptera littoralis_, an insect which is insensitive to most commonly known _B.t._ strains.

On the basis of insecticidal spectrum and size of the toxic polypeptide fragment, we presently defined 3 types of lepidopteran-active B.t. toxins.

Bt2 is an example of a type 1 delta-endotoxin. It has a relatively broad spectrum of insecticidal activity, but is nearly nontoxic to _S. littoralis_. A type 2 toxin is found in pure form in the crystals of _B.t. thuringiensis_ strain 4412 and is toxic to _Pieris brassicae_ but not to _M. sexta_ and _S. littoralis_. An example of a type 3 toxin is present together with a type 1 toxin in _B.t. aizawai_ and is characterized by high toxicity to _S. littoralis_.

Specific monoclonal antibodies were generated against these three types of delta-endotoxins. These antibodies provide an efficient tool for very rapid immunological screening of large numbers of B.t. strains.
In a random set of 28 B.t. strains tested with monoclonal antibodies specific for type 3 toxin, only 8 showed positive reaction. In bioassay only the same 8 strains, and only these, exhibited toxicity to S. littoralis. Thus a perfect correlation exists between the antigenic structure of B.t. toxins and their insecticidal spectrum.

This approach is now further refined and will be extended to the selection and cloning of various new B.t. genes. This will be an essential step in the process towards the engineering of resistance against whole insect complexes in agriculturally important crop species.

Literature Cited

1. Dulmage, H.T. Insecticidal activity of HD-1, a new isolate of Bacillus thuringiensis var. alesti. J. Invertebr. Pathol. 1970, 15, 232-239.
2. Goldberg, L.J. and Margalit, J. A bacterial spore demonstrating rapid larvicidal activity against Anopheles sergentii, Uranotaenia unguiculata, Culex univitattus, Aedes aegypti, Culex pipiens. Mosquito News., 1977, 37, 355-358
3. Krieg, A.; Huger, A.M.; Langenbruch, G.A.; Schnetter, W. Bacillus thuringiensis var. tenebrionis : ein neuer gegenüber Larven von Coleopteren wirksamer Pathotyp, Z. Ang. Ent., 1983, 96, 500-508.
4. Vaeck, M.; Reynaerts, A.; Höfte, H.; Janssens, S.; De Beuckeleer, M.; Dean, C.; Zabeau, M.; Van Montagu, M.; Leemans, J. Transgenic plants protected from insect attack, Nature, 1987, 327, 33-37.

RECEIVED April 1, 1988

Chapter 21

Development of Genetically Engineered Microbial Biocontrol Agents

Frederick J. Perlak, Mark G. Obukowicz, Lidia S. Watrud, and Robert J. Kaufman[1]

Monsanto Company, 700 Chesterfield Village Parkway, St. Louis, MO 63198

Genetically engineered microbial biocontrol agents were developed by the insertion of a gene that produces a insecticidal protein, into two strains of plant-colonizing bacteria. The emphasis of the design of these biological agents were specificity for target insects, protection of the plants colonized by the engineered bacteria, and a low probability to transfer the engineered genetic information to other organisms in the environment. The procedures and technologies used to produce these microbial control agents are reviewed.

During the 1960's concern over widespread use of pesticidal chemicals in the environment spurred an increase in the use of biological control agents in agriculture. The perceived advantages of biological control are human safety, high pesticidal activity coupled with safety for beneficial organisms and a reduction in the chemical load on the environment. Biocontrol agents meet these requirements for selectivity and specificity. Bacteria and viruses pathogenic for insects are used as natural pesticides, although only fourteen have been registered for pesticidal use in the United States and only one, *Bacillus thuringiensis*, has achieved modest commercial usage (1). The reasons for this lack of success are as varied as the organisms tested but historically the chronic drawbacks have been: (a) a lack of unit activity requiring high doses or combinations with chemicals, (b) too great a degree of specificity necessitating mixtures with other insecticides and (c) insufficient residence time of the microbial at the site of the pest problem resulting in the need for multiple applications.

We have applied the principles and technologies of molecular biology to the development of a model microbial pesticide to overcome some of the shortcomings of current biological control agents. Since plants are covered with bacteria, (e.g. *Pseudomonas*, and *Erwinia* species), genetic engineering of pesticidal traits into these organisms could protect the plant by solving the localization, unit activity and persistence problems of the classical biocontrol agents.

[1]Current address: Hemagen Associates, 11444 Lackland Avenue, St. Louis, MO 63146

0097–6156/88/0379–0284$06.00/0

In this paper, we describe the reasoning and methods used in the development of a model genetically engineered microbial biocontrol agent. The evolution of this insecticidal biocontrol agent is followed from the introduction of a cloned insecticidal protein gene into a carefully selected commensal bacteria indigenous to the environment. Methods are described which stably integrated the cloned gene into the chromosomes of these micro-organisms in a manner which minimized the potential for genetic exchange with other microbes in the environment.

Components of a Genetically Engineered Microbial Biocontrol System.

The genetic engineering of a microbial insecticide requires three components: a gene that encodes an insecticide, a plant colonizing microbe, and the technology to transfer and express the gene in the recipient colonizing microbe. First, a gene from a lepidopteran-active strain of *Bacillus thuringiensis* var. *kurstaki* HD-1 (*B.t.*) was isolated. *B.t.* is a spore-forming soil bacterium that produces parasporal crystalline inclusions toxic to lepidopteran larvae (caterpillars of butterflies and moths; 2). The insecticidal activity of this particular *Bacillus thuringiensis* strain is characterized by its specificity and potency for lepidopterans. Second, several plant colonizing bacteria were identified as suitable hosts for the *B.t.* gene. Finally, the technology was developed to insert the gene into the chromosome of the plant colonizing microbes. This minimized the potential for genetic exchange of the engineered gene from the plant colonizing microbe into other bacteria in the environment while maintaining a level of expression sufficient to provide insect pest control.

Gene Selection and Isolation. The gene chosen for the development of a model microbial pesticide system was the gene from *B.t.* encoding the delta-endotoxin (*B.t.* protein). This protein is the active insecticidal ingredient in *B.t.* and is produced during sporulation. It forms a parasporal crystal and is released into the medium upon cell lysis. The parasporal crystal has no biological activity until it is ingested, solubilized and activated by gut enzymes of sensitive lepidopteran larvae to the ultimate toxin. Once activated, the toxin protein destroys cells of the gut epithelium causing the cessation of feeding, paralysis and death (2).

Several groups (3-5) have cloned *B.t.* protein genes from a variety of lepidopteran-active species of *B. thuringiensis*. The gene described in this study was isolated from a library of *Bam*HI digested *B.t.* plasmid DNA inserted into pBR328. Southern hybridization with several synthetic oligonucleotides based on published sequence (3) identified a clone containing a 16 kilobase pair (kb) fragment. This gene has been sequenced (6) and is very similar to a number of reported genes encoding lepidopteran control proteins (7-9). *E. coli* cells containing plasmids carrying the gene produced a high molecular weight protein (134 kD), cross-reactive with antibody made to *B.t.* protein. Insect bioassays using *Manduca sexta* showed that the product of the cloned gene retained a high level of insect toxicity.

Identification of Plant Colonizing Microorganisms. Several criteria were established for the selection of a plant colonizing microbe as an appropriate host for the expression of the insecticidal gene. Essential characteristics include a lack of pathogenicity to plants and humans, effective association with plant tissues, the ability to colonize new tissue during plant growth and development, a lack of indigenous plasmids, sensitivity to antibiotics, a limited potential for genetic exchange and limited environmental persistence after the growing season.

Two strains of *P. fluorescens*, 112-12 and Ps3732, which were isolated from the roots of corn, met the above criteria. To mark these strains, mutants with spontaneous resistance to rifampicin and nalidixic acid were selected in the laboratory (112-12RN and Ps3732RN). These bacteria showed rapid root colonization of corn plants grown in field soil in growth chambers to levels as high as 10^7 CFU (colony forming units) per gram fresh weight of corn roots. These results were repeated when these micro-organisms were tested for colonization under field conditions. An area of farm soil was sprayed with these bacteria (approximately 10^6 CFU per square foot), planted with corn, and the roots of the plants were analyzed for the number of marked bacteria found in the corn rhizoplane. The same rhizoplane analysis was done on plants in the field grown from seeds which had been coated with either 112-12RN or 3732RN (10^6 bacteria per seed). In both experiments, the micro-organisms colonized the corn roots to the same high levels seen in the growth chamber experiments. The level of these bacteria fell, after overwintering, below detectable levels, (20 CFU/g root tissue), on corn planted in the same areas (Unpublished results; M. Miller-Wideman, 3 seasons experiments).

P. fluorescens is a species of bacteria not considered to be a vertebrate pathogen. This was confirmed by subsequent toxicological testing. These isolates lack indigenous plasmids and do not support conjugative plasmids limiting the potential of these strains to transfer and spread engineered traits into other soil or root colonizing microorganisms. Based on these criteria these isolates were ideal choices for genetic engineering.

Introduction of the *B.t.* Insecticidal Protein Gene into Ps3732RN and 112-12RN. Conjugative plasmids of several broad host range incompatibility groups (RP4,IncP; N3, IncN; and pSa,IncW) were found not to stably replicate in Ps3732RN and 112-12RN. Plasmids of the IncQ incompatibility group did replicate in these strains and offered the most expedient approach to introduce the gene. A plasmid developed at Monsanto, pMON5008, (D. J. Drahos, Biological Sciences Staff), was used as the vector. A derivative of pKT230 (10), it contains a gene for resistance to kanamycin (Km^r). In the laboratory, it can be mobilized by a special helper plasmid, pRK2013 (11), into many different bacterial hosts and it contains a *Bgl*II restriction site in a non-essential region. A 4.6 kb *Hpa*I-*Pst*I DNA fragment containing the *B.t.* insecticidal protein gene (*B.t.* gene) was altered to contain flanking *Bam*HI sites (pMAP8;5). A *Bam*HI fragment of pMAP8 containing the *B.t.* gene was inserted into the *Bgl*II site of pMON5008 making use of the ease of ligating *Bam*HI and *Bgl*II cleaved DNA to each other. The pMON5008 derivative with the *B.t.* gene was designated pMAP15.

Triparental matings (11) were used to introduce the plasmids containing the gene into the two selected root colonizers. In this system, 3 strains of bacteria are mixed, concentrated on a filter and incubated on a non-selective agar medium for a period of time suitable for genetic exchange to occur via bacterial conjugation. One was an *E. coli* strain containing the mobilizing plasmid pRK2013, a plasmid with IncP transfer functions. This plasmid can move the IncQ plasmid containing the gene, pMAP15, from another *E. coli* strain into a third non- *E. coli* strain. In this case, the third strain was the recipient, either Ps3732RN or 112-12RN. The desired product of these matings, strains Ps3732RN and 112-12RN harboring pMAP15, were distinguished by selection for the characteristics of the both the donor (Km^r) and the recipient (rif^r), which would only be found in the desired transconjugants. The presence of pMAP15 in Ps3732RN and 112-12RN was determined by plasmid analysis and expression of the *B.t.* gene was confirmed by western blots (immunoassay) and by insect bio-assays.

Although the plasmid containing the *B.t.* gene, pMAP15, provided an efficient means for introducing the gene into these Pseudomonas strains for expression in the laboratory, certain features made it less suitable for pesticidal use. This plasmid and other similar plasmids were unstable in the absence of antibiotic selection, particularly in strain Ps3732RN (about 1% plasmid-free segregants per cell per generation). Such instability is undesirable for field testing and eventual widespread commercial use. In addition, these plasmids could potentially be mobilized by conjugative plasmids likely to be present in the environment. To overcome the problem of plasmid instability and increase containment, procedures were developed for insertion of the gene into the chromosome of our root colonizing bacteria.

Chromosomal Integration of the *B.t.* Gene Using Transposon Tn*5*.

Transposons are ideal vectors to introduce heterologous genes into the chromosome of a soil isolate. It was important to determine if the expression of the *B.t.* gene integrated into the chromosome of a soil isolate would be stable and at high enough levels to retain insecticidal activity. Tn*5* was chosen for the initial insertion of the *B.t.* gene into the chromosome of 112-12RN and Ps3732RN. It can insert into numerous sites in the chromosomes of a wide range of bacterial species (12). Resistance genes for 3 antibiotics (kanamycin, bleomycin and streptomycin) are found in a central region of Tn*5*, flanked by insertion sequences IS50L and IS50R (Figure 1).

The *B.t.* gene (from pMAP8) was inserted into the *Bam*HI site in the central region of Tn*5* creating Tn*5*-*B.t.* (13). Earlier gene mapping had shown that cloning into this site would not impair transposition (a frequency of about 10^{-5} with overnight matings into Ps3732RN and 112-12RN) or the expression of Km^r. Experiments showed that the Tn*5*-*B.t.* construct transposed from the delivery vector into Ps3732RN and 112-12RN at a frequency of 10^{-5} after overnight matings. Southern blot hybridizations established that each Km^r transconjugant contained just one copy of Tn*5*-*B.t.*, that many different sites of

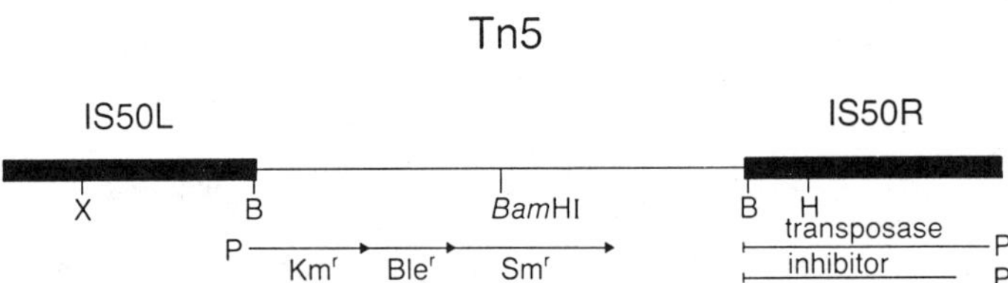

Figure 1. Map of the transposon Tn*5*. The insertion sequences IS50L and IS50R flank the central region which contains the genes for resistance to kanamycin (Km^r), bleomycin (Ble^r), and streptomycin (Sm^r). The P refers to the position of the promoters for expression of antibiotic resistance, transposase and transpoase inhibitor (on IS50R). The restriction sites are marked X=*Xho*I, B=*Bgl*II, and H=*Hind*III.

insertion had been used, and that all other sequences of the delivery vector plasmid had been lost.

Quantitative immunological analysis indicated that the insecticidal protein toxin constituted approximately 1% of the total protein in 112-12RN and Ps3732RN when the *B.t.* gene was in the same orientation as the Km^r gene. Placement of the *B.t.* gene in the opposite orientation to the Km^r gene resulted in lower levels of protein production, estimated to constitute 0.1% of the total protein of the cell.

Table I. Levels of the *B.t.* protein produced in several strains of bacteria with different constructs. Amounts reported are estimates based in several different quantitative immunological techniques

Construct	Strain	*B.t.* Protein as % of Total Protein
pMAP15	*E. coli* JM101	10
	112-12RN	2.0
	Ps3732RN	0.5
Tn5-*B.t*.[a]	112-12RN	1.4
	Ps3732RN	1.0
IS50L-*B.t*.[a]	112-12RN	2.0
	Ps3732RN	2.0
Tn5-*B.t.*[a]*	112-12RN	1.0
	Ps3732RN	1.0

[a] Refers to constructs inserted into the chromosome
* Constructs inserted by homologous recombination (see text)

Elimination of the Transposase Functions of Integrated Tn5-*B.t.* Elements. In principle, the Tn5-*B.t.* element inserted into the chromosome has retained the ability to transpose to other sites in the chromosome or to an incoming plasmid. This concern was lessened by findings that the root colonizing bacteria described here do not maintain plasmids of the most common broad host range incompatibility groups (see Table I). Two different approaches were used to engineer these bacteria to reduce the possibility of transposition dependent spread of the inserted *B.t.* gene. The first approach utilized the transposition of a portion of Tn5 (IS50L) with the *B.t.* gene, without inserting the entire transposon. The second approach utilized homologous recombination between a Tn5-*B.t.* element with a substantial deletion in the transposase and a Tn5 previously inserted in the chromosome.

The first approach depended on the fact that Tn5, depicted in Figure 1, is a composite element whose terminal inverted repeats, IS50L and IS50R, are insertion sequences. Each one is capable of transposition as a separate entity as well as in unison. The movement of a single IS50 element entails the recognition of segments about 19bp long at each end of IS50 by the transposition machinery; these sites are designated O and I to reflect their positions at the outside and inside ends of each IS50 element in Tn*5* (14). Transposition also requires transposase, a 475 amino acid protein encoded by IS50R. The IS50L element differs from IS50R at one position resulting in an ochre stop codon in IS50L (UAC→UAA;15). This terminates translation prematurely resulting in an inactive form of the transposase. The IS50L was modified to insure that the ochre codon of IS50L will not revert or be suppressed. A +4 base pair insertion causing a +1 frame shift was introduced in the defective transposase gene of IS50L at the *Xho*I site. IS50L can transpose independently because the transposase defect of IS50L is complemented by the transposase encoded by the nearby IS50R element. Transposition is inefficient when the transposase is supplied *in trans* (on another plasmid for example) or when the transposase gene *in cis* is separated from IS50L by long distances (12). Insertion of the *B.t.* gene into the Pseudomonas genome as part of IS50L effectively eliminates transposition ($<10^{-12}$) even in the event of invasion by a plasmid or temperate phage containing Tn*5*-related sequences.

A number of recombinant DNA manipulations were required to prepare the IS50L element as the appropriate delivery vector for insertion of the *B.t.* gene (IS50L-*B.t.*) into the chromosome of the two plant-colonizing bacteria. In addition to the *B.t.* gene, IS50L was engineered to carry the Tn*5*-encoded Km^r gene within it's boundaries. The resulting plasmid of these manipulations, pMAP517 (16), is an 18 kb derivative of the suicide vector pSUP1011 (17). The only copy of the Km^r gene is located within IS50L and was utilized to track the transposition of IS50L. Transposase was provided by an intact IS50R element in pMAP517.

Transposition of IS50L-*B.t.*, depicted in Figure 2, was selected by Km^r. Southern hybridizations were used to eliminate Km^r pseudomonads resulting from other possible transposition events (16,18). Approximately, 5% of the Km^r transconjugants corresponded to the desired transposition products; insertions of single copies of the transposase defective IS50L-*B.t.* element into a variety of sites in the pseudomonad genomes. Immunological analysis indicated that a *B.t.* protein of the expected 134 kD size was produced as approximately 1% of the total soluble cell protein in strains 112-12RN and Ps3732RN.

Integration of the *B.t.* Gene by Homologous Recombination. The direct transposition of the IS50L-*B.t.* element into the chromosome of the root-colonizers is a convenient approach for the chromosomal integration of the *B.t.* gene or any other gene. Its limitations include the requirement for a selectable marker such as Km^r and the inability to predict where the element will be inserted. The many different transcription units into which insertion can occur complicates comparisons among various promoters in the evaluation and manipulation of *B.t.* gene expression. These limitations were overcome using homologous recombination between a remnant of Tn*5* flanking the *B.t.* gene and a previously characterized Tn*5* insertion in the chromosomes of Ps3732RN and 112-12RN (19). A substantial deletion (3kb; from the *Bgl*II site in IS50L to the *Hind*III site in IS50R) of Tn*5* was generated which eliminated the drug resistance genes and 324 bp of the

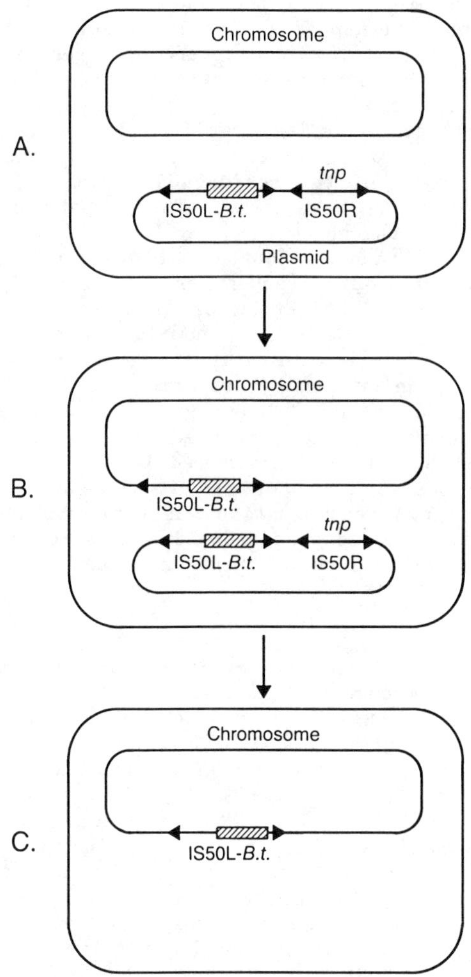

Figure 2. Integration of the IS50L-*B.t.* element into the bacterial chromosome. A. A plasmid carrying IS50L-*B.t.* and IS50R, enters the cell via conjugation. B. The IS50L-*B.t.* element transposes into the chromosome by *in cis* complementation mediated by IS50R. C. Only the IS50L-*B.t.* remains in the cell.

transposase gene of IS50R (19). The *B.t.* gene was cloned into the remaining segment of Tn*5*. Homologous recombination between the segments of Tn*5* flanking *B.t.* and the Tn*5* in the chromosome involved three steps. I. Transposition of Tn*5* (containing a frame shift in IS50L) to the chromosomes of Ps3732RN and 112-12RN and the identification of prototrophic transposition derivatives. II. Introduction by conjugation of pMAP111 (plasmid with the *B.t.* gene flanked by IS50 fragments) into the the strains created in step I. This plasmid cannot transpose on its own and carries a gene for tetracycline resistance. By selecting for tetracycline (Tc) resistant transconjugants an expected result is a single reciprocal crossover between one of the IS50 elements of the resident Tn5 element, and remaining IS50 sequences flanking the *B.t.* gene. This would cause addition of the entire plasmid into the chromosome at the site of the preexisting Tn5 element (Figure 3;A,B). Southern blots confirmed that the co-integrates, recovered at frequencies of about 10-6, had the expected structure. III. Selection of Tc sensitive, Km sensitive derivatives resulting from a second homologous crossover in IS50 sequences, and loss of the vector plus the central region and functional transposase gene of Tn*5* (resolution of the co-integrate). These resolution products were recovered at frequencies of 10^{-3} to 10^{-4} in young cultures, and the expected structure was verified by appropriate Southern blot hybridizations.

Expression of the *B.t.* protein in strains with the *B.t.* gene integrated by homologous recombination was confirmed by Western blot analysis. The strains tested produced a 134 kD protein which retained toxicity to susceptible insects (*Manduca sexta*).

Discussion

The choice of the active insecticidal ingredient in this genetically engineered microbial biocontrol agent was carefully considered. The *B.t.* gene of *B. thuringiensis* was incorporated as part of the system because the parent organism (*B.t.*) is currently used as a natural microbial biocontrol agent, the protein toxin has a high unit of activity requiring very little material for toxicity, and because the *B.t.* protein is specific to lepidopteran insects with a lack of toxicity to other organisms.

A second key aspect of our strategy was the choice of the organism to be used as a recipient of our insecticidal gene. The organisms selected were isolated from corn plants, lacked pathogenicity to humans, plants, and insects and are common in the corn root environment. Both organisms, 112-12 and Ps3732, were characterized as *Pseudomonas fluorescens* by standard criteria. This group of bacteria are not considered to be pathogenic. Our toxicology and plant hypersensitivity testing confirmed the absence of pathogenicity of these strains for humans, plants and other organisms tested.

The concern for the potential of genetic exchange from the microbial pesticide to other organisms was reflected in the evolution of our approaches in introducing the *B.t.* gene into the commensal root-colonizing organisms. Initial prototypes had the *B.t.* gene inserted within a plasmid, which could be mobilized by bacterial conjugation into other organisms. In addition, the plasmids containing the *B.t.* gene were shown to be unstable. The practical conse-

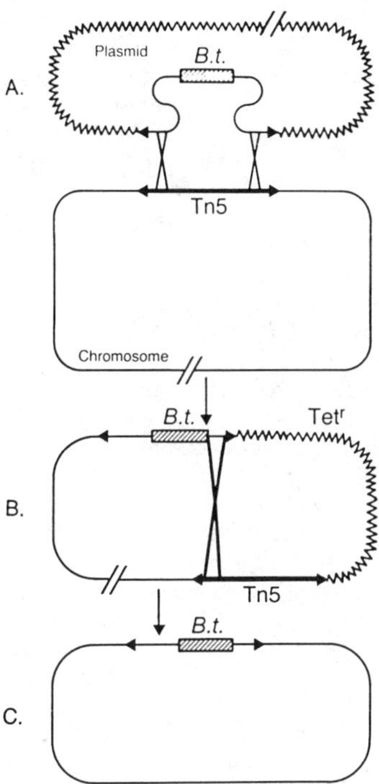

Figure 3. Integration of the *B.t.* gene by homologous recombination. A. Integration of the plasmid carrying *B.t.* by a single reciprocal crossover between the integrated Tn5 (chromosome) and the region of Tn5 flanking the *B.t.* gene. B. Second homologous crossover in the IS50 sequences recombining out the plasmid sequences. C. The resolution of the homologous recombination resulting in *B.t.* integration.

quence of plasmid stability would be decreased protection of the target area resulting in decreasing efficacy as an insecticide. The next generation of microbial pesticides contained the *B.t.* gene, inserted via Tn*5*, into the chromosome of our root-colonizers. Integration stabilized expression and dramatically reduced the potential for transfer of the *B.t.* gene to other micro-organisms in the environment. However, the presence of an active transposase retains the potential of movement of the gene into other organisms.

Elimination of the transposase functions by the use of the IS50L-*B.t.* system or the homologous recombination system reduces the possibility of the transposition of the *B.t.* gene to such a low level that it is non-detectable in the laboratory (<10-12). Additional manipulations such as the frame-shift of the IS50L reading frame effectively eliminate the possibility of a single mutation resulting in a active transposase.

The IS50L-*B.t.* transposition system has two minor disadvantages. Since each gene integration would be an independent event, integration could occur at numerous sites in the chromosome. Second, this system also requires the use of a directly selectable marker such as Km^r for the efficient identification of IS50L transposition events. However, use of this system is straight forward, is fairly efficient and eliminates concerns of gene movement to other organisms in the environment. An added advantage of this system is that Tn*5* has a wide host range among gram negative bacteria. IS50L mediated gene insertion should be possible among the wide variety of microorganisms which can serve as a host for Tn*5*.

The homologous recombination system offers the most advantages for the stable insertion and expression of the *B.t.* gene into the chromosome of the corn root colonizing micro-organisms. Insertion of the *B.t.* gene by this method did not require the permanent presence of an integrated antibiotic resistance gene. The option of linking the *B.t.* gene with an antibiotic resistance gene or some other selective characteristic (such as β-galactosidase) may be an attractive means of monitoring the movement, persistence and inheritance of the *B.t.* gene in the microbial population.

The use of fragments of Tn*5* flanking the *B.t.* gene in the homologous recombination system was a method of expediency rather than necessity. Theoretically, any region of the bacterial chromosome could be cloned, placed on flanking sides of the gene to be integrated and used to provide a region of homology for recombination into the chromosome. Tn*5* was used because the sequence was known and the appropriate fragments were readily available. The use of Tn*5* does have the added advantage of using the system on a wide variety of hosts. Tn*5* has a wide host range and could readily provide homology for homologous recombination. Cloning a fragment from 112-12RN, for example, to provide homology for recombination in 112-12 may not be suitable for recombination in 3732RN.

Other methods of genetic exchange such as natural DNA transformation, generalized transduction, and conjugative chromosomal mobilization were considered as possibilities for mediating the movement of the *B.t.* gene. These have not been demonstrated to work in organisms characterized as *P. fluorescens* but do occur in other organisms at low frequencies and rely upon homology for success. Assuming that one of the above mentioned mechanisms could theoretically work, the gene would be transferred to a closely related organism in the same environment.

Conclusions

The research presented above describes the first steps in developing a genetically engineered insecticide as a commercial product. The cloning of the *B.t.* gene, introducing the gene into carefully selected commensal bacteria indigenous to the environment which they are intended to colonize and demonstrating their toxicity to specific insects are only the start in our program. To be efficient, these living pesticides must be noninjurious to the host plants and must stably maintain and express the pesticidal gene without transmitting it to other bacteria. We have demonstrated the practical use of the transposable element Tn5 to integrate the *B.t.* gene into the chromosomes of two root colonizing psuedomonads. The use of the Tn5 system is theoretically applicable to other genes in a wide variety of gram negative bacteria. Integration increases the stability of the gene reducing the probability of genetic exchange below a level which can be demonstrated in the laboratory (<10-12). Much developmental research and genetic manipulation are required before the concept of a genetically engineered microbial pesticide will be ready for the marketplace. Different pesticidal genes, systems of genetic introduction and expression of genes may be required. It is important that there is continued development of procedures and techniques for genetically engineering microbes to produce microbial insecticides which are highly contained, easily studied, and applicable to a variety of organisms and situations. These systems will allow researchers the opportunity to confirm the high level of safety and containment indicated by the laboratory studies which have already been conducted.

Acknowledgments

We wish to thank P. Lavrik and B. Reich for protein biochemistry and immunological assay support. Thanks are also extended to T. Stone and P. Marrone for insect bioassays, and to E. Mayer, S. Bolten, and K. Kusano-Kretzmer for technical support.

Literature Cited

1. Klausner, A., Bio/Technology 1984, 2, 408-419.
2. Dulmage, H. T. In, Microbial Control of Pests and Plant Diseases Burges, H. D., Ed.; Academic Press: New York, 1964; pp 193-222.
3. Schnepf, H. E.; Whiteley, H. R. Proc. Natl. Acad. Sci. USA 1981, 78, 2893-2897.
4. Kleir, A.; Fargette, F.; Ribier, J.; Rapaport, G. EMBO J. 1982, 1, 791-799.
5. Watrud, L. S.; Perlak, F. J.; Tran, M.-T.; Kusano, K.; Mayer, E. J.; Miller-Wideman, M. A.; Obukowicz, M. G.; Nelson, D. R.; Kreitinger, J. P.; Kaufman, R. J. In, Engineered Organisms in the Environment Halvorson, H. O.; Pramer, D.; Rogul, M., Eds.; American Society for Microbiology: Washington, 1985; pp 40-46.

6. Fischhoff, D. A.; Bowdish, K. S.; Perlak, F. J.; Marrone, P. G.; McCormick S. M.; Niedermeyer, J. G.; Dean, D. A.; Kusano-Kretzmer, K.; Mayer, E. J.; Rochester, D. E.; Rogers, S. G.; Fraley, R. T. Bio/Technology 1987 (August).
7. Geiser, M.; Schweitzer, S.; Grimm, C. Gene 1986, 48, 109-118.
8. Wabiko, H.; Raymond, K. C.; and Bulla, L. A. DNA 1986 5, 305-314.
9. Hofte, H.; de Greve, H.; Seurinck, J.; Jansens, S.; Mahillon, J.; Ampe, C.; Vandekerchhove, J.; Vanderbruggen, H.; van Montagu, M., Zabeau, M.; Vaeck, M. Eur. J. Biochem. 1986, 161, 273-280.
10. Bagdasarian, M.; Lurz, R.; Ruckert, B.; Franklin, F. C. H.; Bagdasarian, M. M.; Frey, J.; Timmis, K. N. Gene 1981, 16, 237-247.
11. Ditta, G.; Stanfield, S.; Corbin, D.; Helinski, D. R. Proc. Natl. Acad. Sci. USA 1980, 77, 7347-7351.
12. Berg, D. E.; Berg, C. M. Bio/Technology 1983, 1, 417-435.
13. Obukowicz, M. G.; Perlak, F. J.; Kusano-Kretzmer, K.; Mayer, E. J.; Watrud, L. S. Gene 1986, 45, 327-331.
14. Sasakawa, C.; Carle, G. F.; Berg, D. E. Proc. Natl. Acad. Sci. USA 1983, 80, 7293-7297.
15. Rothstein, S. J.; Reznikoff, W. S. Cell 1981, 23, 191-199.
16. Obukowicz, M. G.; Perlak, F. J.; Bolten, S. L.; Kusano-Kretzmer, K.; Mayer, E. J.; Watrud, L. S. Gene 1987, 51, 91-96.
17. Simon, R.; Priefer, U.; Puhler, A.; Bio/Technology 1983, 1, 784-791.
18. Nag, D. K.; Das Gupta, U.; Adelt, G.; Berg, D. E. Gene 1985, 34, 17-26.
19. Obukowicz, M. G.; Perlak, F. J.; Kusano-Kretzmer, K.; Mayer, E. J.; Bolten, S. L.; Watrud, L. S. J. Bacteriol. 1986, 168, 982-989.

RECEIVED May 4, 1988

IMMUNOCHEMICAL APPLICATIONS

Chapter 22

Applications of Immunochemistry in Crop Protection and Biotechnology

Bruce D. Hammock

Departments of Entomology and Environmental Toxicology, University of California, Davis, CA 95616

It is very appropriate that a symposium on immunochemical applications be included in a text on Biotechnology in Crop Protection for a number of reasons. Immunoassays are physical assays yet clearly represent a biotechnology themselves which will play an increasing role in many aspects of crop protection. In addition, immunochemical methods are central to the discovery and analysis of other biotechnology products (1). Rather than repeat other texts in describing how immunochemical applications can be carried out, this symposium was designed to provide a sampling of the great range of immunochemical applications in crop protection today and in the foreseeable future. These applications range from the very practical work of plant diagnostics which is well accepted in many countries, to analysis of classical and genetically engineered pest control agents which has aroused wide interest in industrial and regulatory circles, to the use of immunochemistry in the fundamental research efforts which will lead to new generations of crop protection agents. In many ways it is a pity that immunochemical assays were not extensively utilized in the crop protection field over a decade before. New assay formats have made assays a little easier to use, but the same techniques available in 1970 still provide excellent assays today. However, the final article of this section introduces the concept that there will be a revolution in immunochemical technology as proteins such as antibodies are coupled with solid state electronic and optical devices.

In 1980 an article appeared on the Potential of Immunochemical Technology for Pesticide Residue Analysis (2). As evidenced by the articles in this section, that potential is beginning to be realized. However, of greater importance, we can envision far more applications in the future.

0097–6156/88/0379–0298$06.00/0

Misconceptions, Jargon, and Definitions

A common misconception is that one must be an immunologist to utilize immunochemical technology. This is no more true than the assumption that one must be a physicist to use NMR. Like any other biotechnology, however, immunochemical methods usually are surrounded by layer upon layer of jargon. This verbal insulation, in part, is a natural evolution in any field. However, there also is a commercial reason for the confusing jargon in the immunoassay field. A single antibody pool can be used in a number of different assay formats, each with its own relative advantages with regard to sensitivity, speed, ease of use and other parameters. Usually, development of an antibody pool occurs with the use of well established technology so that one does not have patent protection on the antibody. However, it is possible to patent some assay formats. Thus there is a reward in the commercial sector to make one's assay format (whether patented or not) appear as unique as possible.

Certain assay formats offer clear advantages for a particular application, however all immunoassays are based on the reversible but very specific interaction of a protein antibody with an antigen. This interaction can be described by the law of mass action. To deal with the chapters in this section, one only needs the definition of three specialized terms. First, antibodies are a group of serum proteins that react specifically with an antigen. An antigen is a large molecule (usually a protein) which reacts with an antibody and also is capable of inducing antibody production when injected. Finally, haptens are molecules which can react with antibodies, but which are too small to elicit antibody production unless coupled to a protein carrier.

To use immunoassays one normally employs a reporter substance or label. This label can be one of many materials including a radiochemical, a heavy metal, or commonly, an enzyme. A widely used format is the enzyme linked immunosorbent assay or ELISA. This technology is discussed in the chapters by Cheung, Harrison and Van Vuurde. There are numerous variations on a theme as this assay is adjusted to fit a particular analytical need. However, as an example of the complexity of terminology, one can examine the application of the ELISA system in slightly different formats. For instance when the ELISA is carried out on the surface of a nitrocellulose membrane, it has the clever name of dot blot since the end results appear as tiny dots on a piece of paper. If antigens are separated by electrophoresis before the dot blot is run, one has a Western blot. If the antigen is stained *in situ* one might term it enzyme amplified immunohistochemistry even though the biochemical steps are the same in each case. In the commercial sector this same technology has dozens of names. Thus, when one understands the principle of one immunoassay most others are seen as a simple variation on a theme.

Use of Immunoassay in Plant Diagnostics

Immunoassays have been used in research in the plant diagnostic field for over 25 years (3). Kits for the detection of a variety of plant diseases now are appearing on the market, but regulatory agencies in this country have not placed an emphasis on the use of immunoassays for plant diagnostics or quarantine. As indicated in the article by Van Vuurde and co-workers, immunochemical diagnostics are a major component of the plant protection effort in the Netherlands and in many other European countries. Considering the labor intensive alternatives often in wide use in this country, we certainly could benefit from the experience of our European colleagues with immunodiagnostics. The major impact of immunodiagnostics for plant pathogens in Europe also indicates that the technology is of sufficient maturity to apply to diagnosis of pesticides and to other problems in the crop protection area. A variety of immunochemical diagnostics are used in Europe, but ELISA based assays are of increasing importance. Considering that over 10^6 ELISA's were used in the Netherlands for regulatory detection of plant viruses alone in 1986, this format seems sufficiently mature for it to be used in other crop protection areas. Van Vuurde *et. al.* also bring up a 'low technology' approach to bacterial identification where antibodies are used to enrich a bacterial population before plating. This innovative approach of immunoisolation illustrates another power of antibodies in enriching or purifying a molecule or even an organism by affinity chromatography (4).

Libraries of antibodies are excellent tools for the identification of any organism whether it is a resistant insect, a crop pest, or even the meal of a predator used in biological control. Immunodiagnostic systems are perfect for integration into expert systems for diagnostic applications. Immunodiagnostics and expert systems will greatly improve the quality of routine diagnostic work and will prove a valuable time saver even to the expert. However, immunodiagnostic systems need to be checked by experts familiar with both the pathogen and host. Poorly defined diagnostic kits, especially if not used by trained individuals, could prove misleading as discussed below.

Use of Immunoassay in Pesticide Residue Analysis

The potential of immunochemical technology in pesticide residue analysis has been clear for many years (2), but its actual application is just being realized. The number of laboratories reporting on the technology at the recent IUPAC Congress of Pesticide Chemistry attests to the fact that widespread use of the technology soon will be a reality (5,6,7,8). Of greater importance are the large number of agricultural chemical companies that have major in house efforts. As with HPLC technology only after registrations are granted based on immunoassay will acceptance of the technology become widespread.

Of equal importance are the in house efforts in regulatory agencies in both the classical chemical and biotechnology field. As regulatory agencies develop the in house expertise to evaluate the technology, industry will feel more comfortable in advancing it.

The manuscript by Harrison *et. al.* provides some examples of immunochemical applications to classical residue analysis. It summarizes some of the advantages and limitations of the technology as it applies to the field and provides an outline for development of the technology in house.

Use of Immunoassay in Detection of Genetically Engineered Materials

Most attempts to develop genetically engineered products have concentrated on the production of single gene products which by necessity are peptides and proteins. If these products are to be used in crop protection one needs analytical methods to answer regulatory questions as well as for quality control. There is some question whether one should monitor the gene, the message or the translated product. The answer is simple, at least in early stages of the technology, in that one should have techniques available for all three. Fortunately, the probes for the analytical methods needed probably were already developed during the research leading to the product. A very common way to isolate a message is to use an antibody to the desired protein product to screen an expression library. This screening procedure is another adaptation of the basic ELISA format. For instance, in our laboratory the isolation of the message for insect juvenile hormone esterase by Hanzlik and co-workers was accomplished by immunochemical screening of an expression library (9). The same antibody used to isolate the message can then be used to monitor the protein produced in an expression system as it is developed for pest control. This principle is amply illustrated in numerous chapters of this volume.

A potential problem is that, in an industrial setting, the probes for the analytical method normally would be developed in a molecular biology laboratory. The developers of the technology probably will lack the analytical skills, time and interest needed to reduce immunochemical and hybridization techniques to reliable analytical methods to be used with a variety of matrices. Unless there is a level of appreciation among analytical chemists on how to use these technologies, they are likely to be lost or the analytical duties thrown on molecular biologists rather than placed in an analytical group where they belong. The manuscript by Cheung *et. al.* addresses the problem of immunochemical detection of biologicals using the *Bacillus thuringiensis* species as an example. The exciting developments reported in this text regarding the production of genetically engineered plants as well as microbial insecticides indicate that there will be an immediate need for immunochemical detection of genetically engineered crop protection agents.

Use of Immunoassay in Fundamental Research and Product Discovery

As has been discussed by many workers, we have reached a point of diminishing returns with regard to random discovery of crop protection agents by classical screening methods (10-13). Immunoassays probably will be introduced into the industrial setting for short term goals in biotechnology and analytical chemistry. However, it is likely that the major impact of the technology will be in fundamental research on crop protection.

A number of simplistic applications could result from assays to a group of candidate pesticides early in development. Such assays would be very cost effective in monitoring penetration and translocation. If a molecule is large, there is hope that immunochemical techniques could even be used to localize the compound in the target species and possibly to purify the target.

Longer term work will involve the use of immunochemical probes to explore the comparative biochemistry of a variety of target and nontarget species. Excellent examples of this work are seen in the insect molecular biology section of this book. A clear example has been the use of antibodies raised against mammalian neurohormones as a lead in the isolation of invertebrate hormones. These materials are potential targets to use as insect control agents when produced in inappropriate levels in insects following infection with an appropriate expression vector. The peptide structures also may provide leads for the development of inhibitors of processing enzymes. Industrial research on insect neurohormones is totally justified even if one never develops an insect control agent based on them. Just as probes derived from the structures of mammalian neurohormones have proven useful in work on insect neuroendocrinology (14), invertebrate neurohormones are likely to provide immunochemical and oligonucleic acid probes for use in investigating peptide neurotransmitters in the human central nervous system leading to new generations of pharmaceuticals.

Antibodies to candidate herbicides can aid in translocation studies, and for large molecules can even assist in localizing binding to receptors. Ayers and coworkers in this section provide an excellent illustration of the application of both classical and innovative immunochemical technology in approaching problems with plant disease resistance. Although this work is fundamental, there is a clear path towards exploitation of such research in crop protection. This study clearly illustrates how important immunochemical technology is to the modern biochemist, and analogous examples appear in each section of this text.

Potential of New Assay Formats

Several of the manuscripts in this section demonstrate that even mature immunoassay formats in common use over a decade ago still provide very valuable analytical data. However, it is clear that competing analytical technologies, especially in the mass spectral area, have advanced dramatically. Thus, some of the great advantages over chromatographic assays offered by immunoassay are not as dramatic when compared to modern physical methods as they were previously. In the next decade we will see antibodies and other biological molecules combined with solid state electronic and optical devices in many ways to yield hybrid biological-physical sensors. These so called biosensors are likely to have a dramatic impact on analytical technology.

There can be very simple ways to combine immunochemistry with classical residue methodology. Possibly the simplest example would be to analyze fractions from an HPLC run by immunoassay. However, direct combinations of immunochemical and microelectronic systems offer the possible advantages of real time analysis, greater linearity, and increased sensitivity due to smaller sample sizes. The article by Stanbro *et. al.* explains the operation of one such evolving biosensor technology. This article is exciting in its own right and even more exciting in prophesizing things to come.

Perils of a Successful Technology

Clinical diagnostics have matured to the point where people ask how well an assay performs rather than worrying about the chemical basis of the assay. By contrast in the environmental field there is a concentration on the technique used rather than the results it produces. The environmental field must mature to the point where there are clear criteria for acceptable assay performance regardless of the technology upon which that assay is based.

Immunochemical technology was virtually ignored for many years, but now is being widely advanced as a panacea. Possibly this notoriety is due to pressures on analysts to accept biotechnology approaches or possibly it is our perpetual desire for an easy solution. Recent attempts to legislate the type of assays used by analytical chemists indicate that our field may face a severe problem.

There is no question that immunochemistry will make a major contribution to many aspects of environmental analysis, but there is a severe danger that it will be viewed as a panacea. Many authors have listed the numerous advantages of immunodiagnostics. It is important to realize that some of these advantages may be mutually exclusive. Even though the same antibody pool is used, one format may sacrifice speed for accuracy and another sacrifice some sensitivity for cost. If the concept is advanced that every assay will have all benefits, people will be greatly disappointed with the technology.

As mentioned earlier, immunochemistry is applicable to a great many structures, however, in this case as well, the promise of immunoassay can be oversold. The reversible binding of antibodies to a molecule is based on the summation of a number of weak molecular interactions. Hydrogen bonds are very important since they provide a great deal of binding energy as well as directional and distance specificity. Many of the other bonds formed are even more dependent upon close fit. These binding energies however do indicate that there is a lower limit to the size of the molecule to which one can expect to raise good antibodies. Although it is possible to obtain antibodies to molecules which have only three or four carbons, the likelihood of generating high affinity and very specific antibodies is not great. Since immunoassays must be run in a predominantly aqueous system, lipophicity of the target compound also is important. Solubility of the target compound in an aqueous system seldom is a problem since the compound can be presented in a water soluble cosolvent or as a micelle. However, separating the target compound from a lipophilic matrix can be a nightmare. Such procedures can be particularly intimidating to an immunochemist unfamiliar with handling of lipophilic materials. If one must perform many clean up steps prior to the assay to partially purify a lipophilic compound, then the advantages of immunoassay over chromatographic methods are lost. Thus, highly lipophilic molecules may not be optimum targets for immunoassays. Similarly one should avoid highly symmetrical, water unstable or volatile compounds.

It is not impossible to develop a successful immunoassay to molecules that do not seem to lend themselves to immunoassay. However, as the number of contraindications increase, development of the assay will become increasingly difficult. One would not want to try assay development for a small, lipophilic, volatile, symmetrical molecule as an introduction to the technology. The field faces a problem in that there is regulatory pressure to apply simplistic immunochemistry to molecules that are better analyzed by other technologies. In many cases we will be able to develop assays for these materials. However, the great relative advantages of immunochemical approaches will be lost.

The development of the technology is at a critical stage in that it should be advanced rapidly, but if it is oversold the technology could be seen as failing in key applications. Possibly the best strategy is to treat immunochemistry as simply another analytical method which must be carefully validated before use.

Commercial Exploitation

If immunochemical technology in crop protection is to reach its full potential, it must be advanced in the commercial as well as other sectors (15). The greatest commercial benefit will be in money saved by large chemical and biotechnology companies, money made by products which reach the market faster thanks to immunochemical support, and new products generated in part with immunochemical methods incorporated into research programs.

Considering only immunodiagnostics, it will be important for companies to enter the field where profits will be tied to sales of immunochemical technologies. The overhead involved with large companies probably will be too great for them to make a realistic profit from agricultural immunodiagnostics in the short term and small companies or subsidiaries will pioneer the field. For such biotechnology companies there are several clear markets. Probably the best market is in providing a technological service to major companies who want in house methods capable of monitoring product quality, worker exposure, waste disposal, and other housekeeping chores. There is an existing small market in end user assays for many compounds, but only those where many users will be involved offer an immediate profit. Examples of this would be in cases where a farmer would need to know herbicide levels before plant back. When assays are approved for routine use in plant certification and residue analysis, large markets for many assays suddenly will appear. Since assay validation is being pursued on several fronts, it is likely that numerous markets will develop in the near future. Especially at an early stage in the acceptance of the technology, it is critical that well characterized assays be developed for clearly defined goals. Immunochemical technology is certain to represent a major market in the plant protection area. However, the speed with which the market develops depends to a great extent on the quality of the assays used to pioneer the technology.

Acknowledgments

This work was supported in part by NIEHS Superfund Grant PHS ES04699-01, EPA Cooperative Agreement CR-814709-01-0, and a grant from the California Department of Food and Agriculture. BDH is a Burroughs Wellcome Scholar in Toxicology.

Literature Cited

1. Cheung, P. Y. K; Gee, S. J.; Hammock, B. D. In The Impact of Chemistry on Biotechnology: Multidisciplinary Discussions; Phillips, M. P.; Shoemaker, S. P., Eds.; Vol. 362; American Chemical Society: Washington, DC, 1988; p 217-229.
2. Hammock, B. D.; Mumma, R. O. In Recent Advances in Pesticide Analytical Methodology; Harvey, J., Jr.; Zweig, G., Eds.; ACS Publications: Washington, DC, 1980; p 321-352.
3. Clark, M. F. Ann. Rev. Plant Pathol. 1981, 19, 83-106.
4. Van Vuurde, J. W. L. Bulletin OEPP/EPPO 1987, 17, 139-148.
5. Hammock, B. D.; Gee, S. J.; Cheung, P. Y. K.; Miyamoto, T.; Goodrow, M. H.; Van Emon, J.; Seiber, J. N. In Pesticide Science and Biotechnology; Greenhalgh, R.; Roberts, T. R., Eds.; Blackwell Scientific Publications: Oxford, 1987; p 309-316.
6. Mumma, R. O.; Brady, J. F. In Pesticide Science and Biotechnology; Greenhalgh, R.; Roberts, T. R., Eds.; Blackwell Scientific Publications: Oxford, 1987; p 341-8.
7. Newsome, W. H. In Pesticide Science and Biotechnology; Greenhalgh, R.; Roberts, T. R., Eds.; Blackwell Scientific Publications: Oxford, 1987; p 349-352.
8. Vanderlaan, M.; Van Emon, J.; Watkins, B.; Stanker, L. In Pesticide Science and Biotechnology; Greenhalgh, R.; Roberts, T. R., Eds.; Blackwell Scientific Publications: Oxford, 1987; p 597-602.
9. Hammock, B. D.; Harshman, L. G.; Philpott, M. L.; Székács, A.; Ottea, J. A.; Newitt, R. A.; Wroblewski, V. J.; Halarnkar, P. P. In USDA Beltsville Symposia Proceedings in Agricultural Research, 1988, in press.
10. Metcalf, R. L. Ann. Rev. Entomol. 1980, 25, 219-256.
11. Hollingworth, R. M. In Pesticides: Minimizing the Risks; Ragsdale, N. N.; Kuhr, R. J., Eds.; Vol. 336; American Chemical Society: Washington, DC, 1987; p 54-76.
12. Hammock, B. D.; Soderlund, D. M. In Pesticide Resistance: Strategies and Tactics for Management; Glass, E., Ed.; National Academy Press: Washington DC, 1986; p 111-129.
13. Hammock, B. D.; Abdel-Aal, Y. A. I.; Ashour, M.; Buhler, A.; Hanzlik, T. N. ; Newitt, R.; Sparks, T. C. In Human Welfare and Pest Control Chemicals-Approaches to Safe and Effective Control of Medical and Agricultural Pests; Sasa, M.; Matsunaka, S.; Yamamoto, I.; Ohsawa, K., Eds.; Pesticide Science Society of Japan: Tokyo, Japan, 1986; p 53-72.
14. Scharrer, B. Ann. Rev. Entomol. 1987, 32, 1-16.
15. Vanderlaan, M.; Watkins, B. E.; Stanker, L. Environ. Sci. Technol. 1988, 22, 247-254.

RECEIVED May 4, 1988

Chapter 23

Immunological Approaches to the Study of Plant Disease Resistance

Arthur R. Ayers[1], Keith L. Wycoff[2,3], Patricia Loomis[1], Christine Sharetzsky[1], and Misty Bender[1]

[1]Genetic Engineering Technology Program, Cedar Crest College, Allentown, PA 18104
[2]Cellular and Developmental Biology Department, Harvard University, Cambridge, MA 02138

Study of complex biological phenomena is often inhibited by lack of information on the identity of the molecules that mediate critical events. An example is disease resistance, in which plants recognize and respond to pathogens. Traditional immunochemical approaches using polyclonal antisera and, more recently, monoclonal antibodies, have been inadequate to elucidate pathogen antigens recognized by disease resistance gene products of the plant host. To overcome problems of tolerance and immunodominance, we have examined other more powerful immunological approaches including chemical immunosuppression (both general and specific), *in vitro* immunization and neonatal tolerization.

Disease resistance in plants is a complex phenomenon requiring control of hundreds of genes. The expression of genes involved in the accumulation of toxic secondary metabolites (phytoalexins), inhibitors of pathogen enzymes (protease and pectinase inhibitors), enzymes to degrade pathogen structural components (chitinase and endoglucanase) and plant wall components, is dependent on either direct recognition of pathogen constituents or action of these constituents on plant tissues (1). Thus, elaboration of plant defenses is preceded by pathogen detection.

Plants that succumb to pathogens are not deficient in defenses, but rather are defective in their recognition ability. The genes commonly identified by field tests as providing

[3]Current address: Plant Biology Laboratory, Salk Institute for Biological Studies, P.O. Box 85800, San Diego, CA 92138

0097-6156/88/0379-0307$06.00/0

protection against a pathogen permit existing adequate defenses to be mobilized in the presence of the pathogen of interest. These disease resistance genes are identified by their ability to interface with existing defenses and thereby confer pathogen-specific resistance.

Genetics of Host-Pathogen Interactions

The genetic relationship between hosts and pathogens is provocative in both its simplicity and its molecular ramifications. Each gene for disease resistance in a plant is paired with a complementary gene for avirulence in the pathogen. Only a single pair of corresponding genes among many is required to trigger the plant's defenses. The generality of the "Gene-for-Gene" relationship is thought to extend from viruses and bacteria to include fungal pathogens, nematodes and insect pests of plants (4). One molecular interpretation of this profound genetic relationship is that plants produce receptors (disease resistance gene products) that bind pathogen molecules (direct or indirect products of avirulence genes) and initiate the suite of biochemical responses observed as resistance (1).

Function of Resistance Gene Products

Knowledge of the molecular action of disease resistance gene products will determine the potential application of isolated genes in transgenic plants. The existence of disease resistance genes can now only be detected by the interaction of plants with appropriate pathogens and thus limits identification of needed resistance genes to host plants. New genes for resistance to corn pathogens, for example, will therefore, only be found in corn, based on these types of infection assays. This approach is of limited utility, because breeders are already very successful at exploiting existing sources of genetic variation. What is needed is an understanding of the function of the products of disease resistance genes, so that the combined world plant germ plasm can be exploited as a source for new resistance genes.

Limitations of Studies of Gene Structure. It is likely that within the next year or two the first plant disease resistance gene will be cloned and sequenced. This molecular knowledge will quickly lead to the sequence of the protein gene product and information on the localization of this protein, presumably on the cytoplasmic membrane. It may also be possible to begin to piece together the interactions with other plant components that lead to elicitation of plant defenses. But a major piece of this puzzle will still be lacking; what pathogen components interact with this putative membrane receptor? It may be possible for us to have very detailed information about a resistance gene and its primary gene product, but still not be able to use this gene outside of the species from which it was isolated.

Avirulence genes have been isolated and characterized (9). Gene products have been identified, but are not found outside of the pathogen. It is likely that the gene product has enzymatic activity and yields an indirect product that actually mediates pathogen recognition by interaction with resistance gene products. But here too, in-depth knowledge of the gene and gene product does not lead directly to a strategy for implementing resistance genes. What is needed is the identity of the pathogen molecule, potentially the indirect product of pathogen avirulence genes, that mediates recognition by resistance gene products.

Carbohydrates May Mediate Recognition. Carbohydrates have been postulated to mediate pathogen recognition and plants have been demonstrated to be exquisitely sensitive to and specific in their binding of pathogen wall fragments, such as the glucan elicitor (1). The structure of the glucan elicitor is too simple to account for the numerous avirulence genes possessed by fungal pathogens, so the elicitor is obviously not associated directly with the actions of avirulence genes. Other carbohydrates, such as the complex oligosaccharides that decorate the surfaces of extracellular fungal glycoproteins, are better candidates as the mediators of resistance. One possibility is that avirulence genes code for glycosyltransferases that specifically transfer carbohydrate residues to non-reducing terminal glycosyl residues of glycoproteins. Glycoproteins would then serve as carriers reflecting the expression of the complete array of avirulence genes of the pathogen (1).

Function May Lead to Broad Application in Transgenic Crop Plants. If plant resistance genes serve as receptors for pathogen glycosyl residues, knowledge of this function is vital to the deployment of resistance genes in new transgenic crop varieties. Sources of resistance could be characterized by receptors that would bind to pathogen glycoproteins. Because it is likely that the interface between resistance genes and defenses is highly conserved, it will be possible to use resistance genes from any plant as long as the desired binding function is present. Thus, based on function, it will be possible to use resistance genes identified in soybeans or potatoes, for example, to provide needed resistance to corn pathogens.

Immunological Identification of Antigens Associated with Avirulence Genes

Antibodies provide the potential analytical sensitivity and discrimination needed to identify the subtle biochemical differences between pathogen isolates possessing different avirulence genes, i.e. different pathogen races. Since avirulence genes are obligatorily linked to resistance genes, it follows that the structures associated with particular avirulence genes are likely to be those bound to the putative resistance gene receptor.

If these assumptions are correct, it should be possible to produce antibodies specific for each antigen associated with an avirulence gene and the specificity of these antibodies may mimic that of resistance gene products.

Attempts to Produce Monospecific Polyclonal Antisera. Early characterization of antisera of mice and rabbits challenged with mycelial extracts and extracellular glyoproteins of *Phytophthora megasperma* f.sp. *glycinea* (*Pmg*), a fungal pathogen, showed that these components were very immunogenic, but antibodies raised to one race bound just as readily to antigens from another (5). Attempts to create antisera specific for the immunizing race by absorption to antigens from another race led to a complete loss of binding activity -- all of the antibodies raised to one race bound to antigens from another. This was our first hint that a few antigens common to all races of the fungus were dominating the immunological response. Many other antigens were simply not yielding antibodies in the presence of the dominant antigens.

Monoclonal Antibodies from Murine Hybridomas. Murine hybridomas are hybrid cells formed by the fusion of a mouse cancer cell with lymphocytes committed to the secretion of antibodies. The lymphocytes are obtained from the spleens of mice immunized with antigens of interest, e.g. extracellular glycoproteins from a fungal plant pathogen, and under normal circumstances would produce antibodies, but would be unable to divide. Fusion with cancer cells yields hybrids that continue to divide, as well as secrete antibodies. Thus, from a population of lymphocytes from the spleen of an immunized mouse it is possible to generate numerous clones of hybridomas and each clone will yield a single antibody, i.e. monoclonal antibodies.

A library of more that 50 monoclonal antibodies (mAbs) was created for the study of antigens potentially associated with the avirulence genes of *Pmg* (10). These antibodies were placed in six groups based on binding patterns in Western blots of glycoproteins. In all but one case the antibodies bound to complex patterns of glycoproteins, suggesting an interaction with the oligosaccharides common to various groups of glycoproteins. The carbohydrate nature of the dominant antigenic domains was further confirmed by the sensitivity of antibody binding to reagents and enzymes that modify the carbohydrate residues and not the protein component of glycoproteins. There was only a single example of a protein antigen for which mAbs were isolated. These studies and further work evaluating the ability of pairs of mAbs to compete for the same antigenic determinants (K.L. Wycoff and A.R. Ayers, unpublished results), indicates that the extracellular glycoproteins of this fungus are decorated with only a small group of different oligosaccharides. Moreover, these oligosaccharides are the dominant antigenic determinants in both extracellular culture filtrates and purified mycelial walls.

None of the mAbs in our library showed qualitative differences in binding to antigens from different races. The patterns of binding on Western blots of glycoproteins from different races are different, but probably reflect random differences in the proteins of various isolates, rather than differences associated with particular avirulence genes. We think that it is likely that the antigenic determinants that make up the oligosaccharides of the extracellular glycoproteins are in antigenic competition. As a consequence of this competition, antigenic determinants common to all races yield antibodies, whereas the determinants associated with avirulence genes do not.

Deficiencies of Immunological Approaches

Antibodies and mAbs in particular are powerful tools for analysis of molecular structures, but complete exploitation of potential immunological approaches requires an assessment of the limitations, as well as the strengths of the techniques. It is often assumed that any antigen injected into a mouse will yield antibodies. This is not the case. Thus, an obvious strategy would be to start with polyclonal antisera, absorb out antibodies recognizing antigens common to two different races of pathogen, and yield monospecific antisera. But, this approach is dependent on the production of antibodies for each of the antigens injected. Uniform production of antibodies to each determinant in a mixture of inject antigens does not occur for several reasons.

Immunodominance. Antigen presentation to the immune system is a complex phenomenon, many steps and cell types are involved before antibody production begins. Antigens can in essence compete for the limited presentation system and some antigens can dominate the system to such an extent that other antigens fail to generate an immune response. In order to observe responses to all determinants in a mixture, response to the dominant antigens must be controlled.

Tolerance. Mice may also fail to respond to particular antigenic determinants, because those determinants are recognized as self. It would not be surprising to find that many of the carbohydrate antigenic determinants of fungal glycoproteins are also present in mice. We have already observed that some of our mAbs specific for pathogen carbohydrate residues bind to mouse immunoglobulins (K.L. Wycoff and A.R. Ayers unpublished results). If antigenic determinants are recognized as self, no antibodies are produced and the mice are said to be "tolerant". We expect that some avirulence genes might be cryptic because of tolerance, but we expect the majority of the antigenic determinants to be identifiable through antibody production.

New Immunological Approaches

General Immunosuppression. Antibody production by B lymphocytes requires exposure to antigen and subsequent cell division. Reagents, such as alkylating drugs, that interfere with cell division, eliminate antigen-stimulated antibody production. These reagents can function as general immunosuppressants. With a general immunosuppressant, such as cyclophosphamide, it is possible to make a mouse tolerant to an injected antigen, so that subsequent exposure to the same antigen will not result in antibody production (8).

We have tested this approach using pairs of proteins with and without fluorescein isothiocyanate (FITC) as an added antigenic determinant. In two separate series of experiments, we used either bovine serum albumin or casein in initial injections followed by cyclophosphamide treatment. After two weeks, the mice received immunizing injections of either the same protein, or the initial protein derivitized with FITC. A second identical immunizing injection was given two weeks later and sera were evaluated by enzyme immunoassay after two more weeks. Control mice receiving no pretreatment prior to the immunizing injections, produced antibodies that bound to protein with and without FITC derivitization. Response to FITC on heterologous proteins was detectable. Antibodies from mice receiving suppression protocols showed greatly reduced binding to the protein used in the suppression treatment, and enhanced binding to FITC protein. Substantial enhancement of binding to FITC on heterologous proteins was observed. These experiments indicate that general immunosuppression can be a very powerful tool for producing tolerance to dominant antigenic determinants and enhancing antibody production to minor antigenic determinants.

We have also attempted to use general immunosuppression in the search for avirulence gene-associated antigens (2). In these experiments antigens from one race of Pmg were used for suppression with cyclophosphamide, and antigens from a second race were used for immunization. Antisera showed a clear enhancement for binding to the immunizing race relative to the suppressing race when compared by ELISA. Hybridomas derived from these mice yielded antibodies in the first screening that bound only to the immunizing antigens. The race-specific hybridomas as a group showed similar response levels and were uniformly recalcitrant to cloning -- they could not be maintained in cell culture. Subsequent similar experiments have failed to give as striking results in terms of enhancement of binding to the immunizing race at the antiserum level and attempts to generate race-specific antibodies using this methodology have been unsuccessful.

Specific Immunosuppression. Toxic compounds can be linked to an antigen to yield conjugates that can selectively eliminate immunological response to the antigen. We have used copolymers of D-glutamic acid and D-lysine in conjugates with biotin as a model to test this approach of specific immunosuppression. Mice pretreated with biotin conjugated to the copolymer were tolerant to subsequent immunizing injections of biotinylated BSA -- antibodies from these mice showed no binding to biotinylated plate surfaces in ELISA. Mice without the supressing pretreatment produced antibodies showing substantial biotin binding in the same assay.

Specific immunosuppression shows great potential for eliminating the interference of dominant antigens in immunological studies. For example, we believe it will be possible to use conjugates of glycopeptides with the copolymer to eliminate the response to the dominant antigens common to several different races of Pmg, so that the mice can respond to other antigens, presumably including avirulence gene-associated antigens.

In Vitro Immunization. The constraints placed on immunization by the complex presentation system of the animal can be avoided by exposure of lymphocytes to antigens in culture (3). We have performed preliminary experiments (K.L. Wycoff and A.R. Ayers, unpublished results) using Pmg antigens (extracellular glycoproteins and purified mycelial walls) for in vitro immunization prior to hybridoma production. Antibodies from the hybridomas derived from mice challenged with one race of Pmg were evaluated for binding to antigens from other races. None of the antibodies was found to be race-specific. Examples of several of the mAbs from these experiments are being evaluated for comparison to mAbs produced by immunization in vivo.

In additional experiments, in vitro immunization was used to produce monoclonal antibodies to glucomannans purified from Pmg. Glucomannans have been suggested to have race-specific characteristics (7). As part of this project, assays for measuring binding of antibodies to glucomannans were developed using either immobilization of glucommanans found attached to proteins in standard enzyme immunoassays or immobilization through reducing termini available on some glucommannans after oxidation with iodine/potassium iodide treatment under alkaline conditions and linkage of the resulting carboxyl group to Immulon C plates (Dynatech).

Another assay approach tested was biotinylation of glucomannans following periodate oxidation. The oxidized glucomannans were biotinylated with biotinhydrazide. Binding of antibodies to biotinylated glucomannan was quantitated by binding of the test antibodies to immobilized goat anti-mouse Ig, addition of the biotinylated glucomannan, and final addition of streptavidin-alkaline phosphatase conjugate.

These experiments showed dramatic differences (10-20 fold) in the relative binding of some monoclonal antibodies in our Pmg library (10) to glucomannan from different races of Pmg. For example, while antibody KW2 showed an equivalent binding to glucomannans from Pmg races 1 and 7, MF7 showed a 20-fold preference for binding to glucomannan from race 7 relative to race 1. Antibodies generated by in vitro immunization with Pmg glucomannan on the other hand, showed no preferences of more than two fold. Similar results were obtained whether the assay used glucomannans immobilized through linked protein, through covalent linkage following oxidation or using biotinylated glucomannans.

Neonatal Tolerization. Another potential strategy to eliminate response to dominant antigens and permit production of antibodies only to determinants that differ in two antigen mixtures is to expose mice to antigens before their immune systems reach maturity (6). Neonatal mice exposed to antigens will be tolerant to the same antigens when challenged as adults. Exposure of neonatal mice to antigens from one race of Pmg should permit production of antibodies specific to a second race of Pmg when injected with antigens from that race at a later time. Another practical application of this strategy is exposure of neonatal mice to antigens from organisms that commonly show crossreactions in programs involved in the development of antibodies diagnostic for particular pathogens. In preliminary experiments (S. Marshall and A.R. Ayers unpublished results) we have observed that the crossreaction of Phythium ultimum antigens with antibodies produced in response to Pmg antigens may be greatly reduced by pretreatment of neonatal mice with P. ultimum.

Conclusion

We have focused on the use of antibodies for identification of the molecules that mediate pathogen recognition and lead to disease resistance in plants. The immunological approaches described here should permit identification of avirulence gene-associated antigens whether those antigens are proteins or carbohydrates. Although the genetic evidence appears to favor carbohydrates as the pathogen components that mediate recognition by plant hosts, we continue to consider protein antigens as candidates. Several very powerful immunological approaches, including general and specific immunosuppression, in vitro immunization, and neonatal tolerization have been identified as being of potential applicability. Our tests of these approaches have revealed that the techniques work well, but they also reveal the complexity of pathogens and the inadequacy of our knowledge.

Examination of Pmg antigens by the production of a library of monoclonal antibodies revealed that less than a dozen antigens account for essentially the total immunological response to a mixture of Pmg extracellular glycoproteins. This response can, however, be modified by immunological techniques that suppress antibody production to particular antigens. Even with these relatively sophisticated tools, it has proven difficult to screen for antibodies that can be used to identify the antigenic determinants associated with particular avirulence genes. Part of the difficulty may be technical. For example, it is difficult to identify antibodies that bind to antigens present as only a small percentage of a mixture of total antigens. Moreover, it may not be possible to remove uninteresting antigens, because of linkage to the antigenic determinants of interest.

Acknowledgment

Supported in part by the United States Department of Agriculture (85-CRCR-1-1632).

Literature Cited

1. Ayers, A.R.; Goodell, J.J.; DeAngelis, P. In The Biochemical interactions of Plants with Other Organisms, Recent Advances in Phytochemistry, Volume 19; Conn, E.E.; Cooper-Driver G; Swain, T., eds.; Plenum Press, 1985, pp. 1-20.
2. Ayers, A.R; Wycoff, K.L. In Molecular Strategies for Crop Protection; Arntzen, C.J.; Ryans C., eds.; Alan R. Liss, Inc.: New York, 1987, pp. 71-81.
3. Boss, B.D. In Immunochemical Techniques, Part I, Hybridoma Technology and Monoclonal Antibodies, Methods in Enzymology, Volume 121; Langone, J.J.; Van Vunakis, H eds.; Academic Press, Inc.: New York, 1986, pp. 27-33.
4. Day, P.R. Genetics of Host-Pathogen Interaction; W.H. Freeman and Company: San Francisco, 1974.
5. Goodell, J.J.; DeAngelis, P.; Ayers, A.R. In Cellular and Molecular Biology of Plant Stress, UCLA Symposia on Molecular and Cellular Biology, New Series, Volume 22; Key, J.L.; Kosuge, T., eds.; Alan R. Liss, Inc.: New York, 1984, pp. 447-457.
6. Hockfield S. Science 1987, 69:67-70.
7. Keen, N.T.; Yoshikawa, Y; Wang, M.C. Plant Physiol. 1983, 71,466-471.
8. Mathew, W.D.; Patterson, P.H. Cold Spring Harbor Symposium on Quantitative Biology. 1983, 48,625-531.
9. Swanson, J; Kearney, B; Dahlbeck, D; Staskawicz, B. Molec. Pl.-Microbe Interact. 1988, 1,5-9.
10. Wycoff, K.L.; Jellison, J.; Ayers, A.R. Plant Physiol. 1987, 85,508-515.

RECEIVED May 31, 1988

Chapter 24

Immunochemical Methods of Pesticide Residue Analysis

Robert O. Harrison, Shirley J. Gee, and Bruce D. Hammock

Departments of Entomology and Environmental Toxicology, University of California, Davis, CA 95616

Immunochemical methods are rapidly gaining acceptance as analytical techniques for pesticide residue analysis. Unlike most quantitative methods for measuring pesticides, they are simple, rapid, precise, cost effective, and adaptable to laboratory or field situations. The technique centers around the development of an antibody for the pesticide or environmental contaminant of interest. The work hinges on the synthesis of a hapten which contains the functional groups necessary for recognition by the antibody. Once this aspect is complete, immunochemical detection methods may take many forms. The enzyme-linked immunosorbent assay (ELISA) is one form that has been found useful in residue applications. This technique will be illustrated by examples from this laboratory, particularly molinate, a thiocarbamate herbicide used in rice culture. Immunoassay development will be traced from hapten synthesis to validation and field testing of the final assay. Emphasis will be placed on the justification of and the resources required for the successful incorporation of immunochemical technology into an existing analytical laboratory. Special attention will be given to aspects of immunochemical and related technology not covered in other recent reviews. Present use of immunoassay for pesticide analysis will be described and future potential applications and problems will be discussed.

Some basic immunology and definition of a few common terms will allow understanding of the central concepts of immunoassay. Antibodies are serum proteins which bind to specific molecules, called antigens, due to a complementarity of chemical structure between antibody and antigen. Immunization with an antigen preferentially induces the production of antibodies specific for

0097–6156/88/0379–0316$06.00/0

that antigen. While most antigens are naturally occurring proteins or complex carbohydrates, the extraordinary genetic flexibility of the immune system allows for the production of antibodies against virtually any chemical structure, including synthetic antigens. Antigens below a poorly defined size threshold of approximately 1000 daltons will not stimulate the production of specific antibodies. However, small molecules, called haptens, can elicit specific antibodies when attached to larger carrier molecules such as proteins. The phenomenon of immune response to haptens has been studied for over fifty years with the successful production of antibodies to an astounding range of natural and synthetic chemical structures. Thus the potential clearly exists for the development of immunoassays for nearly any compound.

An occasional misconception is that immunoassays are bioassays because of their use of biologically derived antibodies. It is important to realize that although immunoassays are dependent upon antibodies, these antibodies obey the Law of Mass Action just as reagents in any other ligand binding assay and are therefore well suited to quantitative analysis. Theoretical understanding of quantitative immunoassays has evolved significantly in the last three decades through extensive clinical and research use (26). An indication of the massive commercial support for immunochemical techniques is the degree of sophistication of data analysis available in numerous commercial software packages which exploit these developments (HP-Genenchem, Molecular Devices, Dynatech, and many others). This support illustrates the maturity of immunochemical techniques and technology, much of which can be directly appropriated for residue analysis by immunoassay.

The development of immunochemical methods for analysis of haptens such as drugs and steroids (4,11) occurred during the dominance of the pesticide market by highly nonpolar organochlorines, which are easily analyzed by gas chromatography with ion selective detectors due to their lipophilicity and halogenation. There is a long history of extensive clinical and research use of immunoassays for analysis of drugs and hormones, including steroids, peptides, and others (4,11,24,25). The reasons for the lack of use of immunochemical technology in pesticide residue analysis seem to be primarily historical. Present technology makes it feasible to analyze for nearly all pesticides in current use by immunochemical methods.

Reasons for Applying Immunochemical Technology to Pesticides

Since the use of immunochemical technology for pesticide residue analysis was first reviewed by Ercegovich in 1971 (9) several helpful reviews of this application have been published (13,14,23,33). Despite the potential demonstrated during this period, few researchers and regulators apply immunochemical technology to their own problems in pesticide residue analysis.

Immunochemical technology for residue analysis offers the existing analytical laboratory many advantages. Sensitivity and specificity are generally comparable to existing techniques with large improvements in speed and cost. The general applicability of immunochemical technology is also a valuable asset. Immunoassays can be modular; by substituting two reagents, analysis for a different analyte can be performed using the same instrumentation. Multianalyte procedures can be developed through parallel processing. The factors limiting the present use of immunochemical technology for residue analysis are primarily lack of acceptance and poor antibody availability rather than the difficulty of developing antibodies to target compounds. This situation appears to be changing as the advantages of immunoassay for environmental analysis are recognized by analytical chemists. One reason for this recognition is that immunoassay is the best choice among present technologies for the analysis of the large number of samples needed for thorough environmental analysis.

It can be shown by sampling theory that a large number of samples must be analyzed to obtain high confidence estimates of low contamination rates (Figure 1). With immunochemical technology available, the analytical chemist no longer need be dismayed by the prospect of analyzing such large sample loads. The technology for automation of some types of immunoassays is advancing rapidly and sample processing rates are increasing dramatically. The application of robotic systems to the residue laboratory has been discussed (22) and an automated ELISA system using commercially available robots and automated extraction systems has been described which can analyze 10,000 or more seed and plant samples per day for viral and bacterial diseases (34).

Applications of immunoassay to pesticide chemistry have been described which address some difficult problems in analysis by classical methods. These include stereospecific analysis of optically active compounds such as pyrethroids (38), analysis of protein toxins from *Bacillus thuringiensis* (5,37), and compounds difficult to analyze by existing methods, such as diflubenzuron (35) and maleic hydrazide (15; also Harrison, R.O.; Brimfield, A.A.; Hunter, K.W.,Jr.; Nelson, J.O. *J. Agric. Food Chem*, submitted). An example of the excellent specificity possible is seen in assays for parathion (10) and its active form paraoxon (3). Some immunoassays can be used directly for analysis without extensive sample extraction or cleanup, dramatically reducing the work needed in typical residue analysis. An example of this is given in Figures 2 and 3, comparing the direct ELISA analysis of molinate in rice paddy water to the extraction required before GC analysis.

While the use of immunoassay for residue analysis should continue to expand, it is not the answer to all problems in environmental analysis. Immunochemical technology should serve best as a complement to existing methods rather than a replacement for them. It is especially important to recognize the potential of immunoassay for impact in environmental screening through the developments which will be discussed later in this chapter. It is in this area that we expect immunochemical technology to make its greatest contribution to environmental analysis.

Resources Required for Immunoassay Development

Perhaps the most important distinction to be drawn in this chapter is that between the development of immunochemical residue methods and their implementation for routine use. The labor and resources required for developing a specific assay are significant, as will be shown in this section. However, the routine use of validated assays

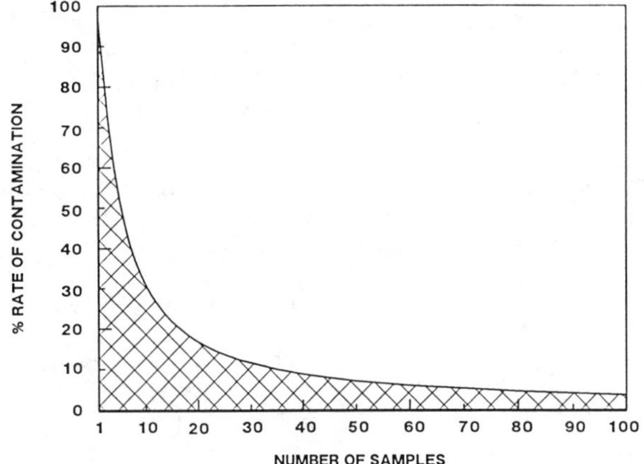

Figure 1. The shaded area represents the 95% confidence interval for true contamination rate, assuming random distribution of contaminated samples and given a constant 0% detected contamination rate (adapted from ref. 1). The upper boundary of the curve represents the upper 95% confidence limit for true contamination rate. Thus, if ten samples were analyzed with no contamination detected, the true contamination rate 95% confidence interval would be between 0% and 30.8%.

requires little expertise, equipment, or supplies not already found in most analytical labs. A summary of the resources needed for immunoassay development is given in Table 1.

TABLE 1 Resources needed for immunoassay development

Organic chemistry: hapten synthesis and conjugation to carriers
Biology: antibody production
Immunochemistry: immunoassay design and optimization
Analytical and residue chemistry: sample preparation, data analysis, assay validation and application

This table illustrates one of the major impediments to the rapid assimilation of immunochemical technology into pesticide residue analysis labs. Because of the amount and variety of work involved, new method development costs may be high when compared to routine chromatographic methods. However, the low cost per run allows for rapid recovery of the initial investment with sufficiently high sample loads. For example, the cost of reagents and supplies for an ELISA for diflubenzuron was estimated to be $0.20/sample as compared with $4 for HPLC or $11 for GC (35). In addition to the lower reagent and supply costs, the major economic advantage of immunoassay is the dramatic decrease in labor costs.

Steps in the Development of Pesticide Immunoassays

The steps involved in the development of pesticide immunoassays have been described previously (14) and numerous reports of pesticide immunoassay development now illustrate this process (3,10,12,15,19,32,35,36,38). These steps will be reiterated here to reinforce salient points and to emphasize the distinction between the development and implementation of immunoassays.

Hapten synthesis. The general synthetic approach chosen will be dictated by the desired assay specificity. Compound or class specific antibodies can be produced depending on which part of the molecule is used for conjugation. Antibody specificity is generally highest for the part of the molecule furthest from the carrier. The importance of carefully planned and rational hapten synthesis cannot be overemphasized. The chosen synthetic route should preserve as much structure as possible of the molecule to be analyzed and must provide a functional group for conjugation to a carrier molecule (usually protein). The optimum hapten structure for antibody production may however lack an apparently essential moiety and this is a tricky area for the non-immunochemist to predict. For example, preservation of the nitro group of parathion by conjugation through the normally unsubstituted aromatic ring positions led to the production of antisera which did not recognize free parathion (31), while parathion specific antisera were produced against conjugates of aminoparathion (10). Conjugation position was also important in the production of antibodies to maleic hydrazide (15; also Harrison, R.O.; Brimfield, A.A.; Hunter, K.W.,Jr.; Nelson, J.O. *J. Agric. Food Chem*, submitted). Immunization with N-conjugates of maleic hydrazide led to maleic hydrazide specific and hapten specific antibodies, while immunization with O-conjugates of maleic hydrazide led to only hapten specific antibodies. There is experimental evidence that haptens are more immunogenic when separated from the carrier by a spacer arm. Presentation of the hapten to receptors on cells of the immune system is less likely to be sterically hindered if the hapten is on a spacer arm several atoms long.

The production of haptens for conjugation is the last point in the assay development process where rigorous structural analysis is possible. Because the assay specificity is determined by the structure(s) of the conjugated haptens, it is critically important for these compounds to be pure and structurally well characterized before conjugation renders this impossible.

Conjugation to carriers. A wide range of proven methods are available for conjugation of haptens to their carriers, most of them using common commercially available reagents. These have been summarized previously (11,14) and many examples of their use in pesticide immunoassay development exist (3,10,12,15,19,32,35,36,38).

Verification of conjugation. As noted above, structural analysis is difficult or impossible after conjugation, but it is possible to verify conjugation even though the structure of the conjugated molecules is unknown. If the hapten has an appropriate UV absorption spectrum, difference spectra between conjugates and original carriers can be used for both qualitative demonstration of conjugation and estimation of the molar ratio of hapten to carrier, often called the hapten density. This method has been applied to both UV absorbing parent compounds (15) and haptens containing a UV absorbing spacer group (3). Alternative methods include quantitation of free primary amino groups and conjugations using radiolabelled haptens.

Antibody production. Antibodies used for immunoassay can be either polyclonal, contained in serum from rabbits or other animals, or monoclonal, produced by mouse lymphocytes immortalized for *in vitro* culture (called hybridomas). The advantages of each have only recently begun to be examined systematically in the context of quantitative hapten analysis (2). Both monoclonal (3,15) and polyclonal (5,10,12,19,32,35-38) antibodies have proven effective for pesticide analysis, although most reported immunoassays have used rabbit sera. Production of monoclonal antibodies is labor intensive and may not warrant the extra effort for the planned application. For standardized methods which will be widely used, monoclonal antibodies may be superior to polyclonal sera because of their biochemical uniformity and defined affinity and specificity. Stable hybridoma cell lines provide a continuous source of their monoclonal antibodies as long as they can be maintained in culture. Monoclonal antibodies will also be useful in the discrimination of related multivalent antigens such as those produced by related strains such as *Bacillus thuringiensis israelensis* and *Bacillus thuringiensis kurstaki* or genetically engineered organisms. The specificity of monoclonal antibodies for single protein determinants will likely also prove useful for distinguishing active and inactive forms of protein toxins, such as from *Bacillus thuringiensis*.

Antibody characterization. A useful description of some important considerations in antibody characterization has been given by Hammock and Mumma (14). Specificity is generally tested using some form of competitive binding, such as a competitive ELISA. An estimation of the average affinity constant may be useful in the comparison of different antibodies and in later optimization of the assay. This value can be conveniently estimated from antiserum dilution curves (27) for selection of the best antibodies in a screening process. This requires the choice of an assay format (see later section on assay formats), such as ELISA, preferably using a different carrier to eliminate carrier crossreactivity. It is important that specificity and sensitivity be evaluated at this point and be deemed acceptable before investing further resources in assays based on inadequate antibodies.

Assay optimization. An optimization step not always taken, but nonetheless important to the success of any immunochemical method of analysis is the selection of materials, such as test tube or plastic plates, which maximize assay performance. An example of this is the selection of 96-well microtiter plates for enzyme-linked immunosorbent assay (ELISA) which give maximum protein binding capacity and minimum interwell variability (28,30; also Harrison, R.O.; Nelson, J.O. J. Immunoassay, submitted). The type of microtiter plate may be the most important single determinant of ELISA performance and this selection should not be made carelessly. Significant error may also occur in the reading of assays performed in 96-well microtiter plates, due to alignment errors of automatic plate readers undetected in normal use (Harrison, R.O.; Nelson, J.O. J. Immunoassay, submitted; Harrison, R.O.; Hammock, B.D. J. Assoc. Off. Anal. Chem., submitted), and a plate reader test should allow further reduction of error. These steps should be taken before a major investment in time, effort, or money is made in an assay system which may later be found to be less than acceptable.

Once developed, any assay system should be optimized to provide the best possible sensitivity. One important variable affecting ELISA performance is the selection of coating antigens. Wie and Hammock (36) showed that prudent choice of coating antigens can significantly improve sensitivity. A useful empirical approach to optimization is described by Hunter and Bosworth (17), consisting mainly of identifying reagent concentrations which provide the highest signal to noise ratio below binding saturation.

Assay application. At this point major differences appear between the historical use of clinical immunoassays and the potential applications of environmental and pesticide immunoassays. Most clinical assays have been applied to simple or well defined and consistent matrices such as urine or serum. In contrast, most matrices likely to be analyzed for pesticides are more complex, less well defined, and more variable. The potential for serious problems with matrix effects in the environmental field is far greater than most clinical immunoassays have encountered. The application of immunoassays to environmental analysis requires sampling strategies, cleanup procedures, and data handling fundamentally similar to those presently in use in any good analytical lab. The critical factor in the success of immunochemical technology will likely be competence in analytical and residue chemistry rather than immunochemical expertise. There are several considerations unique to immunoassay, such as the tolerance of antibodies for organic solvents or variations in pH or ionic strength, and stability of biological reagents in storage. Sample presentation to the reagent antibody is an important consideration. Immunoassay of lipophilic materials in a lipophilic environment is not likely to suceed without some means of presenting the lipophilic analyte to the hydrophilic antibody. This can be done in most cases using water miscible organic cosolvents or detergents to induce micelle formation. Additionally, the effect of the matrix being analyzed on the antibody used for analysis must be evaluated. If the matrix effects are minimal with little or no sample cleanup, then analysis can be done directly. In addition to the analysis of molinate in rice field water (Figures 2 & 3), immunoassays for several other compounds have been reported which require little or no sample cleanup, including parathion (10), chlorsulfuron (19), paraquat (32), and maleic hydrazide (15).

The sample workup necessary for pesticide residue analysis will vary with each combination of analyte and antibody, each of which may have a different tolerance for the matrix and other factors. The effects of these factors must be considered as with the development of any other analytical technique. Matrix effects for one ELISA system are summarized in Figure 4. While the effect of the matrix on the antibodies in Figure 4 is different for each antibody-solvent-matrix combination, the competitive ELISA standard curves for most of these combinations are similar when expressed as percent of the appropriate control. Some systems may not require extensive adjustment, but this must be tested with each individual system. For example, our molinate assay performs equally well in a variety of water types at high concentrations of molinate (Figure 5). The small difference seen between the buffer and water standard curves in Figure 5 was eliminated by the addition of small amounts of concentrated buffer to water samples to equalize them to the buffer composition.

Assay validation. The validation procedure consists primarily of comparison of the immunoassay method to an existing method and verification of the statistical reliability of the new method according to well established principles in analytical chemistry. A general approach to the problem of validation has been summarized in a useful form by Horwitz (16). Participation of groups such as the Association of Official Analytical Chemists (AOAC) and the U.S. Environmental Protection Agency (EPA) in the official validation process should lead to well defined validation protocols suitable for general immunoassay use. These protocols must deal realistically with matrix effects such as those shown in Figure 4. Demonstration of standard curves in the matrix of interest (such as shown in Figure 5) is important, but is not sufficient to predict success of the method. Biotransformation and variability of matrix effects among samples must also be evaluated by the analysis of field treated samples.

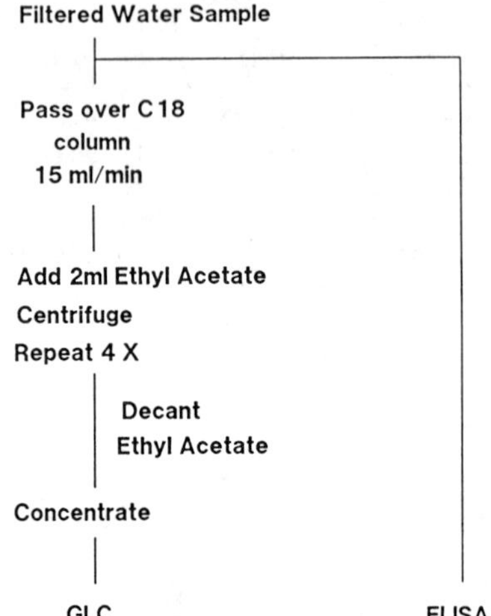

Figure 2. Procedure for extraction and analysis of molinate from water.

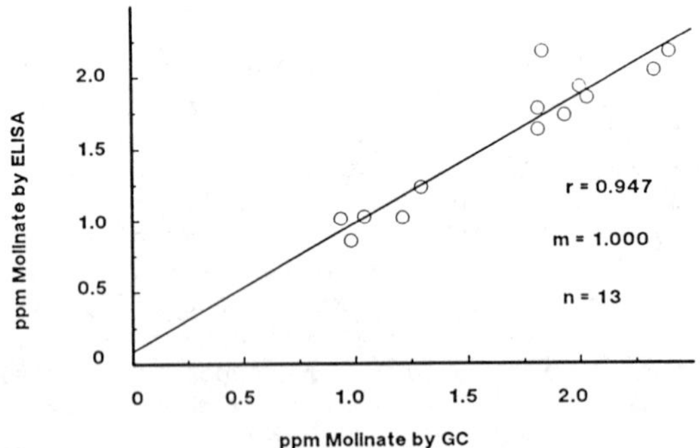

Figure 3. GC vs. ELISA correlation for analysis of molinate field water samples.

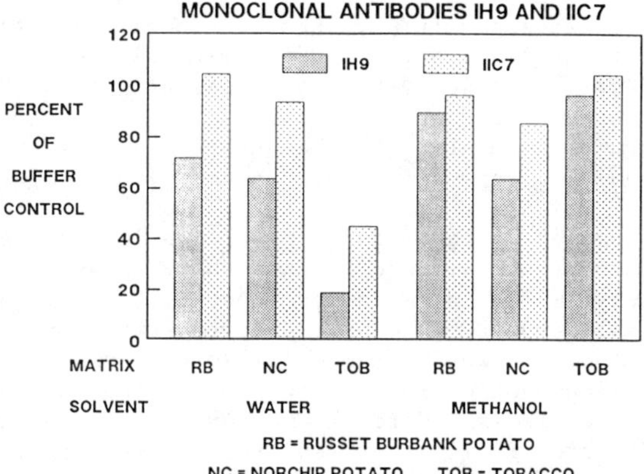

Figure 4. Matrix effects in ELISA of aqueous and methanolic extracts of potato and tobacco using purified anti-maleic hydrazide monoclonal antibodies (modified from ref. 15).

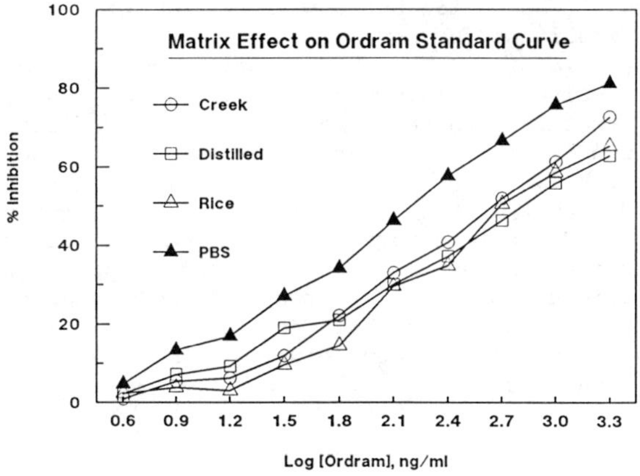

Figure 5. ELISA standard curves for molinate in water from different sources.

Assay formats

Haptens and anti-hapten antibodies are reagents which can be used in a wide variety of assay formats. One fundamental division among immunoassay formats is based on the type of label or tracer which is ultimately detected. Two of the most common types of label are radioisotopes and enzymes, used respectively in radioimmunoassay (RIA) and enzyme immunoassay (EIA). RIA is the older of the two and remains very important and widely used in clinical and research situations. However, the use of radioisotopes carries with it possible health hazards, the need for special handling precautions, and mandatory regulatory oversight. Because EIA avoids these problems, it is gaining popularity rapidly among users of immunoassays.

The format presently favored for pesticide immunoassay is the competitive ELISA or enzyme-linked immunosorbent assay. This is a heterogeneous assay, which is based upon competition between an unknown and variable amount of soluble analyte (the sample) and a small fixed amount of the analyte immobilized on the solid phase, for binding a small amount of soluble antibody. The concentration of analyte in the sample is indirectly measured by the quantitation of bound antibody after it is separated from the free antibody. In contrast, homogeneous assays require no separation step and so can be performed in a single step in one tube. An example of this as applied to pesticides is the Emit analysis of dieldrin and 2,4-D (12). Both heterogeneous and homogeneous assay systems have been used extensively in the clinical field (24,25). Useful comprehensive reviews of the various assay formats have been published (21,24). The choice between these two major classes of assay format is important for several reasons. Homogeneous formats are generally less sensitive, but have fewer steps and are easier, thus lending themselves to field use. Heterogeneous assays are becoming more versatile with the development of many different choices of equipment. There are many different format and equipment possibilities for ELISA alone and selection from among these depends on the needs and budget of the user. Tests can be performed in single tubes or in 96 well microtiter plates using a variety of manual or semiautomatic single-tip or multichannel pipettors. Manual or automatic microtiter plate readers may be used or fully automated systems can be set up which can process thousands of samples per day. A variety of equipment is available for any desired degree of automation.

Assay implementation

The list above is not intended to be comprehensive, but it will serve as a general guide to immunoassay development. We must emphasize again the difference between development and implementation. Assays which work in the laboratory generally require modification before they can be used to analyze field samples. Implementation of assays that have been fully validated for field samples may require little additional commitment by the user, other than analyst training. For the near future, there may be considerable pressure to transfer immunoassay methods to the analytical lab as soon as possible after development, and the analytical chemist may need to do additional work to adapt and fully validate many assays. This will demand that the analytical chemist develop some immunochemical expertise. Many of the implementation problems likely to be encountered will best be solved with traditional analytical chemistry expertise tempered by an understanding of practical immunochemistry.

Present Use

The use of immunochemical technology for pesticide residue analysis has now advanced beyond the potential stage. Some groups are now using ELISA for real residue data. Dupont has submitted cyanazine (Bladex) ELISA data to EPA for product registration (Sharp, J., Dupont, personal communication, 1987) and the California Department of Food and Agriculture is initiating a program of monitoring rice paddy water for molinate and thiobencarb by ELISA.

Some commercial kits are available now, including ELISA tests for atrazine, simazine, and propazine, and for carbofuran, sold by ImmunoSystems, Inc. for $15 per test in a package of ten tests. These kits offer cost and sensitivity competitive with other methods and can be used quantitatively in the field with a portable spectrophotometer. Other companies have immunoassay kits in various stages of development (20,23).

Several recent events indicate the intense interest in immunoassay for analysis of pesticides and other environmental contaminants. Special symposia on immunochemical technology have been or are being presented by the Association of Official Analytical Chemists at their 1986, 1987, and 1988 annual meetings and by the International Union of Pure and Applied Chemistry at the quadrennial Pesticide Chemistry Congress in Ottawa in 1986 (13,23). Several government agencies have recently solicited proposals for development of immunoassay methods for product or environmental analysis, including the U.S. Food and Drug Administration (8), the Food Safety and Inspection Service of the U.S. Department of Agriculture (7), and the U.S. Army (6). The U.S. Environmental Protection Agency is performing a validation study of an immunoassay method for analysis of pentachlorophenol in water, developed by Westinghouse Bioanalytic Systems of Rockville, MD (18).

Future use

One estimate of the size of the market for non-traditional immunoassays is $24 million for plant diagnostics, $180 million for hazardous chemicals, and $126 million for food testing by the year 2000, totaling $330 million (20). For certain pesticides and hazardous chemicals, the market for analysis may be larger than the market for use. Assays for new compounds will continue to be developed, especially including pesticides which are difficult to analyze by traditional methods. Other environmental contaminants and biotechnology products, including genetically engineered pesticides, will also be excellent candidates for immunoassay. We expect that even if limited to only the present level of technology, immunoassays in some form will soon have a large impact on the analysis of pesticides, hazardous chemicals, industrial by-products, and natural toxins in products and the environment. Novel assay formats will be developed which offer greater opportunity for field testing, including rapid or instantaneous procedures, such as described by Stanbro et al. (29). Such new technology has great potential for development of multianalyte methods by merely combining multiple antibodies and haptens on a single probe. The use of monoclonal antibodies for these and other immunoassay applications will increase because of specificity, regulatory, patent, or economic considerations. Increasing automation of the entire immunoassay process will occur even in the absence of radically new technology, due to the increasing availability and sophistication of laboratory robots and dedicated automatic immunoassay aids.

Acceptance of immunochemical technology for pesticide residue analysis will become more widespread as more commercial immunoassay products reach the market and more programs utilizing immunochemical technology begin producing tangible results. Increasing realization of the need for high sample loads for thorough environmental analysis will increase the demand for low cost methods of analysis. Official validation of immunoassay methods will accelerate as analytical chemists become familiar with immunoassay and gain experience with these methods in the validation process. This new experience will undoubtedly be accompanied by better control of assay sources of error and better instrumentation specifically designed for quantitative immunoassay. Greater scientific understanding of underlying principles will be gained as a direct result of the need to deal with matrix effects in the analysis of complex samples and this understanding will facilitate the application of immunochemical technology to new analytical problems. All immunoassays, no matter how novel, will continue to depend on the production of specific antibodies as decribed above. This process in turn depends upon rational antigen synthesis, from hapten design and synthesis through conjugation, to novel methods of antibody production, as outlined in this chapter. Regardless of the exact configuration of the final immunoassay, the principles outlined above will continue to be important for the forseeable future.

Acknowledgments

This work was supported in part by NIEHS Superfund Grant PHS ES04699-01, EPA Cooperative Agreement CR-814709-01-0, and a grant from the California Department of Food and Agriculture. BDH is a Burroughs Wellcome Scholar in Toxicology.

Literature Cited

1. Beyer, W.H., Ed. CRC Handbook of Probability and Statistics CRC Press: Cleveland, 1968; pp 219-237.
2. Bjercke, R.J.; Cook, G.; Langone, J.J. J. Immunol. Meth. 1987, 96, 239-246.
3. Brimfield, A.A.; Lenz, D.E.; Graham, C.; Hunter, K.W.,Jr. J. Agric. Food Chem., 1985, 33, 1237-1242.
4. Butler, V.P.,Jr.; Beiser, S.M. Adv. Immunol. 1973, 17, 255-310.
5. Cheung, P.Y.K.; Hammock, B.D. Curr. Microbiol. 1985, 12, 121-126.

6. Commerce Business Daily July 22, 1985.
7. Commerce Business Daily June 8, 1987.
8. Commerce Business Daily July 14, 1987.
9. Ercegovich, C.D. In Pesticide Identification at the Residue Level; Gould, R., Ed.; Advances in Chemistry Series No. 104; American Chemical Society: Washington, DC, 1971; pp 162-177.
10. Ercegovich, C.D.; Vallejo, R.P.; Gettig, R.R.; Woods, L.; Bogus, E.R.; Mumma, R.O. J. Agric. Food Chem. 1981, 29, 559-563.
11. Erlanger, B.F. Pharmacol. Rev. 1973, 25, 271-280.
12. Ferguson, B.S. Sixth International Congress of Pesticide Chemistry, Ottawa, 1986, abstract 5C-09.
13. Hammock, B.D.; Gee, S.J.; Cheung, P.Y.K.; Miyamoto, T.; Goodrow, M.H.; Van Emon, J.; Seiber, J.N. In Pesticide Science and Biotechnology; Greenhalgh, R. and Roberts, T.R., Eds.; Blackwell: Oxford, 1986; p 309.
14. Hammock, B.D.; Mumma, R.O. In Pesticide Analytical Methodology; Zweig, G., Ed.; ACS Symposium Series No. 136; American Chemical Society: Washington, DC, 1980; pp 321-352.
15. Harrison, R.O. Ph.D. Thesis, University of Maryland, College Park, 1987.
16. Horwitz, W. J.Assoc.Off.Anal.Chem. 1984, 67, 432-440.
17. Hunter, K.W.,Jr.; Bosworth, J.M.,Jr. Meth. Enzymol. 1986, 121, 541-547.
18. Hunter, K.W.,Jr.; Schuman, R.F.; Williams, L.R.; Soileau, S.D.; Chiang, T.C. Association of Official Analytical Chemists 101st Annual Meeting, San Francisco, CA ,1987, abstract 361.
19. Kelley, M.M.; Zahnow, E.W.; Petersen, W.C.; Toy, S.T. J. Agric. Food Chem. 1985, 33, 962-965.
20. Klausner, A. Bio/Technology 1987, 5, 551-556.
21. Langan, J.; Clapp, J.J., Ligand Assay. Analysis of International Developments on Isotopic and Nonisotopic Immunoassay; Masson: New York, 1981
22. McKinney, W. In Pesticide Science and Biotechnology; Greenhalgh, R. and Roberts, T.R., Eds.; Blackwell: Oxford, 1986; p 317.
23. Mumma, R.O.; Brady, J.F. In Pesticide Science and Biotechnology; Greenhalgh, R. and Roberts, T.R., Eds.; Blackwell: Oxford, 1986; p 341.
24. Nakamura, R.M.; Dito, W.R.; Tucker, E.S. Immunoassays: Clinical Laboratory Techniques for the 1980's; Alan R. Liss: New York, 1980.
25. Oellerich, M. J.Clin.Chem.Clin.Biochem. 1980, 18, 197-208.
26. Rodbard, D. In Ligand Assay. Analysis of International Developments on Isotopic and Nonisotopic Immunoassay; Langan, J.; Clapp, J.J., Eds.; Masson: New York, 1981; Chapter 3.
27. Roulston, J.E. J. Immunol. Meth. 1983, 63, 133-138.
28. Shekarchi, I.C.; Sever, J.L.; Lee, Y.J.; Castellano, G.; Madden, D.L. J. Clin. Microbiol. 1984, 19, 89-96.
29. Stanbro, W.D.; Newman, A.L.; Hunter, K.W., Jr. Chapter 25, In Biotechnology for Crop Protection; Hedin, P.A., Menn, J.J., Hollingworth, R.M., Eds.; ACS Symposium Series No. 379; American Chemical Society: Washington, DC, 1988.
30. Stemshorn, B.W.; Buckley, D.J.; St. Amour, G.; Lin, C.S.; Duncan, J.R. J. Immunol. Meth. 1983, 61, 367-375.
31. Vallejo, R.P.; Bogus, E.R.; Mumma, R.O. J. Agric. Food Chem. 1982, 30, 572-580.

32. Van Emon, J.; Hammock, B.D.; Seiber, J.N. Anal.Chem. 1986, 58, 1866-1873.
33. Van Emon, J.M.; Seiber, J.N.; Hammock, B.D. In Bioregulators for Pest Control; Hedin, P., Ed.; ACS Symposium Series No. 276; American Chemical Society: Washington, DC, 1985; pp 307-316.
34. van Vuurde, J.W.L.; Maat, D.Z.; Franken, A.A.J.M., Chapter 26, In Biotechnology for Crop Protection; Hedin, P.A., Menn, J.J., Hollingworth, R.M., Eds.; ACS Symposium Series No. 379; American Chemical Society: Washington, DC, 1988.
35. Wie, S.I.; Hammock, B.D. J. Agric. Food Chem. 1982, 30, 949-957.
36. Wie, S.I.; Hammock, B.D. J. Agric. Food Chem. 1984, 32, 1294-1301.
37. Wie, S.I.; Hammock, B.D.; Gill, S.S.; Grate, E.; Andrews, R.E.; Faust, R.M.; Bulla, L.A.; Schaefer, C.H. J. Appl. Bacteriol. 1984, 57, 447-454.
38. Wing, K.D.; Hammock, B.D.; Wustner, D.A. J. Agric. Food Chem. 1978, 26, 1328-1333.

RECEIVED April 1, 1988

Chapter 25

Applying Biotechnology and Microelectronics for Environmental Analysis

William D. Stanbro, Arnold L. Newman, and Kenneth W. Hunter, Jr.

Biotronic Systems Corporation, 15225 Shady Grove Road, Suite 306, Rockville, MD 20850

A concept for a generic biosensor is introduced that is capable of measuring small molecules in environmental matrices. This sensor, the capacitive affinity sensor, makes use of a combination of antibody and microelectronic technologies. Sensor operation is demonstrated using sensors for hydrocortisone and pentachlorophenol as examples. Because of the mature nature of the critical technologies, sensors based on this design should be commercially available in the near future.

Much of the current agricultural abundance of the United States is due to the availabilty of chemical means of pest control. However, pesticides also represent a considerable threat to the environment when they are used improperly. For this reason the ability to measure pesticide residues at low concentrations in environmental matrices such as surface and groundwaters and soils is of great importance. In this paper we will describe a concept for a generic biosensor, the capacitive affinity sensor, capable of rapidly determining the concentration of many types of small molecules in the environment.

Although several definitions of biosensor exist, we will use the word to mean a microelectronic device that measures the interaction of an analyte with a biologically produced molecule as part of the measurement system. Figure 1 is a block diagram of a generalized biosensor. The most critical element of the sensor is the box marked Transducer; this is where the information about the analyte (i.e. the

0097–6156/88/0379–0331$06.00/0

concentration) is transformed into an electrical signal. A biosensor seeks to exploit the rapidity and specificity of biomolecular reactions, the consequences of which result in a change in the electrical or optical properties that can be transformed into a change in the measured voltage. Once a voltage change occurs it can be further processed to improve the signal to noise ratio and correlate the changes with other sensors. The processed signal can also be used to change the transducer characteristics to improve its performance in a particular situation. The ability to process the transducer output in these ways is one of the great incentives for developing sensor technology.

Our goal is the development of a "user friendly" biosensor for small molecules such as pesticides. To reach this goal the sensor must have a number of characteristics. These include specificity, sensitivity, accuracy, precision, ruggedness and manufacturability. While they are for the most part self-explanatory, the characteristic of manufacturabilty deserves further comment. The best sensor is of little use if it cannot be mass produced at a reasonable cost. For this reason our search for a transduction mechanism for a biosensor has concentrated where possible on well proven technologies that lend themselves to mass production. The biosensor we present here combines two well established technologies; antibodies and microelectronics.

Both polyclonal antibodies and monoclonal antibodies prepared by hybridoma technology, are available commercially for a wide variety of small molecules. The processes for mass producing antibodies at a reasonable cost are well understood (1). Antibodies are also attractive because of their relative stabilty when compared with other proteins such as enzymes (1). Microelectronic device fabrication is a standard technology that can rapidly produce intricate structures on the millimeter to micron scale at very modest cost per unit (2).

Sensor Theory

As discussed above the key to any biosensor is in the transduction mechanism. The mechanism of the capacitive affinity sensor is shown in Figure 2. The sensor consists of a planar capacitor composed of interdigitated metal fingers on an insulating substrate. This structure is then coated with a thin layer of a passivation material. The role of the passivation material is to protect the metal structure from deterioration from contact with the solution. A sample of the analyte or an analog retaining the ability to bind an antibody to the analyte is covalently bound to the

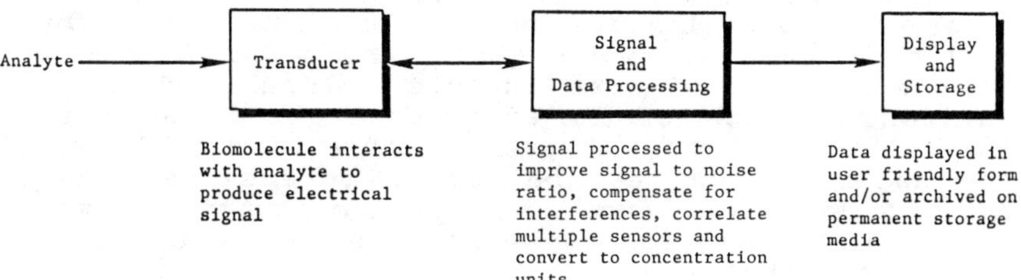

Figure 1. Block diagram of a generalized biosensor.

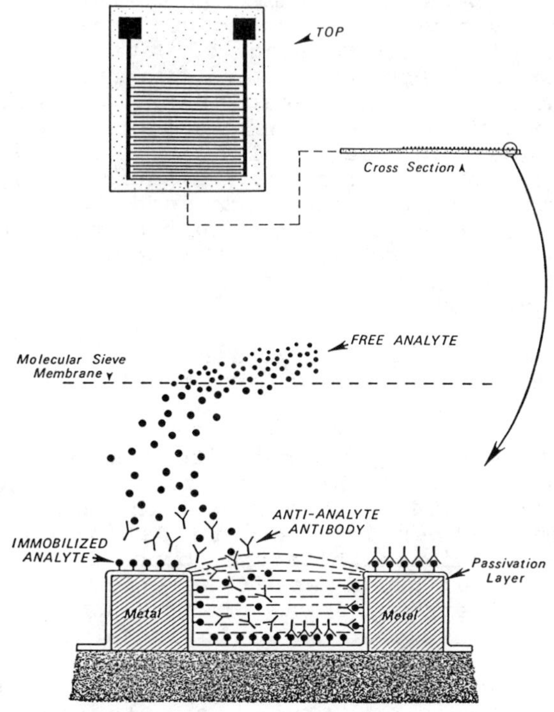

Figure 2. Transduction mechanism of the capacitive affinity sensor.

sensor surface. The sensor is then preloaded with anti-analyte antibodies which bind to the surface bound analyte. Finally, the assembly is covered with a size-selective membrane that retains the antibodies in the sensor, but allows the small analyte molecules to enter or leave. The biosensor is now ready to perform real-time, on-line monitoring of the selected analyte. In the absence of free analyte molecules from the environment most of the antibodies are bound to the immobilized analyte. However, because the bond between the antibody and the analyte is reversible, there exists a state of dynamic equilibrium governed by mass action law kinetics. The capacitor structure is so engineered that the bound antibodies are within the electic field of the capacitor. When free analyte enters the system, it competes with the immobilized analyte for the antibodies, resulting in a displacement of antibodies bound to the immobilized analyte. Since the antibodies have a low dielectric constant (3) relative to the dielectric constant of water, the change in the dielectric constant between two capacitor plates results in a change in capacitance. This change in capacitance can be conveniently and precisely measured with a number of electronic circuits. Since the membrane prevents the loss of the antibodies, when the concentration of analyte decreases in the aqueous environment, the antibodies return to the sensor surface and the capacitance also returns to baseline.

Materials and Methods

Two types of sensors will be described; one to hydrocortisone and one to pentachlorophenol (PCP).

Capacitor Substrate. The capacitors were produced by depositing layers of chromium, copper and gold on an alumina wafer 2.54 cm on a side. The chromium, gold and initial copper layers were deposited by vapor deposition techniques while the copper layer was built up to its final thickness by electroplating. The interdigitated structure was produced using a standard photolithographic process employing a postive photoresist and selective etching to leave the desired structure. The hydrocortisone sensors were then passivated with 2 micrometers of a polymeric material parylene C (a polymerized xylene), followed by 150 nm of sputtered SiO_2 which forms a chemically reactive surface for immobilizing the analyte. The PCP sensor substrate was passivated with a thin layer of polydimethylsiloxane deposited from acetone and allowed to cure overnight at room temperature.

Binding Chemistry. The hydrocortisone sensor surface was created by reacting 3-aminopropyltriethoxysilane (APTS) (Petrarch Sytems) with the surface in dry toluene for two hours at room temperature followed by a 30 minute cure at 60 degrees C. The PCP sensor was coated by dipping in 95% ethanol containing 2% APTS for one minute and curing overnight at room temperature. In both cases the analytes were bound to the amino functional group of the silane through the carbodiimide-catalyzed formation of an amide linkage. The PCP sensor used 2,6-dichlorophenol to which a butanoic acid group was bound at the 4 position of the phenol and the 4 position of the acid (Antech Consultants). This left free the carboxylate group of the acid for reaction with the amine group. The hydrocortisone system used hydrocortisone hemisuccinate (Aldrich Chemical Company), and the coupling reaction was performed in ethanol using 1-ethyl-3-(3,3-dimethyaminopropyl)-carbodiimide (Aldrich Chemical Company).

Antibodies. The anti-PCP were affinity purified monoclonal antibodies obtained from Westinghouse Bio-Analytic Systems Co., Rockville, MD. The anti-hydrocortisone was a commercially available polyclonal antibody purchased from Sigma Chemical Company, St. Louis, MO; and the anti-T-2 toxin antibodies affinity purified monoclonals was obtained from the Uniformed Services University of the Health Sciences, Bethesda, MD.

Capacitance Measurement. Capacitance measurements were made on a GenRad Model 1657 RLC Digibridge. All measurements were made at 1000 Hz.

Measurements of Capacitance Change. All measurements reported here were made in 10 ml phosphate buffered saline (pH 7.4) and the test device had no selective membrane. All experiments were conducted at room temperature.

Results and Discussion. We will present the results of two experiments to demonstrate the ability of the sensor to respond to antibodies binding to a sensor surface and being displaced by free analyte molecules. To illustrate the change in capacitance as antibodies bind to the sensor surface, we have chosen data from the anti-PCP system. The upper curve of Figure 3 shows the change in capacitance (plotted as the negative of the relative change in parts per thousand) as anti-PCP antibodies bind to immobilized analyte on the sensor surface. The bottom curve shows a small change in capacitance when an antibody specific to T-2 toxin is

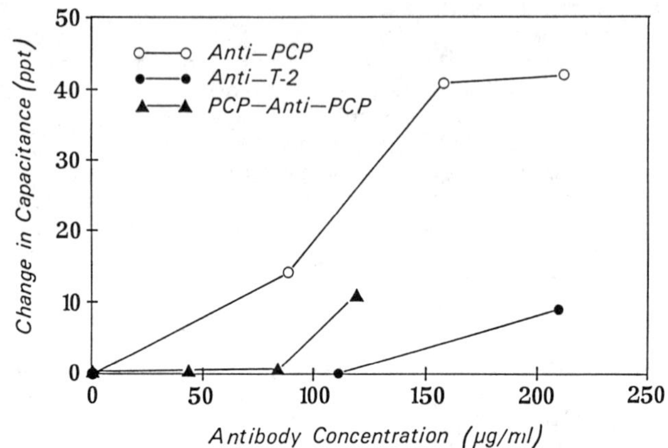

Figure 3. Decrease in capacitance on antibody binding to a pentachlorophenol capacitive affinity sensor.

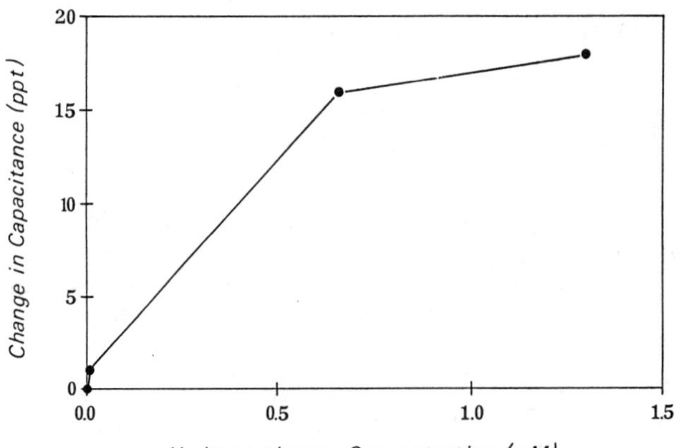

Figure 4. Increase in capacitance when a hydrocortisone capacitive affinity sensor is exposed to hydrocortisone.

added to the PCP system. This provides a control for possible non-specific adsorption of antibodies to the sensor surface. The second control shows the effect of preincubating the anti-PCP antibodies with PCP for several minutes before addition to the sensor. The PCP to anti-PCP molar ratio was 1.5 to 1. Here there is a greatly reduced response which is consistent with the free PCP occupying antibody binding sites and thus reducing their availability to react with the immobilized analyte.

Figure 4 is an example of the sensor response when free hydrocortisone is added to a sensor preloaded with anti-hydrocortisone antibody. The sensor response is expressed as the relative change in the baseline capacitance in parts per thousand. From the results presented here it is apparent that the capacitive affinity sensor is capable of detecting the reversible binding and displacement of antibodies on its surface. It is also apparent that the unbinding reaction can be brought about by the addition of free analyte. These results illustrate the general principles of sensor operation. It should be possible to produce sensors for any analyte against which an antibody has been produced. This of course comprises an enormous array of different compounds, including many agrochemicals such as pesticides. The only components that must be changed in going from one chemical to another are the immobilized analyte and the anti-analyte antibody. In addition, the technologies used in producing the antibodies and microelectronics are well known and amenable to mass production. This means that the time from design to production phase should be rapid, and also suggests that the sensor and its related electronics should be quite inexpensive to manufacture.

In conclusion, we have presented a concept for a generic sensor, the capacitive affinity sensor, based on a marriage of antibodies and microelectronics. Development of this technology is continuing with the expectation of producing commercially viable sensors for a wide variety of analytes in the near future.

Literature Cited

1. Goding, J. W. Monoclonal Antibodies: Principles and Practice; Academic Press: New York, 1983.

2. Gise, P.; Blanchard, R. Modern Semiconductor Fabrication Technology; Prentice-Hall: Englewood Clffs, New Jersey, 1986.

3. Mironov, S. L. Biochemistry and Bioenergetics, 1983, 10, 345-56.

RECEIVED February 12, 1988

Chapter 26

Immunochemical Technology in Indexing Propagative Plant Parts for Viruses and Bacteria in the Netherlands

J. W. L. van Vuurde[1], D. Z. Maat[1], and A. A. J. M. Franken[2]

[1]Research Institute for Plant Protection, P.O. Box 9060, 6700 GW Wageningen, Netherlands
[2]Government Seed Testing Station, P.O. Box 9104, 6700 HE Wageningen, Netherlands

Immunological methods, mainly ELISA, were used in the Netherlands during 1986 for c. 3.2×10^6 tests for quality indexing propagation material for viruses, viz: seed potatoes (46%), ornamentals (30%), bulbs (17%), vegetable seeds (3.5%) and fruit trees (3.5%). Tests are mostly done by inspection services under the supervision of the Ministry of Agriculture (78%). Indexing for plant pathogenic bacteria was done with ELISA for blackleg in seed potatoes (2.5×10^5 tests) and with immunofluorescence microscopy (IF) for bacteria in seeds and potted plants. Immuno-isolation and immunofluorescence colony staining, techniques which combine the advantages of serology and agar plating, show good prospects to overcome problems in the sensitivity and reliability of detection methods for plant pathogenic bacteria.

Reliable and quick detection methods which can be applied to large series of samples at low costs are important tools for quality control of agricultural products and for ecological and environmental studies. Indexing of plant propagation materials for pathogens is an important aspect of the production of high-quality plant products in the Netherlands. This approach has been of great importance in restricting the economic damage caused by viruses and bacteria, pathogens which generally cannot be killed effectively in plants by chemicals or any other method. Quality indexing has also improved the marketing value of export plant propagation materials. Laboratory indexing in addition to field inspection will give extra guarantees with regard to propagative material contaminated with economically important pathogens and organisms mentioned in quarantine lists of importing countries (4).

Serological techniques, such as latex agglutination and micro-precipitation, which were used for routine virus indexing in

0097–6156/88/0379–0338$06.00/0

the Netherlands have been replaced successfully by ELISA. The development of sensitive and reliable serological methods for fungi, nematodes, insects and bacteria is much more complicated than for viruses. This is due to the presence of many non-typical antigenic structures in these organisms which also can be found in non-pathogenic organisms and lead to false positive reactions. In spite of these limitations, reasonably reliable assays were developed to detect some economically important bacteria in seed lots (18, 31).

The aim of this paper is to present the commonly used ELISA procedures for indexing of plant pathogens and their application in the Netherlands. The schemes developed for the detection of seed-borne viruses are used to illustrate how problems in the routine procedures can be solved. In addition, ways to overcome problems in ELISA for other types of plant material and the possibilities for automating ELISA are described. Finally, the present status and strategies for improving the detection of plant pathogenic bacteria are presented.

Application of ELISA in Plant Pathology

Since its introduction in medical research in 1971, various types of enzyme immunoassays have been developed (21). The double antibody sandwich ELISA (DAS-ELISA) introduced in plant pathology in 1976 (23) has become the major test system for plant indexing in the Netherlands. This technique can be applied to a wide range of plant pathogens. Figure 1 shows the principle for the complex antigen situation for detecting bacteria (25).

The major advantages of coating the solid phase with antibodies over the direct coating of the test sample onto the solid phase are a selective concentration of the antigen from the sample onto the solid phase and the washing away of the unbound sample material. Sample compounds bound to the solid phase may increase the background reaction. Virus particles are strongly bound to the solid phase by the coating antibodies, but microscopical investigations at the Research Institute for Plant Protection has shown that the much larger bacterial cells are washed out of the wells. Only small particles such as flagellae, cell wall particles and soluble antigens are involved in the reaction. Enzymes like alkaline phosphatase have linear reaction kinetics, resulting in a better discrimination between antigen-containing and antigen-free samples when the background reaction can be kept sufficiently low (5).

Virus detection with the commonly used immunoglobulin G (IgG) type anti-virus antibodies (MW c. 150,000) conjugated with enzyme (e.g. alkaline phosphatase, MW 80,000-100,000) is more specific than the indirect procedure using non-conjugated anti-virus IgG and alkaline phosphatase-conjugated second antibodies (12, 13). This is explained from steric hindrance of the antibody by the enzyme molecules which reduces the possibilities for coupling to partly homologous antigenic determinants. The best antibody specificity can be expected from test systems using qualified monoclonal antibodies. However, in indexing plant material for viruses in the Netherlands, none of the monoclonal antibodies obtained from various origins gave better or even equally good results so far when compared to testing with polyclonal antisera.

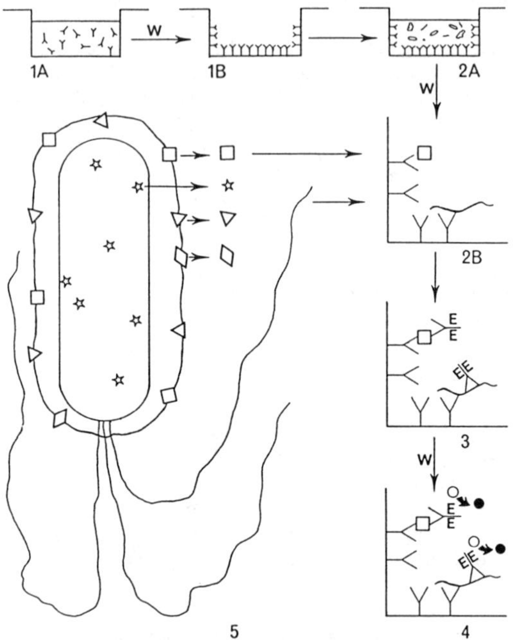

Figure 1. Double antibody sandwich ELISA in a well of a microtitre plate (W, washing with PBS-Tween). 1. Solid-phase coating step: A, incubation with antibodies in high pH buffer; B, solid-phase bound antibodies. 2. Sample step: A, sample incubation in PBS-Tween; B, detail showing antigens trapped by homologous antibodies. 3. Enzyme conjugate step: binding of conjugate molecules to homologous antigens from the sample. 4. Substrate step: color change of the substrate due to enzyme activity of the antigen-bound enzyme conjugate in the case of a positive reaction. 5. Diagram of a bacterium showing various sources of antigen. (Reproduced with permission from ref. 25. Copyright 1988 Akademiai Kiado.)

ELISA in Routine Indexing. Table I summarizes the major types of plant materials and pathogens for which routine quality indexing with ELISA is done in the Netherlands in 1986. Circa 3.2 x 10^6 tests are done for the detection of viruses, mostly in seed potatoes, ornamentals, and flower bulbs. The various viruses involved in seed potato indexing comprise 46% of these. In Table II the viruses for seed potatoes are specified. For bacteria, 0.25 x 10^6 tests were done, and only for seed potatoes. The potato blackleg organism (*Erwinia carotovora* subsp. *atroseptica*), is the only bacterial pathogen for which ELISA is applied on a large scale. Application of ELISA, however, has many limitations for its use in phyto-bacteriology compared to other serological techniques like immuno-fluorescence microscopy and traditional plating techniques (32).

Table I. Indexing propagation material with ELISA in the Netherlands in 1986

Crop	Pathogen	Tests per year[1]	Percentage
Seed potatoes	Viruses	1,465,000	42.5
	Bacteria	250,000	7.2
Bulbs	Viruses	550,000	15.9
Ornamentals	Viruses	965,000	28.0
Vegetable seeds	Viruses	110,000	3.2
Fruit trees	Viruses	110,000	3.2
Total number of tests		3,450,000	

[1] Figures are estimates based on the amounts of antisera supplied.

Table II. Indexing seed potatoes for viruses and bacteria in 1986

Pathogen	Tests per year[1]	Percentage
Potato leafroll virus	205,000	14.0
" virus A	305,000	20.8
" " M	10,000	0.6
" " S	310,000	21.2
" " X	325,000	22.2
" " Y	310,000	21.2
Total number:		
Virus tests	1,465,000	100.0
Bacterial tests (blackleg)	250,000	

[1] Figures are estimates based on the amounts of antisera supplied.

ELISA for the Detection of Seed-borne Viruses

Many aspects involved in ELISA for indexing plant materials can be demonstrated from the procedures for seed testing. For reliable indexing, proper sampling procedures are described in the handbook of the International Seed Testing Association (10). Figure 2 presents two schemes which are developed at the Government Seed Testing Station for seed testing on the base of experience with various types of seeds. The first scheme deals with small seeds which can be easily germinated in large quantities in a limited space at standard conditions adapted from the ISTA handbook (10). The second scheme deals with larger types, like pea and bean seed, for which germination is much more laborious and critical. For these seeds, dry homogenization has shown to be most efficient. The following examples illustrate the procedures and problems in testing small seeds and larger seeds.

Small Seeds. In the Netherlands, lettuce seed lots are certified with regard to lettuce mosaic virus when 2,000 seeds are found free of the agent. ELISA was developed to overcome the sometimes insufficient reliability and the high costs of the growing-on test (17) and the *Chenopodium* test (15) used earlier. Comparison of the results of the growing-on test performed under optimum conditions with those of ELISA, using subsamples containing increasing numbers of seeds, demonstrated that even one slightly infected seed in 99 healthy seeds could be reliably detected with ELISA (28). Therefore, of a seed lot 20 subsamples of 100 seeds each are tested in ELISA. The subsamples are incubated for 4 days at 20°C in a germination box (20x15x2.5 cm) with pleated wet filter paper. One subsample is spread out in one pleat, the maximum capacity per box is 40 subsamples. Prechilling for 2 days at 10°C is used to break dormancy in the seeds and to optimize uniform germination. At the end of the germination period the roots of the seedlings are grown together and form a strip which can be easily taken off the filterpaper. The extraction of the subsample can be done in about 15 seconds by grinding the strip with a power driven crusher (Pollähne, FRG). During pressing, 0.5 ml of extraction buffer is added to the rollers of the press and c. 0.8 ml of extract is collected in a 1 ml polypropylene test tube. The 20 subsample tubes are placed in a tube rack adapted to the use of an eight-channel pipet for routine loading of microtitre plates. The Pollähne crusher contains a vigorous washing system to clean the press in between subsamples in six seconds. The DAS-ELISA is performed according to Clark and Adams (6), but 0.05 M phosphate is used in the buffer for the extraction of the plant sample and for the dilution of the alkaline phosphatase conjugate. The optical density (OD) of the substrates is measured with a photometer at 405 nm. As shown in Figure 2, the results for the test sample are reported negative if all subsamples are negative. The percentage of infected seeds in a positive sample can be estimated with a statistical table from the number of positive subsamples out of the 20 subsamples (28). This test will detect predominantly seed-transmitted virus as it is carried out on germinated seeds. When dry seeds are used, the extraction of the seeds is less consistent, more laborious and false positive results may be obtained with regard to actual seed transmission because

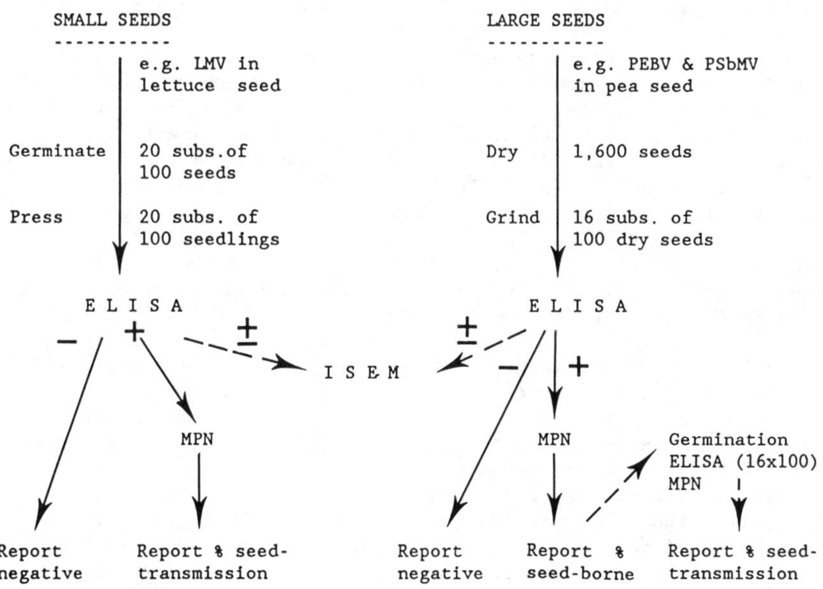

Figure 2. General strategies for ELISA routine testing of seeds for virus, illustrated for lettuce mosaic virus (LMV) in lettuce and pea early-browning virus (PEBV) and pea seed-borne mosaic virus (PSbMV) in pea. Schemes include the estimation of the percentage of infected seeds with the most probable number (MPN) method, confirmation of doubtful results by immunosorbent electron microscopy (ISEM) and check on actual seed transmission of the virus to seedlings.

seed-borne virus from non-embryonic parts of the ungerminated seed may be involved in the reaction (4).

Large Seeds. The difficult handling and the more critical germination of pea seeds has led to a test system based on the grinding of subsamples of 10 to 100 seeds into a fine flour (29). This method is used to eliminate the majority of the negative samples in the most efficient way. The testing scheme for pea seed-borne mosaic virus and for pea early-browning virus is given in Figure 2. In a seed lot, 16 subsamples of 100 seeds are prepared using semi-automated equipment. Each subsample is ground separately in a heavy coffee mill with exchangeable grinder part. A subsample is ground into a fine flour in about 30 seconds. A one ml syringe with a cut tip is used to sample c. 0.20 g of flour from the grinding head by vertical pressing the cut syringe into the flour, with the plunger fixed in the right position, until the cavity is filled and to transfer the syringe into a 1 ml polypropylene tube. After addition of c. 0.7 ml of 0.05 M phosphate extraction buffer with 0.15% Tween 20 and thorough mixing, the same ELISA procedures as used for small seeds are followed. Between a series of subsamples the grinding parts are cleaned. The standard technique can be improved however by combining sample and conjugate incubation. This modification not only reduces the number of steps in the test, but also increases the differences between healthy and virus-containing samples by reducing the background reaction (29). Positive samples in this screening can be further investigated for seed transmission by testing 1,600 seedlings in ELISA. The structure and the humidity of the soil are very critical to obtain consistent germination. The seedlings are extracted in subsamples of 100 with the Pollähne crusher and the sap diluted in extraction buffer (1:5) and tested as described for lettuce mosaic virus. Pea seed-borne mosaic virus almost has the status of a quarantine disease and all positive samples in the ELISA screening are rejected as being potentially dangerous. In case flour testing gives a very weak reaction in ELISA, confirmation is needed especially for quarantine type diseases like pea seed-borne mosaic virus. Immunosorbent electron microscopy proved to be a very sensitive tool for confirmation and enables to use the flour of the test sample that gave a doubtful ELISA result. The model for large seeds presented in Figure 2 was also successful at the Government Seed Testing Station to detect squash mosaic virus in melon seeds.

ELISA for Viruses in Vegetative Plant Materials

In addition to seeds, other plant parts as leaves, stems, bark, roots, bulbs, and tubers are tested. The ELISA procedures used for the various products will be mainly the same. However, sample preparation varies widely among plant species and plant parts. In the Netherlands, the Pollähne press is the most popular tool to prepare extracts. Relatively dry plant parts such as the leaves of some fruit trees, bark material, roots, bulbs and tubers are extracted by adding buffer to the press just above the sample during grinding.

Plant materials which give very slimy extracts, are properly diluted to obtain good results. In case of lily bulb scales, the detection of viruses is enhanced remarkably when the extracts are incubated with enzymes (especially hemicellulase, 2).

Non-specific reactions in ELISA may be caused by antibodies binding to normal plant antigens, or by non-immunological binding of enzyme. The latter can often be inhibited by proper sample dilution and/or by adding relatively high concentrations of Tween 20 (up to 2%), polyvinylpyrrolidon (up to 2%), crude egg albumin (0.1 - 4%) or skimmed milk powder (3%), or combinations of these compounds to the extraction and/or conjugate buffer. Also reducing agents are sometimes applied, e.g. sodium sulphite, 0.01 -0.02 mol/l, is used in buffers for the testing of lettuce, pelargonium, and chrysanthemum. When testing flower bulb materials, non-specific reactions are reduced by increasing the Tween concentration and by adding 1% normal horse serum to the extraction buffer (20). Egg albumin or milk powder proved helpful in reducing background reactions obtained with garlic and shallot (7), whereas egg albumin (1%) together with an increased Tween concentration (1%) were necessary to reliably test leek (Research Institute for Plant Protection).

The reliability of the test is influenced by the season, the plant part tested and its physiological state. Therefore, certain tests cannot be done reliably in summer (e.g. carnations for carnation etched ring virus). In dormant potato tubers, for some viruses reliable detection is possible only after breaking of dormancy (e.g. by rindite treatment, 9). Viruses also are not always evenly spread throughout the plant, requiring careful sampling.

Equipment for Routine Handling and Automation of ELISA

To facilitate and speed up ELISA procedures, besides the Pollähne press, 8-channel pipets and special tube racks mentioned already (Figure 3), other equipment is used or under development.

Based on commercially available multichannel pumps, equipment has been developed to fill series of microtitre plates with coating solutions, enzyme conjugates or substrates. As it is often difficult to completely remove plant extracts from the wells, high-pressure washing devices have been contructed. In many systems results are read with the nake eye. However, more and more this is replaced by (semi-) automatic reading and recording, for instance directly on sample attestation forms, by computerized photometers.

Most handlings are fairly easily mechanized or automated. However, the preparation of the sample demands an individual approach depending on the type of plant material to be tested. Major progress in this respect has been made for the automatic sample handling of bulb scales (20). The scheme of this procedure is given in Figure 4.

Indexing for Plant Pathogenic Bacteria

Besides their suitability for routine application, immunofluorescence (IF) and especially ELISA have some strong limitations for detecting bacteria (31, 32). The detection limit of IF is 10^3 to 10^4 cells per ml, whereas ELISA needs even 10 to 100 times more cells per ml (1, 26). Furthermore, false positive results may be obtained when saprophytic micro-organisms with similar antigenic determinants are present in the test samples. Compared to

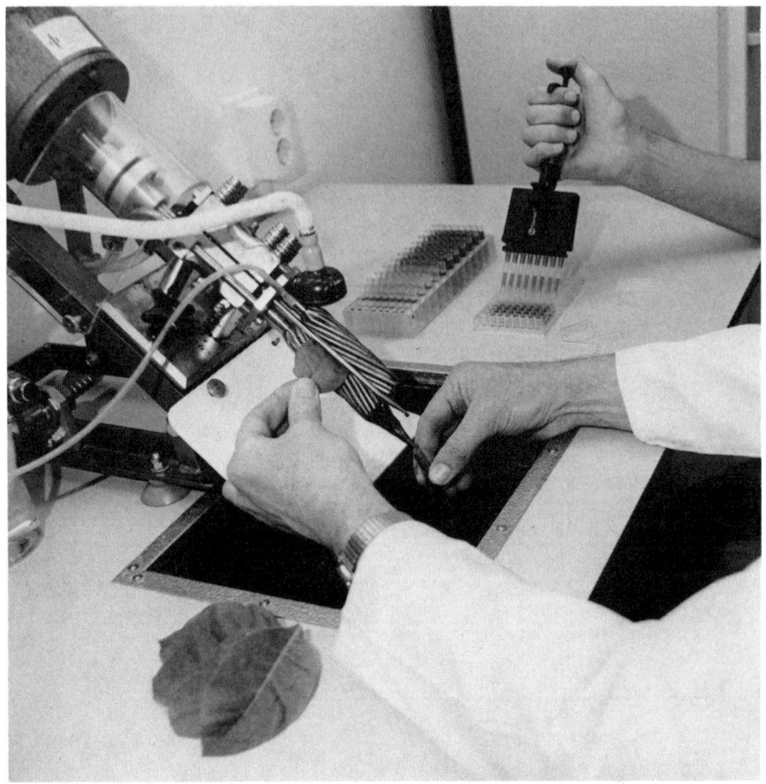

Figure 3. Equipment for routine application of ELISA: extraction of plant material with a Pollähne press and transfer of samples from tubes to corresponding wells in an ELISA microtitre plate with an 8-channel pipet.

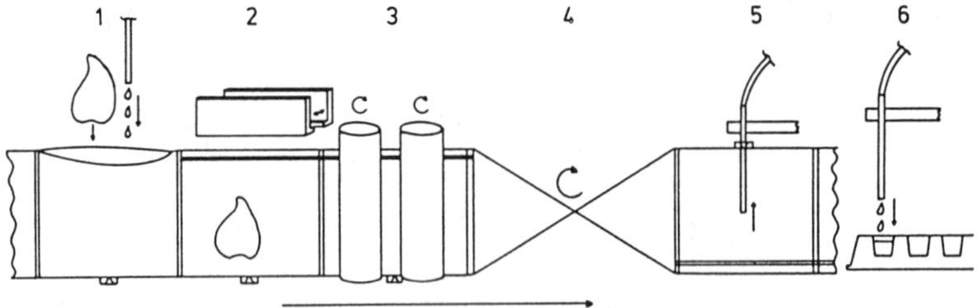

Figure 4. Scheme of the Berleco automated extraction and microtitre plate loading equipment. 1. Filling plastic bag from a belt with plant material (bulb scale) and buffer. 2. Sealing. 3. Pressing. 4. Inversion of the belt. 5. Sampling with syringe through valve. 6. Loading well in microtitre plate. (Reproduced with permission from ref. 20. Copyright 1984 Hofstad Vakpers bv.)

traditional isolation methods, the advantages of IF and ELISA are their consistency, sensitivity, speed and suitability for routine application. The use of antisera, which are checked for specificity with the aid of selected strains of saprophytes isolated from the type of plant material to be tested, limits this risk on false positive reactions (22). If the interference of saprophytes can be limited, isolation on a suitable medium is an important tool to confirm a serological positive sample.

Seed Testing with IF. The following procedure for the detection of *Pseudomonas syringae* pv. *phaseolicola* in bean seed was satisfactory for routine indexing and has been used at the Government Seed Testing Station in the Netherlands since 1982 (26). The test sample is divided into five subsamples of 1,000 seeds. The subsamples are soaked in sterile water in a plastic bag at 6 ^{o}C for 24 h. A 1 ml sample of the soaking solution is taken from each subsample after 6 h to prepare IF slides and the remaining part is stored at -18 ^{o}C. A sample with five IF negative subsamples is reported 'negative' (pathogen free or less than 10^{3} cells/ml). For IF positive subsamples, again 1 ml soaking solvent is taken 24 h after starting the incubation and analyzed by IF and plating on King's medium B. Colonies producing a blue fluorescent pigment under UV-light are biochemically tested and investigated for pathogenicity on bean seedlings. Samples with one or more positive subsamples in IF and from which the pathogen could be isolated are reported as *P.s.* pv. *phaseolicola* contaminated. Samples positive by IF and negative by plating are reported as potentially dangerous when saprophytic bacteria interfered seriously in plating.

Similar procedures are worked out for *Xanthomonas campestris* pv. *campestris* in cabbage seeds (19) *X.c.* pv. *phaseoli* (14, 26), *P.s.* pv. *pisi* (Government Seed Testing Station) and for *Clavibacter michiganensis* subsp. *sepedonicus* for testing seed potatoes in the European Community (11, 16).

Testing Seed Potato Tubers with ELISA. ELISA was first reported for detecting *Erwinia carotovora* subsp. *atroseptica* (blackleg organism) in seed potatoes (24). Starting in 1983 with large scale indexing, 2.5 x 10^{5} tests are now done per year for seed potatoes (Table II) by the General Netherlands Inspection Service for Agricultural Seeds and Seed Potatoes (NAK). Since 1987, seed potatoes are simultaneously tested for *E. chrysanthemi*. Reasonable correlations were obtained between lab testing for *E.c.* subsp. *atroseptica* (50 tubers) and symptom development in the NAK control field (400 plants) in spite of the limitations of ELISA (8).

New Approach in Detecting Bacteria

The sensitivity of serological tests is not influenced by the relative incidence of saprophytes in the sample unless they react with the antiserum. Major critical parameters of ELISA and IF for reliable risk assessment of plant material are the specificity of the serological reaction and the status of the bacterial cells, viz. viable cells, dead cells and/or cell components of the homologous organism. Isolation of the target organism in principle enables a 100% reliable identification, and it allows pathogenicity testing.

The isolation and identification procedures theoretically enable the detection of one or a few colony-forming units in the total test sample if the interference of saprophytic micro-organisms can be eliminated. For some pathogenic bacteria the variable sensitivity of the traditional isolation methods can be partly overcome by the use of selective media.

Immuno-isolation and immunofluorescence colony staining, are designed to combine the advantages of serology viz. selective recognition of the target organism among saprophytes, with those of plating on nutrient media viz. the sensitive isolation of the causal organism for optimum reliability (27, 30, 32).

Immuno-isolation. This technique is based on: (a) the selective trapping of the target organism onto a solid phase coated with antibodies, (b) washing away unbound saprophytes and (c) isolation on a suitable growth medium for identification. Antibody-coated rods, beads, or membranes can be used for this purpose. The antisera for immuno-isolation should be selected or specially prepared against cell wall linked antigens to improve the trapping efficiency and binding strength of homologous bacterial cells. Strong solid phase binding also makes possible selective elution of non-selective and weakly cross-reacting micro-organisms with high-molarity buffers.

Immunofluorescence Colony Staining. This technique is based on: (a) mixing the test sample through a suitable agar medium, (b) incubation of the medium until microscopically visible colonies are formed, and (c) staining of the homologous colonies in the agar with fluorochrome-labeled antibodies for detection with incident UV light microscopy at low magnification.

Agar-mixed sample incubation to obtain micro-colonies gives a 100 to 1000 times higher capacity per plate because of the three dimensional space for colony formation and the reduced biomass production the antagonistic interactions between colonies is reduced in comparison with traditional agar surface plating.

Immunofluorescence colony staining enables the specific recognition of the target colonies. Less than 100 cells of *E. chrysanthemi* per ml can be detected in cattle slurry and potato peel extract at ratios of saprophyte to pathogen colonies > 1,000:1. Isolation of bacteria from IF-positive colonies can be done with a fine needle or glass capillary. The method will be evaluated for other plant pathogenic bacteria.

Optimum sensitivity and specificity can be obtained when immuno-isolation and immunofluorescence colony staining are applied in combination (32). Immunofluorescence colony staining can be adapted into routine procedures, e.g., by using wells of tissue culture plates to replace petri dishes. Reading can be automated using microprocessor controlled video scanning equipment.

Strategies for Improvement of Antiserum Specificity for Bacteria. The possibility of isolating bacteria from IF-stained colonies provides a very efficient tool for investigating the risk on false positive reacting micro-organisms. This is done by testing healthy material of various origins for the presence of bacteria which form colonies that cross-react in IF. Isolation of these cross-reacting strains is of major importance to identify typical antigens in the target bacteria. The knowledge of these antigens will improve the production of more specific polyclonal antisera and of monoclonal antibodies for a more reliable serological detection and identification of bacteria, as was demonstrated for LPS antigens of *Erwinia carotovora* subsp. *atroseptica* (3).

Conclusions

For viruses, over three million tests are done each year to index various types of plant propagative materials with ELISA in the Netherlands. Polyclonal antisera gave good results for all important viruses involved in the testing. The main problem in the development of a test was with optimizing the preparation of the plant sample extracts. In the future, attention will be given to improving the handling of large series by automated procedures.

Compared to viruses, many problems are still associated with the routine indexing of plant materials for pathogenic bacteria. ELISA is less sensitive than immunofluorescence microscopy and both methods involve a risk of false positive reactions due to cross-reacting saprophytic micro-organisms, particularly when antisera are not properly tested.

New methods are presented based on combining immunotrapping, isolation, and identification of colonies directly in agar media. Of these methods, immuno-isolation combined with immunofluorescence colony staining especially has good prospects for detecting low concentrations of the pathogen and for efficient isolation of cross-reacting organisms.

The availability of a representative collection of cross-reacting organisms for a certain type of sample material is a major requirement in the research for typical antigens of the target bacterium, in order to produce more specific monoclonal or polyclonal antisera.

Literature Cited

1. Barzic, M.R.; Trigalet, A. *Agronomie* 1982, *2*, 389-398.
2. Beijersbergen, J.C.M.; Van der Hulst, C.Th.C. *Neth. J. Pl. Path.* 1982, *86*, 277-283.
3. DeBoer, S.H.; McNaughton, M.E. *Phytopathology* 1987, *77*, 828-832.
4. Bos, L. In *Plant Health and Quarantine in International Transfer of Genetic Resources*; Hewitt, W.B.; Chiarappa, L., Eds.; CRC Press Inc.: Cleveland, Ohio, 1977; pp. 39-69.
5. Clark, M.F. *Ann. Rev. of Phytopathol.* 1981, *19*, 83-106.
6. Clark, M.F.; Adams, A.N. *J. gen. Virol.* 1977, *34*, 475-483.
7. Delecolle, H.; Lot, H.; Michel, M.J. *Phytoparasitica* 1985, *13*, 266-267.
8. Endel, E.A.; Van Vuurde, J.W.L. In *EAPR Abstracts of Conference Papers and Posters*; R. Rasmussen: Denmark, 1987; pp. 190-191.
9. Gugerli, P.; Gehriger, W. *Potato Research* 1980, *23*, 353-359.
10. International Seed Testing Assiociation. *Seed Sc. and Techn.* 1985, *13*, 299-355.
11. Janse, J.D.; Van Vaerenbergh, J. *EPPO Bulletin* 1987, *17*, 1-10.
12. Koenig, R. *J. gen. Virol.* 1978, *40*, 309-318.
13. Koenig, R. *J. gen. Virol.* 1981, *55*, 53-62.

14. Malin, M.E.; Roth, D.A.; Belden, E.L. Plant Disease 1983, 67, 645-647.
15. Marrou, J.; Messiaen, C.M. Proceedings of the International Seed Testing Association 1967, 32, 49-57.
16. Miller, H.J. Potato Research 1984, 27, 33-42.
17. Rohloff, I. Proceedings of the International Seed Testing Association 1967, 32, 59-63.
18. Schaad, N.W. Plant Disease 1982, 66, 885-890.
19. Schaad, N.W.; Donaldson, R.C. Seed Sci. and Technol. 1980, 8, 383-391.
20. Van Schadewijk, A.R. LAB/ABC, jaargang 5, november 1984, pp. 30,31.
21. Tijssen, P. Practice and Theory of Enzyme Immunoassays; Elsevier: Amsterdam 1985; pp. 541.
22. Trigalet, A.; Samson, R; Coleno, A. Proc. 4th Int. Conf. Plant Path. Bact. Angers, 1978, pp. 271-288.
23. Voller, A.; Bartlett, A.; Bidwell, D.E.; Clark, M.F.; Adams, A.N. J. gen. Virol. 1976, 33, 165-167.
24. Vruggink, H. Proceedings of the 4th International Conference on Plant Pathogenic Bacteria, Angers, 1978, pp. 307-310.
25. Vruggink, H.; Van Vuurde, J.W.L. In Methods in Phytobacteriology. Klement, Z., Rudolph, K.; Sands, D.C., Eds.; Akademiai Kiado: Budapest 1988.
26. Van Vuurde, J.W.L.; Van den Bovenkamp, G.W.; Birnbaum, Y. Seed Sci. & Technol. 1983, 11, 547-559.
27. Van Vuurde, J.W.L.; Van Henten, C. Seed Sci. and Technol. 1983, 11, 523-533.
28. Van Vuurde, J.W.L.; Maat, D.Z. Seed Sci. & Technol. 1983, 11, 505-513.
29. Van Vuurde, J.W.L.; Maat, D.Z. Neth. J. Pl. Path. 1985, 91, 3-13.
30. Van Vuurde, J.W.L.; Ruissen, M.A.; Vruggink, H. In Proc. 6th Int. Conf. on Plant Path. Bact., 1987, M. Nijhof, Dordrecht, pp. 835-842.
31. Van Vuurde, J.W.L. In Proc. 6th Int. Conf. on Plant Path. Bact., 1987, M. Nijhof, Dordrecht, pp. 799-808.
32. Van Vuurde, J.W.L. EPPO Bulletin 1987, 17, 139-148.

RECEIVED May 4, 1988

MONITORING GENETIC ENGINEERING PRODUCTS IN THE ENVIRONMENT

Chapter 27

Methods Used To Track Introduced Genetically Engineered Organisms

Philip C. Kearney[1] and James M. Tiedje[2]

[1]Natural Resources Institute, U.S. Department of Agriculture, Beltsville, MD 20705
[2]Departments of Crop and Soil Sciences and Departments of Microbiology and Public Health, Michigan State University, East Lansing, MI 48824

Cultural, metabolic, genetic and immunological techniques that could be used to trace genetically engineered microorganisms released into the environment are reviewed. The advantages, disadvantages, and in several cases the sensitivity are considered for each method. Special attention is focused on gene probe technology, since this technique appears to hold considerable promise for monitoring released microorganisms.

Many parallels exist between the current concerns about the intentional release of genetically engineered microorganisms (GEMs) into the environment and concerns expressed about the release of massive amounts of organic pesticides in the environment during the early 1960's. The issues, however, are essentially the same, i.e, the stability, movement, and the safety of altered products. Concurrent with the environmental pressures mounted against the older chlorinated hydrocarbon insecticides was the emerging development of gas chromagraphic (glc) systems that offered increasing levels of sensitivity with a growing number of sophisticated detectors that provided highly specific compound recognition. Unfortunately, as we see in the following section, glc does not offer the selective, automated systems for the rapid or sensitive detection of the altered DNA molecule or the microbial vehicle harboring that DNA. The chlorinated hydrocarbons are stable, lipophilic structures possessing one or more halogens that greatly facilitated detection in the parts per millon range, while modern instrumentation permits even greater sensitivity. By contrast, the DNA molecule is a liable, polar substance in a fragile carrier. More troublesome is the ability of the carrier microorganism to mutate, conjugate, and poliferate in a diverse community. We are faced with a far more complex problem in attempting to monitor the elusive gene in a complex matrix than the stable pesticide with many more easily detectable markers. By comparison, we have only a limited number of methods to detect the specific fragment of DNA of interest in a rather facile delivery system. The object of this section of the symposium is to examine

0097–6156/88/0379–0352$06.00/0

the methods now available to us and the suitability of those methods for potentially vast monitoring programs to allay current apprehensions about the real or preceived damages of intentional release.

CURRENT DETECTION SYSTEMS

A review of currently available detection methods for GEMs has been prepared by Tiedje (1) and McCormick (2). This overview follows these reviews closely. The methods can be conveniently grouped under three major headings of culture and metabolic techniques, genetic techniques and immunological techniques. A detailed, in-depth discussion of each of these options is beyond the scope of this overview paper although gene probe techniques are considered in greater detail due to their promising role in monitoring studies. A few comments about the advantages and disadvantages of these methods is appropriate.

CULTURE AND METABOLIC TECHNIQUES

Plating

Counting organisms on agar plates is a classical technique used extensively by the microbiologists. The advantages are that the method is easy, inexpensive, measures expression, is subject to statistical analysis and can be very sensitive (1). The disadvantages are that the organisms must be culturable, it must possess an inserted marker (e.g., antibiotic resistance), that marker must be stable, and that marker must not disrupt the metabolism or ultimately the survival of that microorganism. For example, rifampin resistance in denitrifier strains has been shown to reduce maximum growth rate (3). The level of sensitivity of plating techniques is in the range of 10-100 cells/gram of soil.

Colormetric Media

This method is based on an obvious color change in the target organism. The advantages are that it is very specific and a good screening tool. Drahos et al. (4) have used this method for monitoring a recombinant *Pseudomonas fluorescens* expressing the *Bacillus thuringenisis* endotoxin. *P. fluorescens* is a naturally occurring root inhabiting microorganism that makes it an ideal candidate as a delivery vehicle for natural pesticides. They developed a sensitive, selectable marker in *P. fluorescens* based on the inability of most fluorescent *Pseudomonas* to use lactose as a sole source of carbon. The *E. coli* lac Y and Z genes were inserted into the engineered *P. fluorescens* chromosome. The Z gene produces B-galactosidase and the lac Y gene encodes for lactose permease. Thus the transformed bacteria can grow on x-gal, a chromogenic substrate. Metabolism produced a characteristic blue-green colony, quite distinct from the pale color of the untransformed cells. In the final analysis, three discreet markers were used to detect the transformed *P. fluorescens* in soils, i.e., B-galactosidase, fluorescent pigments, and rifampicin resistance. These marker systems enabled the detection of lac+ transformants at a sensitivity of 10 cells/gram of soil.

Direct Staining

This is a refinement of the plating technique that gives a total count of living organisms. It overcomes the problem of non-culturability. The addition of naldixic acid to a yeast extract fortified agar plates produces an elongated cell which can be stained with, for example, acridine orange stain, concentrated on a filter disk and counted. Fluorescent monoclonal antibodies can also be added to the plate to give a count of viable organisms specific to an engineered subgroup.

Enriched Media

The addition of soil particles directly in the media has the advantage of being a very classical method for which the soil microbiologist is quite familiar. This approach magnifies the population of interest, but it suffers from a quantitative standpoint, lacks accuracy, and requires a very sensitive screen to be useful.

Spent Media Analysis

This is a system based on the identification of microorganisms by the analysis of the culture medium for unique byproducts of that specific group of organisms. The advantages of this approach is that the sensitive, automated chemical systems, like glc or hplc, could be used. The disadvantages to this approach are obvious, i.e., that a tremendous database on the qualitative profile of those metabolic products would have to be available. Given the complexity of that analysis, it is unlikely that this is a practicable solution.

GLC and HPLC

As discussed in the introduction, these tools have a very limited role in direct microorganism tracking. They are used in situations where microorganisms can be identified by detection of characteristic profiles of fatty acids, lipids and steroid linkages. This would also require a huge database that does not exist.

Protein Product Fingerprinting

This method is based on the identification of microorganisms by analyzing polyacrylamide gel electrophoresis patterns produced by radiolabeled proteins. This again would require a detailed knowledge of those patterns. This method is best used to distinguish between closely related strains.

GENETIC TECHNIQUES

Gene Probes

These techniques rely on the specificity provided by the hybridization between two complementary (or nearly so) sequences of nucleic acids. Sequences can be selected for probing that are very common among organisms, for functional or phylogenic groups, or very specific for only one strain of an organism.

The main advantages of this methodology includes: (1) the ability to detect organisms that cannot be cultured because they are stressed or fastidious, which is a common limitation in the study of the total microbial population in nature, (2) the ability to detect organisms without the necessity of having a specific selectable marker, such as antibiotic resistance or requiring the development of other selective media, (3) the ability to track the gene of interest regardless of whether it is expressed or transferred to another organism, and (4) the potential for higher sensitivity than is possible by any other methods. The disadvantages of the gene probe method are that it is more complex which reduces the number of samples that can be processed, quantitation of the number of gene copies per gram of sample is still difficult and imprecise. The use of the method in microbial ecology is still in its infancy so that problems with reproducibility, adaptability to different soil composition, different organisms and probes will take time to resolve before the method can be used effectively.

A number of research groups are now actively pursuing the gene probe method for use in the environment, and improvements in its sensitivity and ease of use are expected in the near future. The directions taken by many of the U.S. workers have been reviewed by Holben and Tiedje (5), and the recent REGEM Conference Proceedings reports the work of many international workers on this subject (6). The following is a brief summarization of important features and advances in the use of gene probes for detecting microorganisms in nature.

Selection of Probes

There are three basic classes of genes which have been used as probes. Each has been used successfully, and at this stage of knowledge, none can be argued to be generally preferred. These classes are probes targeted for ribosomal RNA (rRNA), for randomly cloned sequences of DNA unique to the organism of interest, and for the engineered sequence itself. If one is tracking a genetically engineered microorganism (GEM) the latter is usually used because the probe is already available and it is this gene that is of greatest interest for GEMs. For non-GEMs, the first two probe types are more likely to be used. Focusing on rRNA has the advantage of potentially providing probes common for several groups or specific for one strain depending on whether one selects a conserved or variable region of the 16S rRNA.

Obtaining Sample DNA

Three approaches have been used to recover DNA from natural samples for probing. The first reported involved cell lysis and extraction of total DNA in sediment by SDS and alkaline treatment (7). This method has the advantage of a more quantitative recovery of DNA from the soil but has the disadvantage of requiring more clean-up of the DNA since other problem contaminants are also extracted.

A second approach is to first elute the microbial cells from soil before lysis thereby protecting the DNA from soil contaminants. This method also has the advantage in that the DNA is retained in large molecular weight fragments (approximately 50 kb) so that restriction digestion, and gel separation yields discrete fragments that allow analysis by Southern blots (8). Both of the above methods give more DNA than can be accounted for in the population detected by plate counting implying that DNA from non-culturable organisms is also being recovered.

The third method is to first foster growth of the population of interest, but after dilution and separation by the most probable number (MPN) protocol. The cultured cells in wells are then lysed and probed with the gene (9). In this case the presence or absence of the target gene is used with a MPN table to estimate the original population density of the organism. The advantage of this method is that the target DNA is amplified and purified by growth. Detection limits of 10 to 100 organisms/g, depending on the particular organisms growth response, were obtained with this method.

The Choice of Detection Method

Once the probe and sample have been obtained there are several choices for detection based on the experimental goals. The detection methods vary in sensitivity, ease of use, stability of probe, and whether information on gene location is desired. The most common analysis format is by dot or slot blot in which the DNA is fixed to a filter, the ^{32}P-labeled probe added, non-hybridized DNA washed away and the amount of radioactivity remaining analyzed usually by autoradiography. Using the format and the M13/^{32}P probe systems of Holben et. al. (8), 1000 organisms/g soil were detected. More information is obtained, however, if the DNA is cut by restriction enzymes, separated according to length in gels, and the location of the different sized fragments detected by autoradiography after Southern transfer to blots. Multiple populations of organisms with the same target can all be detected in one sample and any change in gene position either within its host or if transferred to other organisms can be detected because the target gene would have different flanking regions and thus be in different sized fragments. Using the M13 probe, 10^4 organisms/g could be detected in Southern blots (8).

If the goal for detection is sensitivity, the polymerase chain reaction (PCR) methodology offers exciting new possibilities. With this method a primer unique to the target is added to the environmental DNA sample, and through repeated cycles of synthesis of complementary strands and denaturation, the target is greatly magnified thereby increasing sensitivity. When Steffan and Atlas (10) added a known organism (and target gene) to sediment, they were able to detect one organism/g. A very different approach is to detect the microbial cell itself in a microscopic field by using gene probes to identify the particular cells of interest. This method, developed by Pace's group (11), is based on probes to the 16S rRNA. Because cells have many ribosomes, there is sufficient target in an individual cell so that the signal can be easily recognized by microautoradiography (^{35}S or ^{32}P) probes or visually by fluorescent probes. This method preserves the information on cell morphology and perhaps location in a consortium or on surfaces if that information is important.

All detection so far in microbial ecology has been with ^{32}P-labeled probes. Because of the radiation exposure and short shelf-life of these probes, non-radioactive probes would be of great value. Such probes have been developed, e.g., the biotin-labeled residues which yield a color reaction, but they have not yet proved sensitive or specific enough to generate any interest by microbial ecologists.

Gene Sequencing

At this time, gene sequencing can probably be dismissed as a reasonable monitoring tool because of the high labor input, cost and large database to interpret the results.

IMMUNOLOGICAL TECHNIQUES

Immunofluorescence

This procedure utilizes a species-specific antiserium and fluorescent dye. On examination with an epifluorescene microscope, stained cells appear as a peripheral green band visible beneath the cell wall. The method has been used to study Rhizobium ecology and to monitor the survival of pathogens in an aquatic environment (12). The advantages are speed, and it detects organisms that do not grow. The disadvantages are the small size of the sample analyzed, lack of specificity, and unknown genetic composition. The sensitivity is about 1000 cells/gram of soil. It is not likely to be generally useful for monitoring GEMs because it cannot distinguish the GEM from its parent or close relatives in nature.

Immunoradiography

The same principle applies here as with immunofluorescence, except that the antiserum is coupled with a labeled compound.

Flow Cytometry

This is a cell sorting technique that uses a laser beam and fluorgenic staining techniques to measure cell size, and cellular components in a single cell in a large population. This automated cell-by-cell analytical tool has had limited environmental monitoring application. It's advantages are speed of analysis and sensitivity. The disadvantages are expense and ease of operation.

Affinity Concentration

This is a method of concentrating low populations of released microorganisms from field samples. One option being considered is a selective immunobiological fixation on polystyrene beads or other supports coated with a suitable monoclonal antibody.

Literature Cited

1. Tiedje, J. M. Proc. Pros. Phys. Biol. Contain. Genet. Engin. Organ. ERC-114 1985, 115.
2. McCormick, D. Biotechnology. 1986, 4, 419.
3. Smith, M. S.; Tiedje, J. M. Can. J. Microbiol. 1980, 26, 851.
4. Drahos, D. J.; Hemming, B. C.; McPherson, S. Biotechnology. 1986, 4, 439.
5. Holben, W. E.; Tiedje, J. M. Ecology. 1988, 69, 561.
6. REGEM 1. First International Conference on the Release of Genetically Engineered Microorganisms. Programme and Information. Cardiff Wales, UK. April 5th-8th, 1988.
7. Ogram, A.; Sayler, G. S.; Barkay, T. J. Microbiol. Methods. 1987, 7, 57.
8. Holben, W. E.; Jansson, J. K.; Chelm, B. K.; Tiedje, J. M. Appl. Environ. Microbiol. 1988, 54, 703.
9. Fredrickson, J. K.; Bezdicek, D. F.; Brockman, F. J.; Li, S. W. Appl. Environ. Microbiol. 1988, 54, 446.
10. Steffan, R.; Atlas, R. Appl. Environ. Microbiol. 1988, in press.
11. Giovannoni, S. J.; DeLong, E. F.; Olsen, G. J.; Pace, N. R. J. Bacteriol. 1988, 170, 720.
12. Colwell, R. R.; Brayton, P. R.; Grimes, D. B.; Roszak, S. A.; Huq, S. A.; Palmer, L. M. Biotechnology 1985, 3, 817.

RECEIVED June 19, 1988

Chapter 28

Monitoring *Bacillus thuringiensis* in the Environment with Enzyme-Linked Immunosorbent Assay

Peter Y. K. Cheung[1] and Bruce D. Hammock[2]

[1]Terrapin Diagnostics, 165 Eighth Street, Suite 306, San Francisco, CA 94103

[2]Departments of Entomology and Environmental Toxicology, University of California, Davis, CA 95616

Use of *Bacillus thuringiensis* (BT) in insect control in many ways exemplifies the advantages and problems with genetically engineered pesticides. Being biochemical and/or biological in nature, classical analytical procedures are often not applicable to the monitoring of these agents. A two step competitive enzyme linked immunosorbent assay (ELISA) procedure was developed to the BT *israelensis* δ-endotoxin. Application of the ELISA for monitoring the δ-endotoxin was examined. A direct correlation between immunochemical assay and bioassay was determined under laboratory conditions. The ELISA procedure did not perform well under acidic conditions but worked well at alkaline pH's. Presence of bovine serum albumin improved the ELISA by reducing background. When water samples collected from the field were spiked with the δ-endotoxin and tested, sensitivity and precision of the ELISA procedure decreased regardless of the pH tested. Two procedures, anion exchange chromatography and dot-blotting, were explored for sample clean-up and concentration when sample work-up is needed. In addition, problems with assay development are also discussed. As more biochemical agents are introduced into the agricultural market, derived either from classical or molecular approaches, it is critical that analytical methods are in place for quality control and to monitor the presence of these agents in the environment.

The introduction of molecular genetics into the field of pesticide chemistry, has in a few years, resulted in the development of a new class of pest control agents. These agents have presented a new challenge to the environmental chemist. In many situations, classical GLC and HPLC procedures are no longer applicable to the analyses of these complex biological and/or biochemical materials.

0097–6156/88/0379–0359$06.00/0

Due to this lack of analytical methods, bioassays are often the only viable method for residue analysis. A good example are the larvicidal toxins of *Bacillus thuringiensis* (BT).

The BT endotoxins are large proteinaceous molecules with excellent efficacy as well as host specificities and are easily degraded in the environment (1-8). These ideal traits have made BT toxins among the most intensely studied bio-control agents. Resources are also devoted to improve the performance of these toxins by increased expression and by expression in alternate hosts and plants (9,10; also see papers in this volume by M. Veck and D. Witt). It is apparent that the next generation of biopesticides will not be exclusively fermentation products, but genetically engineered "live" materials as well. Although these biochemical pesticides will less likely be environmental contaminants, it is of great concern to the environmentalist and regulatory agencies that we may be polluting the "wild" gene pool with human-made genetic materials. At the same time, the development of monitoring technology for the implementation and regulatory strategies of these new agents are lagging. However, this does not mean that analytical methods are not available. On the contrary, sophisticated methods such as immunochemical assays and genetic probes are used regularly in the research, production, and quality control of biotechnology products. It is therefore reasonable to adapt these analytical techniques to actual residue procedures. Immunoassays may well be the best analytical tool for monitoring these new pesticides as well as for monitoring the actual expression of these genetic materials in the environment.

Enzyme linked immunosorbent assays (ELISA) have been successfully developed for the detection and quantitation of the BT *kurstaki* and BT *israelensis* toxins (11-14). These ELISA are used routinely to supplement bioassays in monitoring the production and quality control during fermentation process. Reported here are the laboratory studies of the application of ELISA for monitoring the BT toxin in environmental samples.

Antigen Preparation and Antiserum Production

BT *israelensis* was isolated from a commercial preparation provided by Sandoz Inc. Procedures for crystal production, isolation, and dissolution were as described previously (3,13,16,17). The alkaline solubilized δ-endotoxin was dialyzed against a 50 mM sodium acetate buffer (pH 4.5) until a precipitate developed. The precipitate was removed by centrifugation and the supernatant was readjusted to pH 8.5 with a 0.1 M Tris solution. This toxin solution was then fractionated on a DEAE-anion exchange column (30 ml bed volume; DE-52, Whatman) pre-equilibrated with a 50 mM Tris-HCl buffer at pH 8.5. Effluents from the anion exchange chromatography were bioassyed against 4th instar larvae of *Aedes aegypti*. Toxic fractions were pooled and used in the immunization of New Zealand white albino rabbits (Batton and Kingman, Fremont, CA). Rabbits were immunized with three initial injections (sc) in Freund's Complete Adjuvant, and then three booster shots in Freund's Incomplete Adjuvant a month later. Serum was collected by exsanguination in a single bleeding.

Results of DEAE-anion exchange chromatography are shown in Figure 1. Partially purified BT *israelensis* crystal proteins (25 to 28 kilodalton peptides) were used in the immunization of rabbits. Western Blotting (Figure 2) showed that this antiserum reacted strongly with the 25- to 28- kDa peptides. Cross reactivity with the other BT *israelensis* crystal proteins and the BT *kurstaki* toxin was minimal.

ELISA

A two-step competitive ELISA was developed for quantitating the alkaline solubilized BT *israelensis* δ-endotoxin. The ELISA procedure used was a modification of methods described by Voller et al (18) and by Wie et al (11). Samples of dissolved toxin were incubated overnight at 25°C with the anti-BT antiserum in phosphate buffered saline (PBS/Tween; 0.14 M sodium chloride, 1.5 mM dibasic potassium phosphate, 8.3 mM monobasic sodium phosphate, 2.7 mM potassium chloride, 0.02% sodium azide, 0.05% Tween-20, pH 7.2). Microtiter plates were coated overnight at 4°C with a toxin solution (1μg/ml, 200μl/well). After overnight incubation, the microtiter plates were washed 3X with PBS/Tween, 200μl of each reaction mixture was transferred into individual wells and allowed to react for two hours at room temperature. The microtiter plates were then washed 3X with PBS/Tween, 200μl of a goat anti-rabbit IgG alkaline phosphatase conjugate (1:3000, Miles Laboratories Inc.) was added to each well and allowed to react for an additional 2 hours. After three more washings with PBS/Tween, 200μl of substrate solution (0.1% p-nitrophenol phosphate in 10% diethanolamine and 0.1% magnesium chloride, pH 9.8) was added. Enzyme reactions were stopped after 30 min by addition of 50μl of a 3N NaOH solution. The optical density of each reaction was measured by a Titertek EIA reader (Flow Laboratories Inc.). Data were collected by the use of a personal computer and a PC-EIA software package (Dorian Software Co., Maryland), then transferred with a program developed in house (19) and analyzed with a Lotus 1-2-3 software package (Lotus Corporation). A standard ELISA titration of the BT *israelensis* δ-endotoxin is shown in Figure 3. The limits of detection by the ELISA procedure ranged from 15 ng/ml to 1000 ng/ml. No cross reactivity with BT *kurstaki* toxin was detected.

To test the correlation between immunoassay and bioassay, fourteen BT *israelensis* formulations (Biochem Products, US Division) with known biological activities were assayed with ELISA. Particulate matter (including crystals) from the formulations were first collected and washed with distilled water by centrifugation. The pellets were extracted overnight with a 0.5% sodium carbonate plus 0.02% azide solution at 4°C. After alkaline extraction, the pH's of the supernatants were adjusted to 7.5-8.0. These extracts were then diluted with PBS/Tween and assayed by ELISA. Concentration of BT *israelensis* δ-endotoxin in each extraction was determined by comparing with a standard curve (Figure 3). Results of these comparisons are shown in Table I.

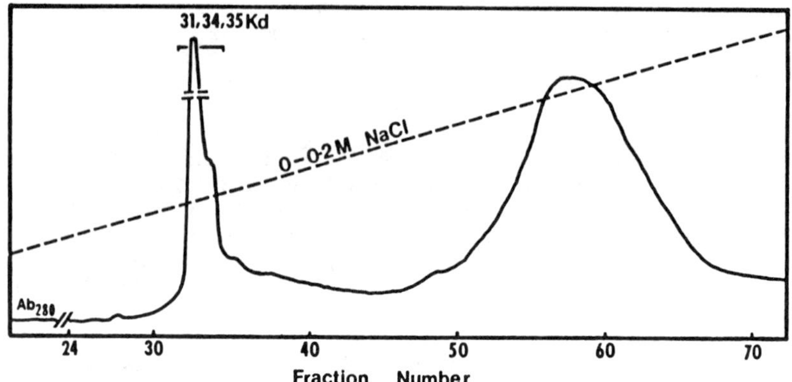

Figure 1. Separation of BT *israelensis* δ-endotoxin by DEAE-anion exchange chromatography. Fractions 56 to 65 were collected and pooled and used for immunizing rabbits.

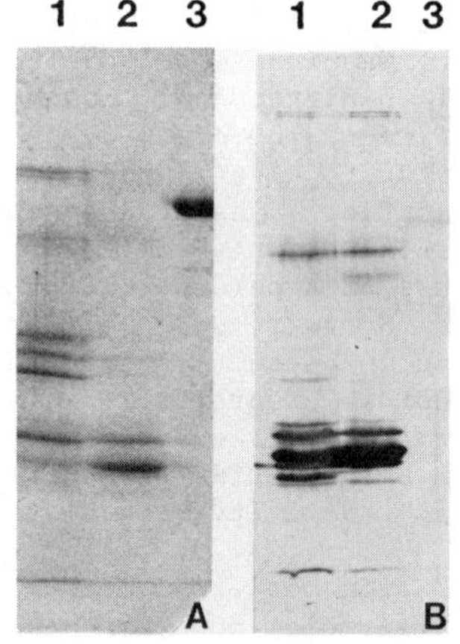

Figure 2. Western Blot of BT *israelensis* and BT *kurstaki* δ-endotoxins. A. SDS-PAGE separation of crystal proteins before transfer onto nitrocellulose membrane. Lane 1, BT *israelensis* protein; Lane 2, Protein used for immunization; Lane 3, BT *kurstaki* toxin. B. Western Blotting of the same proteins using a rabbit antiserum raised against proteins in lane 2.

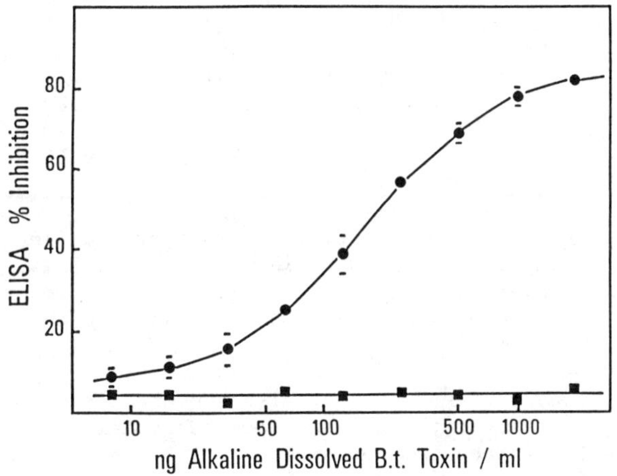

Figure 3. Indirect ELISA inhibition standard curve using the rabbit antiserum to partially purified BT *israelensis* toxin. ●---● *israelensis* proteins; ■---■ *kurstaki* proteins.

Table I. Comparison of ELISA and Bioassays of Commercial Formulations

Formulations	LC_{50} (mg per L)	IPS/78 LC_{50} (mg per L)	IU/mg	μg ELISA/mg Formulation	IU-μg ELISA
SWI 1	0.8581	0.3768	440	13.13	33.51
SWI 2	10.8000	0.5000	46	8.45	5.44
SWI 3	0.2736	0.2650	969	11.96	81.02
SWI 4	0.2815	0.2650	942	/	/
SWI 5	0.3138	0.2650	844	11.41	73.92
SWI 6	0.3820	0.2650	694	9.69	71.62
SWI 7	0.6284	0.3998	648	10.10	73.16
SWI 8	0.6193	0.3768	608	8.85	68.66
SWI 9	0.7810	0.3814	488	5.35	91.20
SWI 10	0.3932	0.2405	612	8.61	71.08
SWI 11	0.4817	0.2405	499	8.44	59.08
SWI 12	0.5876	0.3768	641	9.71	65.28
SWI 13	0.6512	0.3274	503	9.82	51.22
SWI 14			799	10.20	78.33
SWI 15			799	9.61	83.14

If both assays are measuring the toxic element of the δ-endotoxin, numbers in the last column should be identical for all 14 formulations. An average of 69.3 ± 14.9 IU was found to be associated with each microgram of toxin detected with the ELISA procedure.

As in bioassays, immunoassay is also a quantitation of the toxic principle(s) within the BT *israelensis* crystal toxin. Thus it is possible to express immunochemical activities in terms of biological activities or vice versa (ie. number of international units (IU) per μg of ELISA detected protein, or LC_{50} = μg of ELISA detected materials). In Table I, the biological activities of the formulations are expressed in IU/mg formulation (column 4), and the immunological activities are expressed in μg of toxin detected by ELISA per mg of formulation (eg. column 5, μg ELISA/mg formulation). A correlation between bioassay and immunoassay is demonstrated when the biological activities are divided into the immunological activities for each formulation (column 6). The resulting unit will be IU/μg of toxin detected by ELISA (eg. IU/μg ELISA). Theoretically, this unit should be identical for all the formulations when the same strain of BT *israelensis* was used. For the 14 formulations, an average of 69.3 ± 14.9 IU was found to be associated with each microgram of toxin detected using the ELISA procedure. The coefficient of variance for the ELISA procedure (21%) is very close to that of the bioassay. Toxicity of the IPS/78 standard powder has a 22% coefficient of variance (LC_{50} = 0.3585 ± 0.0815 mg/ml).

Environmental Samples

It has been widely observed that field efficacy of BT *israelensis* toxin is short term (7,8,9,23). Biodegradation is often suspected as one of the major causes of lost efficacy, because the proteinaceous crystals are an excellent substrate for microorganisms. To examine how biodegradation affected sample handling and sample work-up, commercial BT *israelensis* material was suspended in distilled water plus 0.02% sodium azide, tap water, and water collected from a road side ditch. The biological and immunochemical activities were monitored for six days. Samples were removed at different time points for bioassay against mosquito larvae or for immunoassays after alkaline extraction. The results are shown in Figure 7. Both biological and immunochemical activities deteriorated rapidly when an anti-microbial agent was not present. It also appeared that the immunochemical activity deteriorated faster than the biological activity. However, presence of immunoassay interfering substances could have contributed to this observation, resulting in an apparent immediate disappearance of immunochemical activity at 1 hr. post inoculation.

Several assay conditions were also tested to determine if the ELISA is applicable to analysis procedures for environmental samples. The assay was performed at pH's 5-9 and also in the presence of 0.02 - 2.0% bovine serum albumin. In addition, tap water, water collected from a rice field, and effluent water from a sewage treatment plant were spiked with alkaline-dissolved toxin and then assayed with the ELISA.

Performing the assay at pH's between 7.2 to 9 did not affect the dose responses of the ELISA (Figure 4). In acidic conditions, the relationship between concentration and immunoactivity of the assay became less well defined. The upper detection limit also was lower. This effect is probably due to the toxin precipitating at acidic pH's. Presence of BSA (0.02-2.0%) reduced background color development and thus gave a better correlation between concentration and immunoactivity (Figure 5).

When water samples (spiked with BT *israelensis*) were titrated with the ELISA procedure, the detection limits were reduced to between 100-1000 ng/ml. This was especially obvious with the rice field water and the sewage effluent water (Figure 6). Adjusting the pH's of the water samples prior to testing did not improve the assay. It is likely that the presence of organic matter and microorganisms in these water samples interfered with the ELISA procedure. Adsorption of the antigen and antibody to organic matter would have affected the precision and accuracy of the assay. In addition, biodegradation by microbes and proteases released from these microorganisms would also interfere with the antibody/antigen reaction.

These results illustrate two important factors in working with biological and biochemical pesticides. First, sample clean-up is probably needed before environmental samples can be applied to an immunoassay procedure. Second, to correlate field and laboratory data, samples should be handled in such a way as to minimize biodegradation.

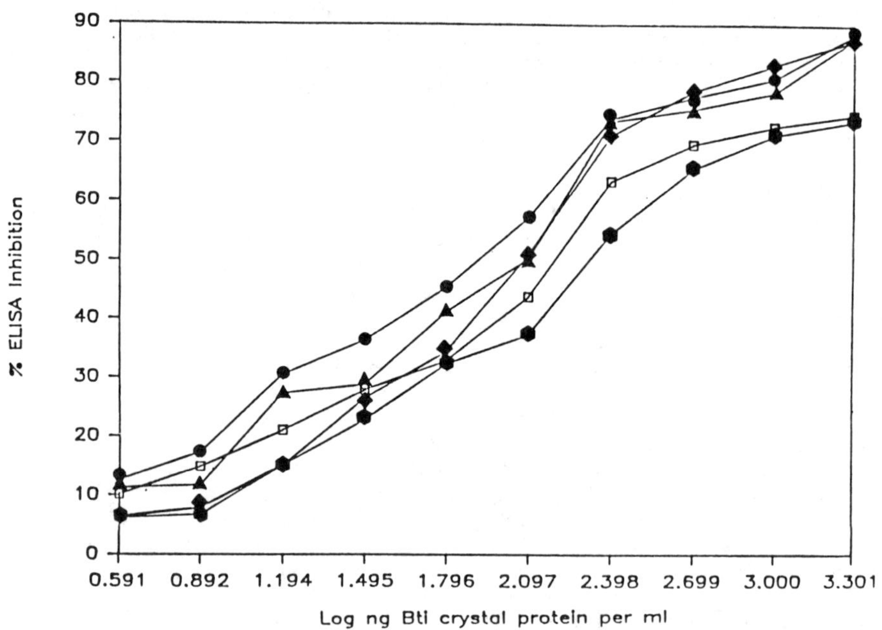

Figure 4. Titration of BT *israelensis* proteins by ELISA at pH 5 --⬢-- , pH 6 --□-- , pH 7.4 --♦-- , pH 8 --▲-- , and pH 9 --●--.

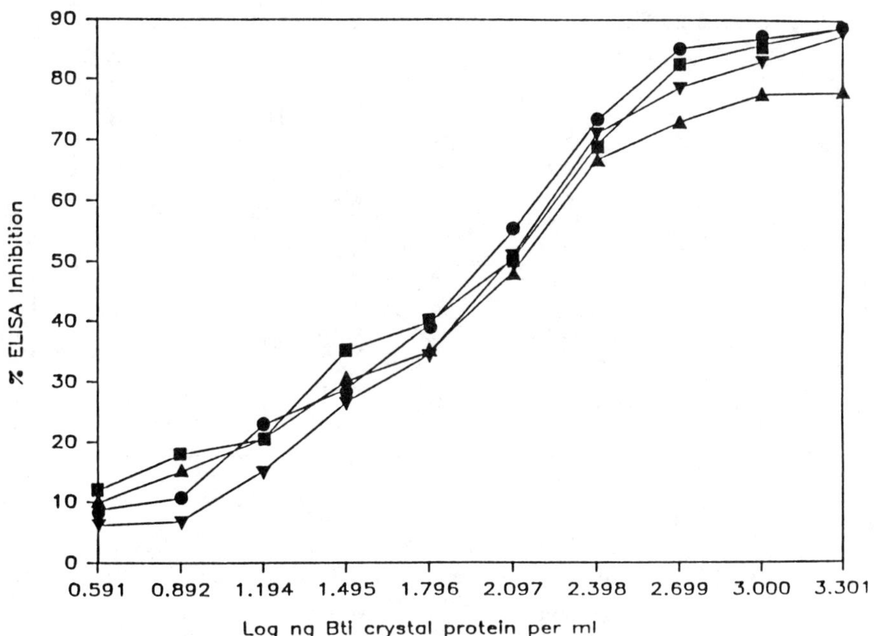

Figure 5. Titration of BT *israelensis* proteins by ELISA in the presence of 0.02% BSA --●-- , 0.2% BSA --▲-- , 2% BSA --■-- , and no BSA --▼--.

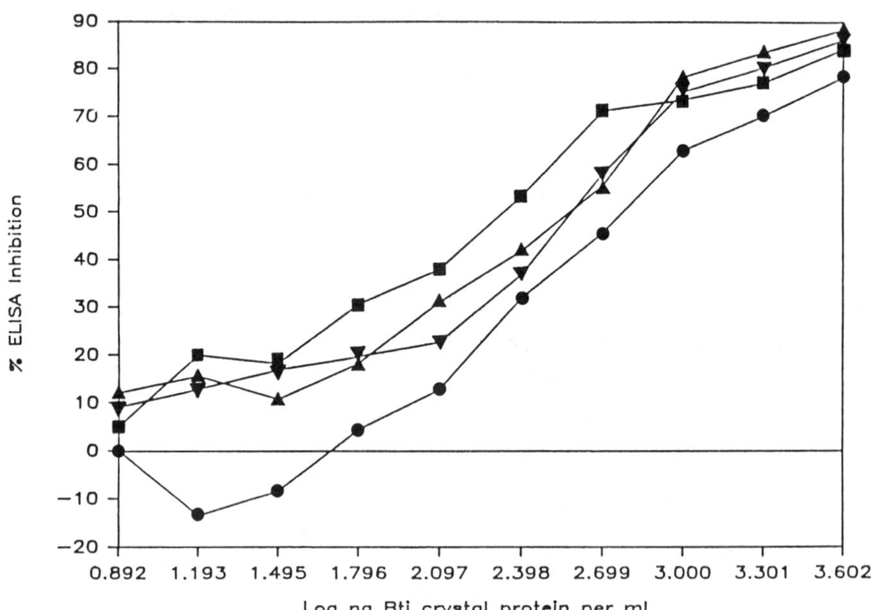

Figure 6. Titration of BT *israelensis* proteins in field collected water samples; buffer control --▼-- , tap water --▲--, rice field water --■-- , and sewage plant effluent water --●--.

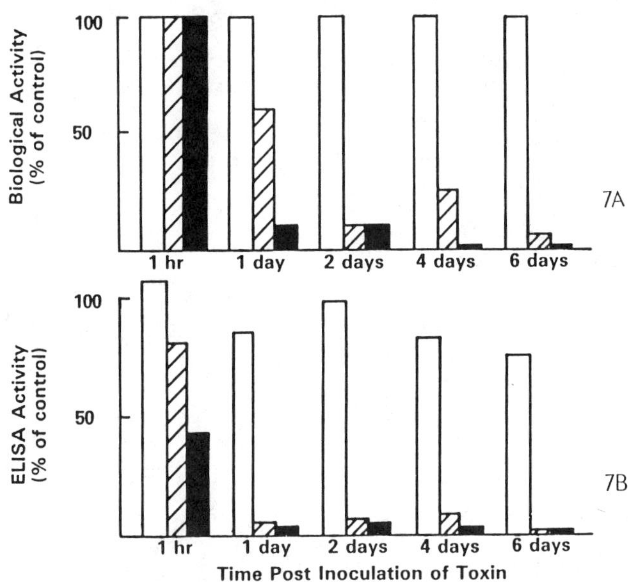

Figure 7. Biodegradation of BT *israelensis* toxin. Commercial formulation of toxin was suspended in water samples. At different times post inoculation the biological activity (A) and the immunochemical activity (B) was determined and expressed as % of activity at time zero. Water samples included: [open bar] distilled water plus 0.04% sodium azide; [hatched bar] tap water; [solid bar] water collected from road side ditches.

Sample Work-up

Bioassays for BT toxin are sensitive because the test animals function as a concentrator and a detector simultaneously. When exposed to very low concentrations, the toxin is accumulated as the animal feeds over a period of days. Compared to immunoassays, bioassays for the BT toxins are generally 50-100 times more sensitive (Table II). As with other residue procedures, sample work-up can be employed to extend the detection limits of an enzyme immunoassay and to remove interfering materials at the same time. Such work up will be essential if immunoassays are to be as sensitive as bioassays using filter feeding organisms.

Table II. Comparison of sensitivities between bioassay and ELISA for the detection of BT *israelensis* δ-endotoxin

BT toxin	Bioassay LC_{50}	ELISA IC_{50}[a]
BT *israelensis*	< 1 ng/ml[b]	100-200 ng/ml
BT *kurstaki*	5.5 - 8.5 ng/cm^2 [c]	10-20 ng/ml

[a] IC_{50}, concentration at which 50% inhibition of ELISA activity occurred.
[b] With purified crystals (2,6)
[c] With purified crystals (2,22); toxicity of BT *kurstaki* is measured in terms of amount of toxin per unit surface area of insect diet.

Immuno-Blotting. Many affinity membranes adsorb molecules such as proteins, nucleic acids, carbohydrates, etc. The Immuno-Blot technique takes advantage of the fact that antigens can be quickly immobilized onto these membranes and then be analyzed with an ELISA procedure. Sample work-up is quickly achieved by repeated sample applications. Antigens are concentrated on the membrane, thus extending the limits of detection. Depending on the binding capacity of the membrane used, the increase in sensitivity can be quantitative. With five sample applications onto a pure nitrocellulose membrane, we were able to extend the ELISA detection limit of BT *israelensis* toxin from 22 ng/ml to 4 ng/ml (20). In addition, immunoblots can be run in a 96-well format. After drying and clarifying the membrane, blots can be read in some standard readers. It remains to be seen if problems with plate variability can be avoided with the immunoblot format.

Solid Phase Extraction. DEAE-anion exchange chromatography was employed to concentrate BT *israelensis* toxin from a large volume of very diluted solution. A liter of rice field water was spiked with 250 ng of the toxin (final concentration 250 pg/ml). This solution was then passed through a 1 ml DEAE-column (Sephacel; Pharmacia). The toxin was eluted with 1 ml of 0.15 M NaCl solution and then assayed with the ELISA and the Immuno-Blot procedures. The assay, in combination with solid phase extraction, can measure BT toxin down to the 100 picogram/ml level.

Conclusion

Application of immunoassay to pesticide residue work has gained support due to its sensitivity, specificity, speed of analysis, and cost effectiveness (20-22). The preliminary data presented in this study address several potential problems when applying enzyme immunoassay to a residue procedure for BT *israelensis* toxin in specific and other biological pesticides in general. These large proteinaceous molecules are usually highly immunogenic and do not require hapten carrier conjugation for antibody production. Procedures for assay development and optimization are well established. However, because environmental samples often contain ELISA interfering materials, sample clean-up or exchanging into a different sample matrix is required to maintain the precision and accuracy of the EIA's. In addition, like many other biological and genetically engineered pesticides, BT *israelensis* toxin is very susceptible to biodegradation. Procedures are needed to deal with sample handling and sample work-up.

Acknowledgments

This study was supported, in part, by the UC Fund for Mosquito Research, NIEHS Grant R01-ES02710-05, the California Department of Food and Agriculture, and by the Western Regional Pesticide Impact Assessment Program. Bruce D. Hammock was supported by NIEHS Research Career Development Award 5 K04ES00107-05 and is a Burroughs Wellcome Scholar in Toxicology. We would also like to thank Biochem Product US-Division, Salzburg Laboratories Inc. for providing us with the BT *israelensis* formulations and the bioassay data for these materials.

Literature Cited

1. Bulla, L. A., Jr.; Kramer, K. J.; Cox, D. J.; Jones, B. L.; Davidson, L.I.; Lookhart, G.L. J. Biol. Chem. 1981, 256, 3000-4.
2. Tyrell, D .J.; Bulla, L. A., Jr.; Andrews, R. E.; Kramer, K. J.; Davidson, L. I.; Nordin, P. J. Bacteriol. 1981, 145, 1052-62.
3. Cheung, P. Y. K.; Roe, R. M.; Hammock, B. D.; Judson, C. L.; Montague, M. A. Pest. Biochem. Physiol. 1985, 23, 85-94.
4. Goldberg, L. J.; Margalit, J. Mosquito News 1977, 37, 355-8.
5. Klowden, M. J.; Held, G. A.; Bulla, L. A., Jr. Appl. Environ. Microbiol. 1983, 46, 312-5.
6. Tyrell, D. J.; Davison, L. I.; Bulla, L. A., Jr.; Ramoska, W. A. Appl. Environ. Microbiol. 1979, 38, 656-8.
7. Gaugler, R.; Finney, J. R. Misc. Publ. Entomol. Soc. Am. 1982, 12, 1-17.
8. Mulligan, F. S.; Schaefer, C. H.; Wilder, W. H. J. Econ. Entomol. 1980, 73, 684-8.
9. Kirschaum, J. B. Ann. Rev. Entomol. 1985, 30, 51-70.

10. Jaworski, E. G. In Pesticide Science and Biotechnology; Greenhalgh, R.; Roberts, T. R., Eds.; Blackwell Scientific Publications, 1987; pp 585-9.
11. Wie, S. I.; Andrews, R. E., Jr.; Hammock, B. D.; Faust, R. M.; Bulla, L. A., Jr. Appl. Environ. Microbiol. 1982, 43, 891-4.
12. Wie, S. I.; Hammock, B. D.; Gill, S. S.; Grate, E.; Andrews, R. E., Jr.; Faust, R. M.; Bulla, L. A., Jr.; Schaefer, C. H. J. Appl. Bacteriol. 1984, 57, 447-457.
13. Cheung, P. Y. K.; Hammock, B. D. Appl. Environ. Microbiol. 1985, 50, 984-8.
14. Smith, R. A.; Ulrich, J. T. Appl. Environ. Microbiol. 1983, 45, 586-590.
15. Winkler, V. W.; Hansen, G. D.; Yoder, J. J. Invert. Pathol. 1971, 18, 378-382.
16. Cheung, P. Y. K.; Hammock, B. D. Curr. Microbiol. 1985, 12, 121-6.
17. Cheung, P. Y. K.; Buster, D.; Hammock, B. D.; Roe, R. M.; Alford, A. R. Pestic. Biochem. Physiol. 1987, 27, 42-9.
18. Voller, A.; Bartlett, A.; Bidwell, D. E. J. Clinical Pathol. 1978, 31, 507-520.
19. Gee, S. J.; Miyamoto, T.; Buster, D.; Hammock, B. D. J. Agri. Food Chem. 1988, in press.
20. Cheung, P. Y. K.; Gee, S. J.; Hammock, B. D. In Impact of Chemistry on Biotechnology: Multidisciplinary Discussions; Phillips, M.; Shoemaker, S. P.; Middlekauff, R. D.; Ottenbrite, R. M., Eds.; American Chemical Society Symposium Series: Washington, DC, 1988; Vol. 364, pp 217-229.
21. Hammock, B. D.; Gee, S. J.; Cheung, P. Y. K; Miyamoto, T.; Goodrow, M. H.; Van Emon, J.; Seiber, J. N. In Pesticide Science and Biotechnology; Greenhalgh, R.; Roberts, T. R., Eds.; Blackwell Scientific Publications, 1987; pp 309-316.
22. Van Emon, J. M.; Seiber, J. N.; Hammock, B. D. In Bioregulators for Pest Control; Hedin, P. A. Ed.; American Chemical Society Symposium Series: Washington, DC, 1985; Vol. 276, pp 307-316.
23. Lacey, L. A.; Undeen, A. H. Ann. Rev. Entomol. 1986, 31, 265-296.

RECEIVED April 1, 1988

Chapter 29

Phylogenetically Based Studies of Microbial Ecosystem Perturbation

David A. Stahl

College of Veterinary Medicine, University of Illinois, Urbana, IL 61801

The use of comparative ribosomal RNA sequencing for studies in microbial ecology and microbial evolution is discussed. A specific study, using the 16S ribosomal RNA to monitor the bovine rumen microbial community, is described. The response of this community to antibiotic addition was evaluated by hybridization of oligonucleotide probes to nucleic acid isolated from the rumen. The application of the ribosomal RNA based measure of microbial communities to general evaluations of community disruption, as might follow the release of genetically engineered microorganisms into the environment, is discussed.

Current concern over the pending and pilot release of genetically engineered microorganisms (GEMS) and viruses into the environment (1,2) is a reflection of our very limited understanding of natural microbial ecosystems. Recent advances in biotechnology have greatly expanded the repertoire of genetic constructions and so, the phenotypic variety available for commercial application. However, unlike other products, microorganisms are self-replicating. Their release must be treated as irreversible. This has served to fuel concern.

Two opposing attitudes concerning "deliberate release" prevail, or at least dominate attention. The conservative viewpoint would ask that any novel organism (either the product of genetic engineering or more conventional genetic manipulations) be closely scrutinized (1). This position holds that we know very little about how microorganisms compete in nature, what constitutes a functionally stable microbial community and how buffered those communities are to insult... so it is essential to assess ecosystem effects. The alternative position claims that the products of recombinant technology differ little (if at all) from those arising from natural processes and those of more traditional genetic selections (2). This controversy has shown how poorly prepared the scientific community is to evaluate natural microbial communities and the impact of engineered microorganisms on these communities. Fortunately, the same technical revolution that

0097-6156/88/0379-0373$06.00/0

spawned genetically engineered microorganisms has also served to provide new tools for the evaluation of natural microbial communities. This chapter describes the application of some of these techniques to the characterization of the most poorly characterized of natural communities. The foundation of this work is the analysis of ribosomal RNAs (rRNAs) or their genes in the environment.

Historical Impediments to the Study of Microbial Ecology

There is little disagreement that the greatest handicap in the study of microbial ecology is characterizing the members of microbial communities. This is a two part problem. The first part is the prerequisite pure culture isolation of the various microorganisms making up a natural community. The pure culture description is the foundation of microbiology and is an essential prelude to the complete description (biochemical, physiological and genetic) of a microorganism. Yet, the greater part of many natural populations cannot readily be grown using standard techniques (3). This problem has often been cited as the greatest blind spot in microbial ecology and in measures of microbial diversity.

The second part of the problem of characterization is nomenclature. Although the necessity of cultivation is perhaps the better appreciated limitation in studies of microbial ecology, the more fundamental limitation is taxonomy. The phenotypic characteristics of a microorganism frequently do not adequately define it. Differences in physiological attributes examined may belie an underlying genetic similarity. Alternatively, similar physiological and morphological attire may obscure genetic diversity (12). There are many examples of each. Classification would therefore seem to be the blind spot in studies of microbial ecology. The study of microbial ecology rests upon a correct and workable taxonomy. The later rests upon an understanding of microbial phylogeny.

Microbial Phylogeny and the Use of Biopolymers as Historical Documents

Historically, the classification of microorganisms has been driven by two considerations. These are the practical need for a useful determinative scheme for identification and, an ideal, that the classification reflect natural relationships. The drive toward a phylogenetically correct taxonomy was early mired in technical and conceptual shortcomings. The ranking by morphology and physiology (e.g. photosynthetic, chemosynthetic and heterotrophic metabolisms) was ultimately shown to be arbitrary (4). An added confusion concerned diversity within designated taxons. The "splitting" or "lumping" of classes of microorganisms often reflected the immediate needs (and resources) of the study, not a uniform classification. Insights into the natural relationships among microorganisms had to await the development of methods for studying their molecular architecture (5).

The genealogy of microorganisms is most clearly recorded in their common biopolymers, the amino acid sequence of homologous proteins and the nucleotide sequence of homologous nucleic acids. When certain conditions are met (below), divergence of sequence can be related to evolutionary divergence of the organism. This can be inferred by several means. Methods commonly used include nucleic acid hybridization, antigenic relatedness, electron microscopy (e.g. of ribosomes), and sequencing. Among these, the least ambiguous is comparative sequencing. Unlike the other methods, sequencing generates a cumulative data collection that can be treated analytically. Thus, the recent revolution in nucleic acid sequencing technology has fostered comparative sequencing studies and is placing microbiology for the first time within an evolutionary framework. Comparative sequence analysis, most importantly of the ribosomal RNAs, has yielded the most complete understanding of microbial phylogeny (4,6). This, and a clear evolutionary perspective has served as the foundation for the explicit characterization of natural microbial communities (7,8,9,10).

The ribosomal RNAs and the Universal Phylogeny

The use of ribosomal RNAs to infer evolutionary relationships is now well documented (4). These ubiquitous biopolymers are ideally suited to studies of microbial phylogeny and to studies in microbial ecology for the following reasons.

1. As integral elements of the protein-synthesizing machinery, they are functionally and evolutionarily homologous in all organisms. Thus they can be used to infer a universal phylogeny of extant life.

2. They are highly conserved in both sequence, size and higher-order structure. Consequently, the rRNAs (or their genes) can be identified in natural samples by both size and sequence.

3. The conservation of nucleotide sequence across the length of the molecules is variable. Some regions of sequence (and structure) are invariant. These can be used to identify these molecules in the environment and also for the sequence alignment necessary for calculations of similarity. The more variable regions are used to infer evolutionary distance and to provide signatures of identify for individual species or strains of microorganisms.

4. The rRNAs provide ample sequence information for statistically significant comparisons.

5. The rRNA genes appear not to be transferred among different microorganisms. The phylogeny of the molecules therefore reflects the phylogeny of the parent organisms.

There are three ribosomal RNAs in bacteria; 5S (ca 120 nucleotides), 16S (ca 1600 nucleotides), and 23S (ca 3,000 nucleotides). The eukaryotic versions of the 16S and 23s rRNAs are somewhat larger. In addition, most eukaryotes have a fourth species, the 5.8S rRNA, that is homologous with the 5' end of the 23S rRNA. For historical reasons, most early analyses used the 5S and 16S species in comparative studies. The smaller species were focused upon mostly because the then-available sequencing technology was not adequate for sequencing the larger 23S rRNA. However, recent developments in sequencing technology have now made all species equally accessible (11). Since the precision of comparative analysis grows with the amount information (sequence) analyzed, greater use of the 23S rRNA is anticipated for the study of both microbial evolution and microbial ecology.

The first inclusive phylogeny (encompassing both eukaryotes and prokaryotes) was derived from the comparative sequencing of 16S rRNAs (6). This study was primarily the effort of Carl Woese and collaborators and has served as the foundation for the studies in microbial ecology discussed here. This earlier work relied upon the sequencing of oligonucleotides (RNA fragments from one to ca 20 nucleotides in length) derived from ribonuclease T_1 digestion of the 16S rRNA. The collection (the catalogue) of oligonucleotide sequences comprises a partial, and discontinuous, sequence of the parent molecule. Similarity between organisms is inferred from the fraction of oligonucleotides common to both and from the limited application of signature sequences (6). The latter are relatively highly conserved oligonucleotides that define major lines of evolutionary descent. Although, much of the existing data collection of 16S rRNA sequence is in the form of oligonucleotide catalogues, the use of rRNA templated reverse transcriptase sequencing is rapidly expanding the collection of complete or nearly complete 16S rRNA sequences (11). Continuous sequences will soon replace the catalogue collection.

It was the analysis of the 16S rRNAs that first revealed the unique evolutionary position of the archaebacteria, and defined the primary evolutionary divisions of life on this planet (4,6). Figure 1 shows an unrooted phylogenetic tree based upon complete 16S rRNA sequences of representatives of the three primary lines of evolutionary descent; the archaebacteria, the eubacteria and the eukaryotes. Within these primary lines of descent other major lines of descent have been delineated. Within the eubacteria, for example, some ten major divisions are now recognized. It has been suggested that these be given a systematic rank equivalent to phylum (12). However, this chapter will not elaborate upon the emerging phylogenetic description of life on this planet. The reader is referred to a recent review for a dedicated treatment of this subject (4,12).

Microbial Ecology and Microbial Phylogeny: Common Ground

Full appreciation of microbial evolution must come from a complete knowledge of extant microbial diversity. As well, full appreciation of microbial diversity can only come from an understanding of their evolution. Although these studies generally have been treated independently, this is mostly because there has been no unifying conceptual and technical framework. The ribosomal RNAs now provide the framework. They therefore offer the criteria to explicitly characterize any organism on this planet and to place it in a phylogenetic context. They can be used to characterize natural communities without the requirements of cultivation or previous pure-culture descriptions of the community members (7,10).

Two general technical approaches to using the ribosomal RNAs for the characterization of natural microbial communities have so far been evaluated (8). The method first explored used the 5S rRNAs to

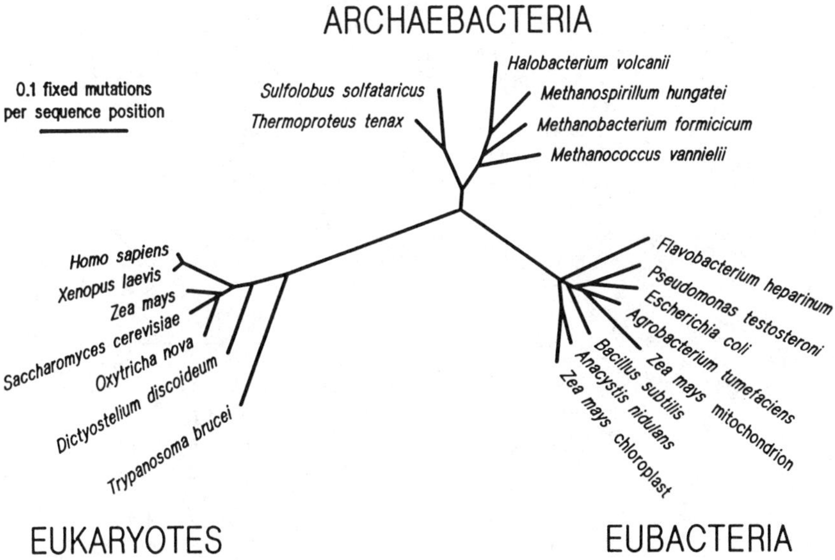

Figure 1. Universal phylogenetic tree based on comparative 16S rRNA sequencing showing the relationship of representatives of the three primary lines of evolutionary descent. Sequence differences through regions of unambiguous alignment (about 950 nucleotides) were corrected for multiple nucleotide substitutions per site and these estimates of sequence divergence (mutations fixed per sequence position) used to infer the phylogenetic tree as previously described (26). (Reproduced with permission from Ref. 23. Copyright 1986 Cell Press.)

characterize natural communities of limited complexity (9,13). In this approach 5S rRNAs are directly isolated from available biomass, separated on high-resolution acrylamide gels, and sequenced. The number of unique 5S rRNAs defines population complexity. Each community member is characterized by a unique sequence that can be related to a reference collection of 5S rRNAs from organisms in pure culture. An identical sequence in the reference collection is not required, since the reference organism rRNA sequence most closely related to the environmentally derived sequence is always known. Because this approach is dependent upon the analytical fractionation of similar molecules, it can not easily resolve the microbial components of very complex communities (> 10-20 species). This is the greatest limitation of the analysis. Also, the 5S rRNA is relatively small (ca 120 nucleotides) and does not provide enough information to discriminate between very closely related organisms or for determining meaningful estimates of greater phylogenetic distances, as separate the various eubacterial phyla.

The second approach is not so limited by population complexity or information content. Here 16S rRNA genes are "shotgun cloned" using DNA purified from available biomass (7,8). In this case, the originating population complexity does not matter since the rRNA genes are clonally isolated from a recombinant library of DNA directly derived from the natural microbial community. The different clones are sorted and sequenced using rapid sequencing techniques. The reader is referred to previous review articles on the ecological use of rRNAs for details of the 5S rRNA based analysis and the clonal isolation of rRNA genes from natural samples (8,9,13).

The interpretation of rRNAs identified in the environment is dependent upon the completeness of the reference collection of sequences. Although there will not soon be a reference collection complete enough to assign exact identities, this should not be perceived as too great a limitation of technique. The more immediate change this new technology will bring is in mindset. Pure culture isolation has so long been a necessary prelude to studies in microbial ecology that this has nearly become an accepted limitation. Now the pure culture representation of a community can be compared to an _explicit_ measure of community composition; rRNA sequences. This alone should greatly alter how natural systems are characterized. Recognition of a dominant community member by sequence but not by culture serves as powerful incentive to pursue the more elusive community member.

A second change this molecular approach brings to studies of microbial ecology is more difficult to succinctly state. This derives from understanding evolutionary relationships and the value of a phylogenetic framework in designing studies of natural systems. As should become apparent, a phylogenetic framework forces integrated studies of natural systems. The basis for the phylogenetic framework (rRNA sequence) provides explicit characteristics (signature sequences) for monitoring microorganisms in any setting. The same measure is applied to every system. For the first time, microbial ecologists working with different microbial systems can cross-talk, using the ribosomal RNAs as a common tongue. To more clearly illustrate the couple between phylogeny and ecology, a specific study will be discussed. This is part of an ongoing study of ruminal microbial ecology (14).

The Rumen as a Model Microbial Ecosystem

The rumen is the modified stomach region of herbivores responsible for fiber digestion. (15,16,17). Having a volume of about 30 gallons in cattle, the rumen is an anaerobic environment harboring large numbers of obligately anaerobic bacteria and eukaryotes. Total microbial numbers average 10^9 to 10^{10} cells/ml. These microbes are the principal agents of fiber breakdown, converting it to the protein and energy sources (primarily volatile fatty acids) used by the animal. The animal is therefore entirely dependent upon these microbial symbionts. We have elected to use this microbial ecosystem for developing some of the molecular techniques described in this chapter.

Why the Rumen?

There are a number of reasons for using the rumen microbial community as a model system.

1. The rumen offers an abundant and complex microbial population. It is composed of representatives of all the primary kingdoms (archaebacteria, eubacteria, and eukaryotes), including anaerobic fungi and protozoa in addition to the better described bacteria.

2. It is well characterized bacteriologically, and therefore offers a basis for judging the fidelity of these molecular descriptions. Although the resident bacteria are quite well described biochemically and physiologically, there has been relatively little work on the ecological relationships between the different groups.

3. It is quite analogous to other systems of anaerobic decomposition (e.g. sediments and sewage digesters), except that fatty acids and lowly substituted benzenoids are not completely degraded (18).

4. It is a contained and easily sampled population. Exogenous microorganisms are easily introduced into the system and the system as a whole readily perturbed, for example, by altering diet, introducing exogenous organisms, or administering antibiotics.

The last point is of particular interest from the standpoint of assessing the risk of releasing genetically engineered microorganisms (GEMS) into the environment. One principal concern of GEM release is their impact on existing microbial communities. But there is little understanding of what constitutes a stable microbial community. Measures of normal population variation and functional redundancy among natural community members are virtually absent in the literature. Yet these must be done to predict risk or to evaluate the ecosystem effects of GEM release. Community perturbation must be defined relative to normal variation in community composition. To more clearly establish these baseline criteria for risk assessment, the rumen stands as a very useful model.

We have initiated a comparative sequencing survey of dominant rumen microbial flora (Montgomery, L.; Flesher, B.; Stahl, D.A. Submitted to Int. J. Syst. Bacteriol.). The goal is to define each organism by 16S rRNA sequence. There are two components to the sequence characterization, with current effort directed to the first. The first is the comparative sequence analysis of rumen microorganisms now in pure culture. This provides a molecular signature (16S sequence) for each and also validates or serves as a basis for restructuring the existing taxonomy. Regions of variable sequence are used to determine phylogenetic relationships and are also used as hybridization targets for synthetic oligonucleotides. The relative amount of hybridization of specific or general oligonucleotide probes to rRNA isolated from the environment offers a rapid and quantitative measure of organism abundance without need for cultivation.

The second component of the characterization is the comparative sequencing of 16S rRNA genes cloned from DNA isolated from total rumen contents. The comparative sequencing of 16S rRNA genes clonally isolated from naturally available DNA offers the most unbiased measure of rumen microbial content. Yet, as already discussed, the clonal analysis of the population cannot stand alone. It must be related to the existing culture collection of rumen microflora and their corresponding 16S rRNA sequences. Although a reference collection large enough to adequately cover the microbial diversity of this planet is not attainable in the foreseeable future, the rumen bacterial community is possibly well enough circumscribed to be reasonably well-defined by both the pure culture and 16S rRNA sequence criteria.

Use of rRNA-Targeted Hybridization Probes for Studies of Microbial Ecology

This section will cover two aspects of the use of DNA oligonucleotides as hybridization probes to monitor environmental populations of microorganisms. The first will discuss the nuts-and-bolts of technique. The second is more conceptual and addresses the design of these probes within a phylogenetic framework.

Overview of Methodology.

An unbiased assessment of microbial makeup based upon nucleic acid hybridization necessitates an unbiased recovery of nucleic acid from the various organisms represented in the environment. This is not straightforward. Cells greatly vary in susceptibility to breakage by enzymatic, chemical and mechanical disruption. In general, however, mechanical disruption offers the most uniform breakage of different cell types. Although mechanical disruption cannot be used to recover high molecular weight DNA, it can be used to recover intact DNA and rRNA suitable for oligonucleotide probe hybridization. In our experience, disruption with glass beads in a reciprocating shaker offers excellent recovery of nucleic acid from a great variety of organisms (Stahl, D.A., unpublished observations). Even the more recalcitrant types (e.g. Gram-positive cocci, mycobacteria) are efficiently broken (unpublished observations).

For the rumen perturbation study, total nucleic acid was recovered from unfractionated rumen contents by mechanical disruption (Mini-beadbeater, Biospec Products, Bartlesville, OK) with glass beads (150-200 microns) and phenol. Phenol was present to minimize the degradation of nucleic acids by nucleases. Following additional extractions with phenol/$CHCl_3$, total nucleic acid (primarily rRNA) was recovered by ethanol precipitation, denatured with 1.5% glutaraldehyde, and spotted on nylon membranes (19). Since there are usually 10,000 to 20,000 copies of the ribosomal RNAs per cell, rRNA-targeted hybridization is a much more sensitive measure of organism abundance than hybridization to DNA sequences. DNA oligonucleotide probes were labeled with ^{32}P using polynucleotide kinase and gamma-labeled ATP (19). Hybridization conditions were optimized for sensitivity and specificity relative to total nucleic acid isolated from various other rumen and non-rumen microorganisms (19,20).

For these studies we have used radioactive probes. This at present remains the most sensitive detection system. Non-radioactive signalling systems (e.g. biotin and alkaline phosphatase tagged oligonucleotides) have yet to demonstrate comparable or better sensitivity. This should change, but discussion of these alternatives is beyond the scope of this chapter. For the rumen perturbation study, hybridization of the various oligonucleotide probes was measured by densitometry of film exposed to the nylon support membrane following hybridization. Film response throughout the range of exposures used was linear. Figure 2 is an example of one hybridization series taken from our rumen study.

Design of Hybridization Probes and their Application to Monitoring Microbial Communities

It is the capacity to define rRNA targeted probe specificity within a phylogenetic framework that lends such versatility to these studies. Figure 3 displays a 16S rRNA folded into a consensus secondary structure with positions shaded according to relative evolutionary conservation among thirty different eubacterial species. Darker shading corresponds to higher conservation. As displayed by this figure, the molecule is a patchwork of relative evolutionary conservation. Some positions and locales are invariant; others are common to a given kingdom but vary between kingdoms; more variable regions contain signatures for the various eubacterial phyla. The most variable domains can be used to discriminate between species or even subspecies of bacteria. Thus the spectrum of organisms addressed by a single hybridization probe varies according to the region of the molecule selected as the hybridization target. The net cast by a given hybridization probe is defined by the needs of the study. The casting of wide or narrow nets in studies of microbial ecology is best illustrated by example. Oligonucleotide probes of "nested" specificity have been used by us to monitor populations of one of the more important cellulolytic bacteria in the rumen, *Bacteroides succinogenes*.

Prior to our characterization of various strains of this rumen bacterium, it was viewed as a coherent species. Now, based upon the marked divergence among the 16S rRNA sequences of strains of this "species", it is now recognized to be made up of a collection of

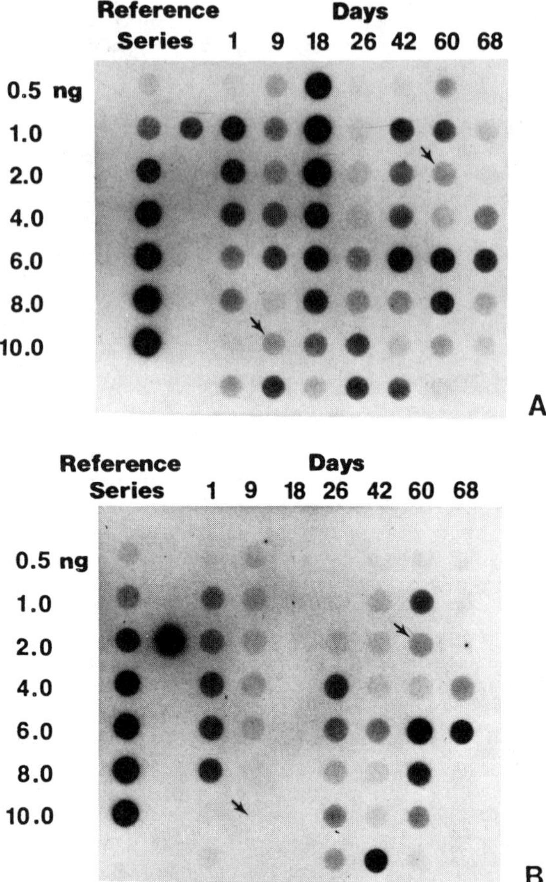

Figure 2. Representative hybridization of the ruminal probe (2A) and cecal probe (2B) to nucleic acid extracted from daily rumen fluid samples and spotted to a nylon membrane. The oligonucleotide probes listed were used in the complete study. The following are complementary to positions 207-226 in the Escherichia coli 16S rRNA numbering: Lachnospira multiparus-specific, CTTATACCACCGGAGTTTTTCA; Bacteroides succinogenes strains NR9- and DR7-specific (cecal probe) CCGCATCGATGAATCTTTCGT ; and B. succinogenes strains S85- and A3c-specific (ruminal probe) CCATACCGATAAATCTCTAGT. The B. succinogenes signature probe AATCGGACGCAAGCTCATCCC is complementary to positions 225-245 in the E. coli numbering. Total 16S rRNA was estimated by hybridization to an oligonucleotide ACGGGCGGTGTGTRC complementary to a region (near position 1400, E. coli numbering) of virtually all 16S-like rRNAs so far characterized (17). Bound probe was quantitated by densitometry relative to reference standards after autoradiography with preflashed film exposed at -85o C. Arrows mark the addition to, and removal from, the feed of the ionophore antibiotic monensin.

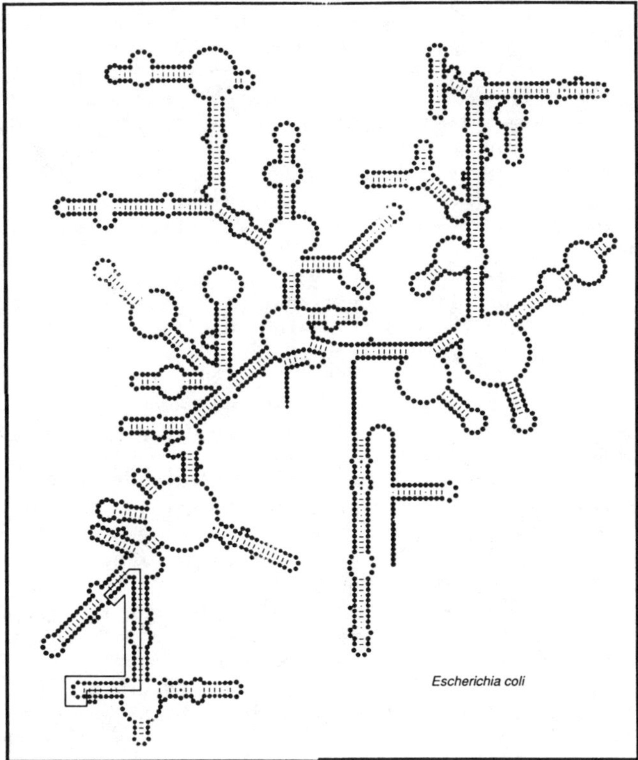

Figure 3. Positional nucleotide conservation in 27 diverse eubacterial 16S ribosomal RNA sequences. Shading intensity increases in proportion with conservation. Invariant positions within this sequence set are black. With the exception of the universal probe, the oligonucleotide probes used in this study hybridize within the boxed region of the 16S rRNA sequence. (Reproduced with permission from Ref. 10. Copyright 1986 Nature Publishing Co.)

genetically distinct bacteria (Montgomery, L.; Flesher, B.; Stahl, D.A. Submitted to Int. J. Syst. Bacteriol.). Although the assemblage of strains is phylogenetically coherent, to the exclusion of other characterized bacteria, the phylogenetic depth among them (ca 90% similarity) is greater than that which separates Proteus vulgaris from Escherichia coli (ca 93% similarity). The phylogenetic depth among representative strains of B. succinogenes is best illustrated by the phylogenetic tree (fig. 4) inferred from 16S rRNA sequence similarity. The depth separating the two major lines of descent with B. succinogenes is deep enough to divide the collection into different species or genera, but there are, as yet, no satisfying phenotypic criteria to base such a division upon; the various strains are now distinguished mainly by vitamin requirements and variations in morphology. As we continue to build our reference collection cryptic diversity is a common theme. A single strain cannot be assumed representative of a species. These observations highlight the limitations of traditional identification schemes for studies in microbial ecology.

To monitor populations of B. succinogenes in the rumen over time, three different oligonucleotide probes were synthesized. The target regions within the 16S rRNA are indicated in figure 3. The "signature" probe is complementary to the 16S rRNA of all but one strain of B. succinogenes (DR7, S85, HM2, REH9-1, and A3c but not NR9) so far characterized by comparative sequencing. Strain NR9 differs in a single A to G transition. The other probes identify either of two natural groups within the larger assemblage. One group (ruminal type) is represented by rumen strains S85 and A3c; the second group (cecal type) is represented by strains NR9 and DR7 isolated from rat and pig ceca, respectively. These probes therefore are of nested specificity. A fourth oligonucleotide probe identified Lachnospira multiparus, a Gram-positive pectinolytic rumen bacterium. One additional oligonucleotide was used to measure the total 16S rRNA content of each nucleic acid sample applied to nylon membranes. This probe is complementary to a sequence element present in all small subunit ribosomal RNAs so far characterized (eukaryotic and prokaryotic) (11). Hybridization to this probe was used to normalize hybridization of each specific probe. Abundance is expressed as the target-group fraction of the total 16S rRNAs in the sample.

Perturbation of the Rumen Microbial Ecology by Monensin

Monensin is a sodium ionophore antibiotic that is routinely included in cattle feedlot diets to improve feed utilization efficiency (21). Increased propionate production, decreased methane production and a protein sparing effect have been suggested to be in part responsible for improved feed conversion (21). Yet, it remains unresolved whether alteration of the rumen population or alteration of the physiology of a relatively unchanged population contributes most to these effects.. Polyether antibiotics, such as monensin and lasalocid, are generally considered to be most active against Gram-positive organisms, such as Lachnospira spp. and Ruminococcus spp., whereas organisms with Gram-negative cell walls (e.g. Selenomonas and Bacteroides spp.) are considered to be relatively resistant (22).

In this study we have used monensin primarily as a means to perturb this ecosystem in order to evaluate our methods for measuring

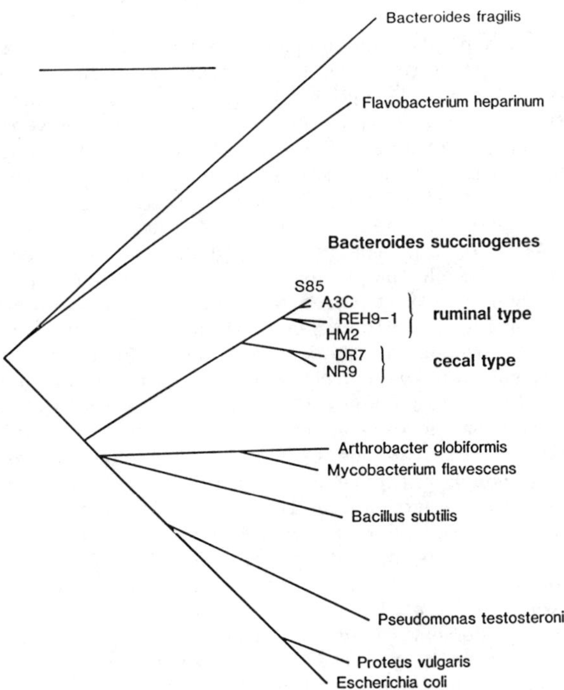

Figure 4. Phylogenetic tree inferred by comparative analysis of 16S rRNAs of various strains of Bacteroides succinogenes and the most closely related eubacterial species (see Fig. 1 and Ref. 20 for descriptions of method).

change in community composition. The experimental animal was a ruminally cannulated, non-lactating Holstein cow (ca 12 yrs) fed a maintenance diet of 1.8 Kg grain mix and 1.8 Kg alfalfa hay at 8 AM and 4 PM each day. The cow was adapted to the diet for 20 days before sampling. The rumen contents were sampled daily for 77 days at 2 PM from the mid rumen region. Monensin was added at 35 mg per Kg of feed from sampling day 16 to day 62, comparable to levels used in feed lot diets. Total nucleic acids were extracted from about 1 mL of each daily sample and hybridized to each of the various oligonucleotide probes described above. The results of this study are in figure 5.

The two groups of B. succinogenes responded very differently to addition of the antibiotic. The proportion (as a percentage of total 16S-like rRNAs) of ruminal-type B. succinogenes increased about five-fold (relative to the baseline period) immediately after addition, whereas the proportion of cecal-type B. succinogenes was depressed. The ruminal type remained elevated (accounting transiently for about 1% of total ribosome numbers) for several days before dropping, over the next two week period, to below baseline proportions. At that time the cecal type gained relative dominance, although present in much lower levels than the earlier peak of the ruminal type. Over the next three week period the ruminal type predominated, showing two clear peaks of several days duration each. Upon withdrawal of monensin, the cecal types reached their highest levels (0.4%), transiently equalling the proportion of the ruminal-type before again dropping to earlier levels (<0.1%). The numbers of L. multiparus were expected to fall with addition of monensin, based on its Gram-positive wall structure and sensitivity to monensin. Although relative numbers in the two to three week period after addition were generally about half of those for the baseline period, the response was not immediate, nor was it sustained. The clearest population response followed removal of the antibiotic from the animal's diet, at which time L. multiparus transiently exceeded 1% of total ribosome numbers.

This work has demonstrated that these techniques are suitable for monitoring these microorganisms in their natural setting. However, more importantly, it has demonstrated the value of formulating the analysis within the phylogenetic framework. This is illustrated in the comparison of the group and signature probe hybridization to the various strains of B. succinogenes. Hybridization to the signature oligonucleotide probe, which detects most strains of B. succinogenes so far characterized by comparative sequencing, was generally consistent with the trends shown by the group-specific probes. There was one notable exception. Between days 24 and 30, the signature probe response greatly exceeded the sum of the ruminal- and cecal-types. Thus, a relatively large proportion of B. succinogenes-like organisms were unaccounted for by the group-specific probes, suggesting the presence of another assemblage of B. succinogenes-like microorganisms that had not been characterized. Thus the analysis has likely identified a previously unrecognized group of B. succinogenes-like bacteria. An integrative analysis of this kind is essential for sorting out the complexities of natural microbial ecosystems.

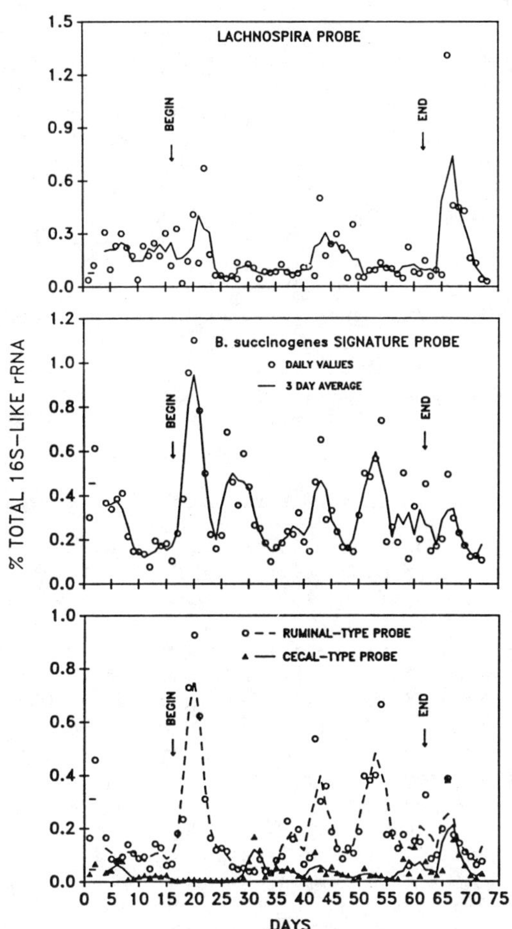

Figure 5. Dot Blot Quantitation. Arrows mark the addition to, and removal from, the feed of the ionophore antibiotic monensin. Points correspond to normalized daily values. The profiles are drawn from a running three day average of the daily values (see text for discussion). (Reproduced with permission from Ref. 14. Copyright 1988 American Society for Microbiology)

The Big Picture: More General Assessments of Microbial Ecosystem Perturbation

An expansion of nested hybridization is the inclusion of phylum specific and kingdom specific probes in general environmental assessments. Probes are now available that specifically hybridize with the 16S rRNA of virtually all members of each of the primary kingdoms (Giovannoni, S.J., Delong, E; Olsen, G.J.; Pace, N.R. J. Bacteriol., in press). Also, although not yet systematically examined, the various major lines of descent within the primary kingdoms (phyla) can be defined by various signature positions (4). Thus the various microbial phyla should also be amenable to identification by one or several signature sequences. We have in our research identified several 16S rRNA target sequences that encompass most sulfate-reducing bacteria available in pure culture (unpublished). A battery of phylum-specific probes could be used to establish diversity indices for natural populations. Change in a general diversity index could prove to be a useful measure of microbial ecosystem perturbation, as might follow the release of genetically engineered microorganisms into the environment. We are currently developing a collection of general probes for use in assessing general changes in microbial ecosystems.

General Caveats of Oligonucleotide Probe-based Identification

A concern regarding microorganism identification by nucleic acid hybridization is target group specificity, which ultimately must be empirically tested. Yet, given that most natural systems remain poorly characterized, this generally is not possible. However, by framing the question of specificity in a phylogenetic context, specificity can be evaluated and the ecosystem can be better defined. For example, the "phylogenetically-nested" hybridization probes for *B. succinogenes* evaluate specificity. Consistency between general probe and group- or strain-specific probe hybridization increases confidence in target group selectivity. Although observations of inconsistency, as observed between the *B. succinogenes* signature- and group-specific probes between days 24 and 30, might indicate a problem with probe specificity, it is also suggestive of unrecognized diversity within the larger assemblage. Thus, the use of probes having nested specificity serves to affirm target group selectivity as well as point to unrecognized diversity within the larger assemblage.

Summary

Studies in microbial ecology must begin with an understanding of natural diversity. Traditional difficulties of assessing this diversity have limited most studies of natural systems. The limitations of culturing are generally well appreciated. Nevertheless, traditional classification schemes probably are the more serious constraint. Available phenotypic criteria often obscure diversity or miss specific relationships. Comparative sequencing of the ribosomal RNAs now provides the criteria necessary to explicitly characterize microorganisms and phylogenetically relate them. The ribosomal RNAs therefore offer the means to assess extant microbial diversity, both within the pure culture collection and the environment. Thus, studies in microbial ecology and microbial evolution are closely allied. And, only through both can there be detailed understanding of general microbial ecosystems and change within those systems, as might follow the introduction of engineered microorganisms.

Acknowledgments

I thank Berdena Flesher for much of the technical work included in this article, George Fox, Lee Krumholz, Berdena Flesher and John Urbance for critical comment and Norman Pace for critical comment and providing the print for Fig. 1. This work was supported by cooperative agreement CR812496-01-3 to D.A. Stahl from the U.S. Environmental Protection Agency.

Literature Cited

1. Sharples, F.E. Science 1987, 235, 1329-1335.
2. Davis, B.D. Science 1987, 235, 1329-1335.
3. Atlas, R.M. In Current Perspectives in Microbial Ecology; M.J. Klug; Reddy, C.A., Eds.; American Society for Microbiology: Washington, D.C. 1983; pp 540-545.
4. Woese, C.R.; Stackebrandt,E.; Macke, T.J.; Fox, G.E. System. Appl. Microbiol. 1985, 6, 143-151.
5. Zuckerkandl, E.; Pauling, P. J. Theor. Biol. 1965, 8, 357-366.
6. Fox, G.E.; Stackebrandt, E,; Hespell, R.B.; Gibson, J.; Maniloff, J.; Dyer, T.A.; Wolfe, R.S.; Balch, W.E.; Tanner, R.S.; Magrum, L.J.; Zablen, L.B.; Blakemore, R.; Gupta, R.; Bonen, L.; Lewis, B.J.; Stahl, D.A.; Luehrson, K.R.; Chen, K.N.; Woese,C.R. Science 1980, 209, 457-463.
7. Pace, N.R., Stahl, D.A.; Lane, D.J.; Olsen, G.J. In Advances in Microbial Ecology, Plenum Press: New York, 1986; Vol. 9, Chapter
8. Olsen, G.J., Lane, D.J.; Giovannoni, S.J.; Pace, N.R.; Stahl, D.A. Ann. Rev. Microbiol. 1986, 40, 337-366.
9. Stahl, D.A.; Lane, D.L.; Olsen, G.J.; Pace, N.R. Appl. Environ. Microbiol. 1985, 49, 1379-1384.
10. Stahl, D.A. Bio/Technology 1986, 4, 623-628.
11. Lane, D.J.; Pace, B.; Olsen, G.J.; Stahl, D.A.; Sogin, M.L.; Pace, N.R. Proc. Natl. Acad. Sci. USA 1985, 82, 6955-6959.
12. Woese, C.R. Microbiol. Rev. 1987, 51, 221-271.
13. Stahl, D.A., Lane, D.J.; Olsen, G.J.; Pace, N.R. Science 1984, 224, 409-411.
14. Stahl, D.A.; Flesher, B.; Mansfield, H.R.; Montgomery, L. Appl. Environ. Microbiol. 1988, 54, 1079-1084.
15. Hobson, P.N.; Wallace, R.J. Critical Reviews of Microbiology 1982 9, 165-225; Critical Reviews of Microbiology 1982, 9, 253-320.
16. Bryant, M.P. In Dukes' physiology of domestic animals, ninth edition; Swenson, M.J., Ed.; Cornell University Press: Ithaca, 1977; Chapter
17. Hungate, R.E. The Rumen and Its Microbes; Academic Press: New York, 1966.
18. McInerney, M.J.; Bryant, M.P. In Biomass Conversion Processes for Energy and Fuels; Sofer, S.S.; Zaborsky, O.R., Eds.; Plenum Publishing Corporation, 1981; pp. 277-296.

19. Berent, S.L.; Mahmoudi, M.; Torczynski, R.M.; Bragg, P.W.; Bollon, A.P. Bio-Techniques 1985, May/June, 208-220.
20. Wallace, R.B.; Shaffer, J.; Murphy, R.F.; Bonner, J; Hirose, T.; Itakura, K. Nucl. Acids Res. 1979, 6, 3543-3557.
21. Bergen, W.G.; Bates, D.B. J. Anim. Sci. 1984, 58, 1465-1527.
22. Dennis, S.M.; Nagaraja, T.G.; Bartley, E.E. J. Anim. Sci. 1981, 52, 418-426.
23. Pace, N.R., Olsen, G.J.; Woese, C.R. Cell 1986, 45, 325-326.

RECEIVED May 25, 1988

REGULATORY CONSIDERATIONS FOR BIOTECHNOLOGY PRODUCTS

Regulation of Biotechnology Products in Advanced Stages of Development

The diversity of crop protection products brought about by modern biotechnology will be limited only by the ingenuity of those seeking to develop new and useful products. The chapter concentrates on regulatory considerations for products in the advanced stages of development and that are close to commercial use. These include several microbial pesticides and other pesticide producing organisms (e.g., plants), as well as herbicide resistant crop plants.

Regulatory oversight of agricultural biotechnology products is administered at the federal level through existing statutory authorities. For example, the U.S. Department of Agriculture's Animal and Plant Health Inspection Service (USDA/APHIS) oversees the movement of potential plant pest organisms under the Plant Pest Act, and the Environmental Protection Agency (EPA) regulates the use of pesticide products under the Federal Insecticide Fungicide and Rodenticide Act (FIFRA). EPA also sets tolerances for pesticide residues in food crops under the authority of the Federal Food Drug and Cosmetic Act (FFDCA). Meaningful implementation of these authorities involves a complex and evolving interaction between the regulatory authority (e.g., EPA), the regulated community, and the general public (Figure 1).

Each of these three participants brings a different perspective to bear on the common objective of bringing safe and useful products to the market place for the benefit of society. Regulatory authorities, such as EPA, are charged by law with ensuring that the benefits of a new product outweigh any human health or environmental risks that may be associated with the use of the product. This requires a careful review of the proposed product and the manner in which it is to be used. Industry comprises a major segment of the regulated community and is similarly concerned that their products are safe and effective. But industry must be cognizant of the need for timely product development and marketing in order to ensure their economic survival. Thus, the costs of regulatory compliance, in terms of time and money, are an essential consideration for successful product development. The public is likely to share interests of both the regulators and the regulated community. Interest in the former stems from the obvious concern for public and environmental health, while interest in the latter relates to the desire for new and improved products, as well as industrial productivity and competitiveness.

The regulated community is represented in this chapter by two companies currently on the forefront in development of inovative microbial pesticide products through the use of classical advanced genetic manipulations. Bruce Carlton will discuss strategies the Ecogen company is using to develop genetically improved strains of *Bacillus thuringiensis* for use as microbial pesticides. David Miller of the Genetics Institute will summarize their thoughts on seeking an experimental use permit for a genetically engineered viral insecticide.

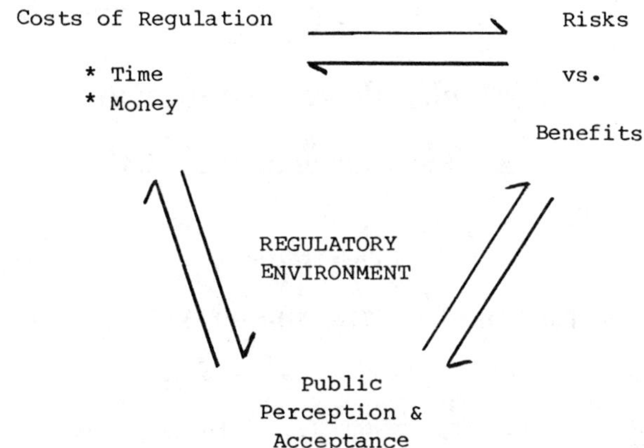

Figure 1. The objectives and responsibilities of regulatory agencies, industry, and the general public, all interact to create the regulatory environment for agricultural biotechnology products.

Academicians may represent interests in all three areas. They may perform basic and applied research, represent concerns on behalf of the public sector, and advise regulatory agencies in their product evaluations. The latter role is discussed in this chapter by James Tiedje who currently serves as a member of EPA's Scientific Advisory Panel and Biotechnology Science Advisory Committee, in addition to actively conducting research in microbial ecology and agricultural biotechnology.

To date only a handful of public interest groups have shown an active interest in biotechnology issues. These groups have articulated a wide range of concerns and used a variety of approaches for participating in the implementation of biotechnology regulations and have input into the regulatory decisionmaking process. Jack Doyle, of the Environmental Policy Institute will discuss an environmentalist's view of regulations in the area of biotechnology.

Regulatory responsibility and authority may reside at the federal, state and local levels. Discussion in this chapter is limited to the activities in two federal agencies. Jerome Miksche will provide an update of the USDA Biotechnology Research Guidelines and Fred Betz will discuss EPA's regulatory program and risk assessment for agricultural biotechnology products.

Frederick S. Betz
U.S. Environmental Protection Agency
(TS–769C), 401 M Street, SW
Washington, DC 20460

RECEIVED May 27, 1988

Chapter 30

Biotechnology Regulation

An Environmentalist's View

Jack Doyle

Environmental Policy Institute, 218 D Street, SE, Washington, DC 20003

Environmentalists do not intend to be the enemy of the biotechnology industry. Rather, they recognize the great potentials and desire to help move biotechnology forward without unduly endangering the environment. Environmentalists contend that regulation of the release of genetically-altered organisms must be given to EPA. They cite illegal release incidents and problems with the management within the coordinated framwork. They believe new legislation is needed to clarify the present situation legally, to protect the environment, and to insure a "local playing field" for all participants.

Despite what many have heard, or are inclined to believe, environmentalists are not shutting down the biotechnology industry. One need only look as far as Wall Street and the recent public offerings by biotechnology companies to know that this industry is robust indeed. Investors continue to be interested, even though the pay-off is years away in many cases. Agracetus has tested genetically engineered tobacco plants in the field; Ciba-Geigy has field tested herbicide tolerant plants; Advanced Genetic Sciences has tested "ice minus" in the field, as has University of California researcher Steven Lindow.

Secondly, it is important to mention the fact that some environmentalists think a great deal about economic impacts as well as ecological concerns. Too often, it seems, environmentalists are thought of as only concerned with the picture-book environment, with the cosmetic side of the outdoors, derisively referred to as "bugs and bunny" people or "tree huggers". However, most environmentalists are very concerned with economic health and, particularly, economic opportunity, fair pricing, and genuine competition in the marketplace.

And third, The Environmental Policy Institute (EPI) does not oppose biotechnology, but rather we believe that these <u>are</u>

0097-6156/88/0379-0394$06.00/0

exciting times for biological research. The revolution that is occurring in the biological sciences at the cellular and molecular levels is awesome and it is astounding. The concerns with biotechnology have more to do with how this new knowledge will be used, what knowledge will be pursued and capitalized, and the economic as well as environmental consequences of using biotechnology merely to continue old and damaging pactices.

Biotechnology As A Knowledge Opportunity

Biotechnology needs to be approached as a knowledge opportunity, not simply as a manufacturing opportunity. What is happening today in the biological sciences at the molecular and cellular levels is truly astounding. We are on the verge of understanding the biological realm in fundamentally new ways; in ways that should help agriculture become profitable as well as environmentally responsible.

But if this new biology is used only to divide up the natural world into its smallest possible commercial parts, rather than improve our ability to work with the biological realm, the result may be further economic and environmental problems for the future.

Government's Role In Biotechnology

Government's role in biotechnology is a multifaceted role, one that is complicated by different tasks and different constituencies. There are at least three distinct roles -- and they are often found at the state level as well. First, there is one role that includes being a funder of biotechnology research, at several levels, such as through the National Institutes of Health, the National Science Foundation, the USDA, and the land grant universities. Second, there is another role that includes being a booster and promoter of biotechnology for the ostensible purpose of maintaining a competitive advantage, or in the case of the states, to attract new industry. And third, there is the role of regulator for reasons of public health and safety, product efficacy, and environmental protection.

It is possible that these three government roles -- funding, promotion and regulation -- may not always receive equal attention by government agencies, and may in fact come into conflict with one another.

EPI is in favor of new legislation for regulating the environmental release of genetically-altered organisms. The current regulatory framework is inadequate, and probably over the long run, unworkable and fraught with legal pitfalls.

On a purely environmental level -- with respect to deliberate release -- the regulatory process, as currently crafted, is lodged in all the wrong places, with the exception of some of EPA's authorities. Biotechnology-derived products will be used for all kinds of purposes in the environment in the years ahead; for toxic waste clean-up, for enhanced oil recover, for cleaning up oil spills, for applications in forestry, for sewage waste treatment, and for a range of uses in agriculture.

It seems reasonable then, when considering the potential environmental effects that engineered organisms could have when released into the air, water and soil, that environmental science and environmental protection be the paramount consideration in any regulatory scheme. But that is not the case with the present regulatory framework.

As early as 1982, concern about, and control over, the regulation of environmental biotechnology began to emerge in the White House Office of Science & Technology Policy (OSTP). This began initially as an inquiry into the U. S. biotechnology industry's competitive international position, and soon included a long list of concerns related to the Administration's interest in competitiveness. In addition to OSTP, other agencies, including the State and Commerce departments, and later OMB, also became involved in the question. Industry representatives were brought into the Administration's process early on. After the completion of this review of the new biotechnology industry, the stated intention of the Administration became clear: maximize the U. S. competitive position in biotechnology, minimize regulation, and deregulate any existing laws thought to encumber the progress of high technology generally -- from drug export laws to antitrust laws.

In October 1982, the White House Office of Science and Technology Policy (OSTP), under the direction of George Keyworth, commissioned an inquiry into the competitiveness of the U. S. biotechnology industry. That effort was organized as a working group headed by William J. Walsh of the State Department. In April 1983, a smaller group of individuals involved with the study met with nine key representatives of biotechnology companies and major corporations at a Virginia retreat for two days to finalize the report's recommendations. Walsh described that draft--which was submitted to OSTP on May 27, 1983-- as "a data base with suggestions on how government agencies should change their ways in such areas as patents, licenses, export controls and FDA regulations, so as not to lend themselves to selling out state-of-the art cheaply, and to unshackle industry in pursuit of biotechnology". The report recommended, for example, removing certain disincentives to biotechnology in the tax, patent, and antitrust area. "The current situation", wrote Walsh in his letter to Keyworth transmitting the study, "does not warrant more restrictive control measures, which could be counter-productive".

Then in June 1983, there came the first signs of government interest in the regulation of genetically engineered organisms proposed for release into the environment. At that time, EPA's Assistant Administrator Don Clay, then appearing before Rep. Albert Gore's House Science & Technology subcommittee, asserted that EPA had authority to regulate certain products of biotechnology under FIFRA and TSCA.

Clay's testimony also raised the first official signs of uncertainty with regard to the available predictive capabilities and data base needed for assessing the risk of genetically engineered organisms in the environment. In his testimony, Clay pointed to the technical problems of scientifically gearing up to do environmental assessments of genetically-engineered organisms. "There are almost no accepted methodologies for evaluating the safety of genetically-engineered products", he explained, adding that the risk-assessment tools and data used for chemical substances could not apply in the case of organisms. Clay then explained that the EPA needed to develop methodologies to evaluate environmental fate, human exposure, and the potential environmental and health hazards of genetically-engineered organisms, but that such tests were "still several years away".

In early 1984, after EPA had circulated a draft biotechnology regulation for publication in the Federal Register with regard to its perceived responsibilities under FIFRA and TSCA, Christopher DeMuth, an OMB official, objected. In a March 12th, 1984 memo drafted for the Cabinet Council on Economic Affairs, DeMuth argued that the council, not EPA, should decide how to regulate biotechnology. In the memo, DeMuth referred to EPA's proposal as a "gambit", and voiced concerns that too much regulation would hurt biotechnology companies, which he viewed as "extraordinarily innovative" but "unusually sensitive to regulatory costs and delays". He also outlined several options: doing nothing, asking the National Academy of Sciences to study the matter, and keeping the status quo intact under NIH, among others. One report at the time charged that OMB and DeMuth were under pressure from the Department of Commerce and Commerce Secretary Malcolm Baldridge, who was concerned that EPA regulation would hinder U.S. biotechnology companies in competition with Japan.

Meanwhile, within the White House Domestic Council, all of this agitating over EPA's move to regulate biotechnology, led to the demise of EPA's and William Ruckelshaus' lead on biotechnology issue. On April 30, 1984, the Working Group on Biotechnology of the Cabinet Council on Natural Resouces was created under the chairmanship of OSTP's George Keyworth. On May 9th, this newly constituted group held its first meeting, with Keyworth as chairman. According to EPA's Jack Moore, at least part of Keyworth's message was: "Let's maintain our competitiveness. Don't unwittingly do anything to stifle the technology". It was this OSTP-led working group that formulated the first coordinated framework approach to federal biotechnology regulation, first published in the Federal Register December 31. 1984.

Throughout early 1985, the regulatory framwork was criticized on a number of fronts, and the issue of environmental risk contined to be discussed.

In February 1985, the Cornell University Ecosystems Research Center issued one of the first comprehensive overviews of what was possible and what was lacking to adequately assess the environmental risks of new biotechnology products. The findings of that report are still quite remarkable, and underscore how little we know about microorganisms--how they survive, how they are dispersed in the environment, why some grow rapidly, and why others do not.

The most important conclusion of the Cornell study is what it said about the ability to make predictions about the behavior of organisms in the environment: "Methods for predicting the likelihood of survial and proliferation of a given organism in the environment are crude... Methods are available for assessing some potential effects, but there are many deficiencies in current knowledge and theory. Generally, we lack any true data base against which to compare test results or predict environmental consequences."

In June 1985, following the Cornell study, in Philadelphia, molecular biologists and ecologists gathered for the first time in a public formum to discuss some of the questions surrounding deliberate release. Not much agreement was reached at that meeting, but more ecologists did become involved.

In November 1985, the Biotechnology Science Coordinating Committee was formed to help guide and coordinate the regulatory process among federal agencies. By this time EPA had announced the approval of Advanced Genetic Sciences' (AGS) field test for "ice minus". Then in December 1985, Senator Durenberger introduced S.1967. The Biosafety Act, aimed at amending TSCA to require that EPA review and approve genetically engineered microorganisms before they could be released into the environment. Hearings were never held, and the bill died in the last Congress.

The first quarter of 1986 was not a good period for the biotechnology industry. On February 11th, the Monterey County (California) Board passed a moratorium on the release of genetically engineered organisms within that local jurisdiction. Monterey County was the site chosen by AGS to field test ice minus.

On February 26th, it was revealed in news accounts that AGS had illegally tested ice-minus in the open air a year earlier without informing EPA. A month later, EPA fined AGS $20,000 and suspended their permit to test ice minus, which the agency had approved just four months earlier.

Less than three weeks after the revelations about AGS' illegal activities, it was subsequently revealed that USDA had approved a genetically engineered vaccine without the review of its own RAC. USDA then suspended the license it issued for the vaccine and began an environmental review.

Meanwhile, back in Congress, the Fuqua/Volkmer bill--The Biotechnology Science Coordination Act of 1986-- was introduced in March 1986. This bill took a more comprehensive approach than did the Durenberger bill, but it died too. One hearing was held.

In June 1986, the final revision of the Administration's Coordinated framework for the regulation of Biotechnology was published in the *Federal Register*. This now constitutes the guidelines but they lack adequate or agreed upon definitions of some very important threshold terms, such as "release into the environment" and "pathogen".

With this new package of regulations concerning biotechnology, there arose some controversy surrounding some exemptions of genetically altered organisms the Administration had proposed. For example, one proposal offered to exempt genetically engineered organisms that had a gene or genes deleted. Some scientists immediately objected to this exemption. For example, MIT biologist Jonathan King stated: "It is a medieval scientific view that a deletion is automatically less risky". The deletion of a gene, he said, can have a wide range of effects. Similarly, Cornell University biochemist Liebe Cavalieri stated that the deletion of DNA segments does not necessarily mean that you have changed just one characteristic. The Administration also proposed exempting organisms that had only changes in regulatory genes; that is, genes that regulate or control the activity of other genes. Some ecologists objected on this point too because the alternation of regulatory sequences can affect the characteristics of organisms in ways that affect their survival, reproduction and effects on ecological systems. Stated Elliott Norse, then representing the Ecological Society of America: Regulatory sequences that change growth or reproductive rates by just a few percent could dramatically alter competitive balances among microorganisms, plants, or animals in nature. In fact, stated Norse, "exempting any of the processes of altering genomes that have developed in the last 10-15 years from rigorous review seems premature".

In December 1986, the "Volkmer report" was issued by the House Science & Technology Committee's Oversight and Investigations Subcommittee, entitled: "Issues in the Federal Regulation of Biotechnology: From Research to Release". This report tracked the handling of the illegal AGS and USDA release incidents, and also explored the problems apparent with the management of the Coordinated Framework.

From an environmentalist's prespective, there are a number of important problems to be considered. First and foremost is the question of legal authority. Is the Coordinated Regulatory Framework for Biotechnology legally sound? Does it provide clear legal authority? Secondly, there is also the critical matter of definitions. It seems that a set of basic, operative, and scientifically sound definitions must be in place before a regulatory system can be established. Yet, the Administration's Coordinated framework is just the reverse. Currently, we are back-tracking on definitions where they exist at all; trying to make them fit the package. In EPA, some key rule-makings needed to make their part of the package work under TSCA are dependent on definitions such as those for "pathogen", "release into the environment", and "containment", none of which yet exist.

Thirdly, the issue of determining risk to the environment from genetically altered organisms is a matter for environmental scientists and ecologists, not molecular biologists and medical scientists. Yet the principal architects of the current framework have been the molecular biologists, medical scientists, and industrial biotechnologists. Ecologists, for the most part, have been excluded from this process.

As the regulatory framwork has emerged under the guidance of the BSCC, there has been a clear bias toward the medical sciences in trying to define the environmental and ecological risks faced with biotechnology. EPI does not believe, for example, that consulting with infectious control officers at hospitals, the National Library of Medicine, or using the TOXLINE and MEDLINE data bases will help significantly with predicting ecological consequences of genetically engineered viruses, bacteria, and insects destined for use in the environment. There may be some parallels, and these data bases may be somewhat helpful, but in the main, disease etiology/epidemiology and environenmtal ecology are different phenomena, with different thresholds and consequences. So there needs to be much more attention paid to the science of ecology in the administration of regulation.

Yet another question is to whether there will be more or less litigation. Based on the intra-agency tug-of-war in the handling of biotechnology regulation so far, it is probable that we will see over the next few years, a never ending tangle of new definitions, amended definitions, ad-hoc rule-makings, BSCC subcommittee meetings, scientific advisory committee meetings, exemption listings and de-listings, appeals, and finally, litigation. Some of this litigation will be based on procedural flaws that are guaranteed to come with a process that attempts to cover five agencies; it can't help but make more gray areas as it moves forward. And some of this litigation may well come from environmental organizations or consumer groups. But some may also come from industries frustrated with the process.

EPI supports the view that new legislation is needed to clarify the present situation legally, to protect the environment, and to insure a level playing field for all participants. This is in industry's interest too, because it will streamline the review process. For both environmentalists and businessmen, there is a need to straighten out this convoluted framework in order to avoid unnecessary bureaucracy and litigation in the years ahead.

EPI has called on Congress to review and evaluate the present regulatory situation, and to clearly articulate what laws should apply, whether those be old laws clarified with new amendments, or the crafting of an entirely new piece of legislation. From time to time in the past, others have also called for Congress to become involved legislatively. Others have recommended legislation on the deliberate release situation as early as 1981. New legislation was one of the options outlined by OTA in its 1981 report Impact of Applied Genetics: Micro-Organisms, Plants, and Animals:

Congress could pass legislation regulating all types and phases of genetic engineering from research through commercial production. This option would deal comprehensively and directly with the risks of novel molecular genetic techniqiues. A specific statute would eliminate the uncertain ties over the extent to which present law covers particular applications.

Rep. John Dingell called for new legislation in a September 1985 letter to the editor of Issues in Science & Technology. "New statutory provisions seem essential to provide a sensible and fair set of rules for all aspects of this new field", he wrote. And in 1986, the background paper for Brookings' Second Annual Conference on Biotechnology included an option for "a new integrated law to regulate biotechnology".

In Congressional testimony in June 1987, and in comments to the BSCC in September 1986, EPI advocated that Congress give EPA the lead in reviewing all proposed releases of genetically-altered organisms, whether for academic test-plot experiments or full-blown commercial releases. It was proposed that a Biotechnology review Board be created within EPA charged with reviewing and approving all applications for outdoor testing. Additionally, a permit system was proposed as an integral part of this system, with EPA having final authority for approving or vetoing individual applications for release on the basis of ecological and/or public health risks. Some legislators are now talking more seriously about the need for legislation.

Recently, Senator Albert Gore, in a May 19, 1987 speech before the Keystone Center Forum on Biotechnology held in Washington, D. C., issued the following excerpted statement:

Looking back over the past five years since I held my first hearing on biotechnology, I'm afraid that many of the questions raised remain unresolved...

Back in 1983, for example, the Subcommittee on Investigations and Oversight concluded that one of the major risks associated with the release of genetically altered organisms was not necessarily the releases themselves, but our inability and failure to make risk assessments. Four years later, we still haven't developed adequate standards for risk assessment...

...I have always felt that government should encourage this new science, not stand in its way. But we did see the need for new input and called for development of a "predictive ecology" to help shed light on some of the nagging uncertainties. Yet those uncertainties have only multiplied...

It is the job of government and the Congress not only to reduce risk, but to reduce fear. Biotechnology will not realize its promise until we have a biotechnology industry that trusts our governmental process and is willing to sit down with the elected representatives of the public, the Congress, to draft comprehensive legislation to dispel the public's fears and thus, insure a solid foundation for its future.

I do not understand the reluctance (of industry to do this). It was Congress that passed the Drug Export Bill, the Patent Term Restoration Bill, and the research and development sections of Superfund. Yet, despite this track record of support, industry continues to react with fear whenever talk of biotechnology legislation arises.

I've been told it's fear of the unknown that leads industry to oppose new biotech legislation. But the industry must realize that a full, vigorous public debate now, when things are going well, will only help in the long-run. Nothing is more dangerous to the future of biotechnology than to pospone the debate until a crisis occurs...

EPI predicts that more members of Congress will call for legislation to clarify the situation.

Another development occurring on the regulatory front is the activity at the state and local level. New Jersey, Texas California, Wisconsin, and North Carolina have all considered new or special approaches to biotechnology regulation, some with new legislation. In New Jersey, a moratorium measure was introduced in 1986, which later, as a compromise bill incorporating a review commission, passed unanimously in the state Senate. That bill is still waiting a vote in the New Jersey assembly.

It is likely that as more and more field tests occur in more and more states, there will be more concern in state legislatures. It should be in the states' interest to begin assembling their regulatory and ecological expertise in these areas, and not to rely on Washington or assume that the current regulatory framwork is going to survive.

It appears that a final resolution will not be soon in coming. In many ways, the debate is just beginning, both around the country and in the U. S. Congress. The advent of biotechnology may provide a fundamental opportunity to phase out chemical pesticide use in agriculture. It may be an opportunity whose time has come, and therefore it needs to be pushed hard, and elevated to a national goal.

Consider for a moment some of the opinions and activities in this country with regard to pesticides:

Public Opinion. A January 1984 consumer survey conducted by the Food Marketing Institute, 77 percent of those polled expressed concern over pesticide and herbicide residues in food, indicating the problem to be a "serious hazard". (Hammonds, 1984).

Public Opinion. A September 1986 Des Moines Register poll taken in Iowa shows that 58% of those surveyed believe farm chemicals are the biggest threat to water quality, and that 78% favor limits on farm chemicals.

Consumer Pressure. In May 1986, the center for Science in the Public Interest began a national campaign called "Americans for Safe Food", with a 5-point plan of action that called for, among other things: laws that require disclosure of pesticides, drugs, and other chemicals used in the production of foods; a ban on pesticides and animal drugs known to pose a serious risk to consumers; and national standards for "organic", "natural", and "pesticide-free" foods.

Business Reaction. In July 1986, Safeway Stores, Inc., the nation's largest grocery chain, announced that it would stop buying apples treated with the chemical growth regulator Alar, despite EPA's decision in January to allow its use while further studies are done. In addition, the state of Maine proposed a non-detectable standard for daminozide to be reached this year, and the state of Massachusetts has enacted regulation to reduce Alar in baby foods and heat-processed foods to a non-detectable level by 1988. (Wall Street Journal, 1986).

Business Reaction. In November 1986, the H. J. Heinz Co. announced that it was planning to restrict the purchase of crops used in the manufacture of baby foods that had been treated with certain pesticides. Heinz listed 12 chemicals: alachlor, aldicarb, captan, captafol, carbofuran, carbon tetrachloride, cyanazine, daminozide, dinocap, ethylene oxide, linuron, and TPTH. Heinz told farmers that it would likely test crops for the absence of these chemicals -- all of which were then still legal, but under review by EPA as possible health hazards. (Meier, 1986).

Farmworker Health. In May 1986, United Farmworkers leader, Caesar Chavez sent out a mass mailing appeal to Americans nationwide, announcing a new grape boycott aimed at eliminating five pesticides that endanger farmworker health. In his appeal, Chavez asked consumers not to buy fresh California table grapes until growers agree to ban the five most dangerous pesticides used in grape production -- captan, dinoseb, parathion, phosdrin, and methyl bromide. (Chavez, 1986).

Farmer & Farm Family Exposure. Last year, the U. S. Environmental Protection Agency warned women of childbearing age to avoid farm fields recently treated with the herbicide dinoseb because the chemical might cause birth defects. Another study, conducted by a team of medical scientists, found that Kansas farmers exposed to a widely-used corn and wheat herbicide for 20 days or more a year had a sixfold increase in a certain kind of lymph cancers compared to non-farmers. And a third study from Wisconsin noted possible immune-system suppression due to aldicarb exposure.

With biotechnology, it may be conceivable to begin a national research program that has as its goal the phasing out of pesticides in agriculture. At the very least, it should be possible to drastically reduce the use of pesticides by "building in" disease and insect resistance into crops and livestock, and making that a major priority in USDA and the land grant system.

In-fact, such a program could be targeted as a national goal--a goal no less important than putting a man on the moon, or NIH's war on cancer. Certainly, eliminating the source of one of our greatest public health threats, as well as a continuing source of groundwater pollution, and a primary source of farmer, and farmworker poisonings--would be a laudable national goal.

Such a program would elicit widespread support and accomplish several things simultaneously. First, it would reduce public, farmer, and farmworker exposure to pesticides. Second, it would reduce the cost of production for farmers and thereby improve farm income and profitability. Farmers currently spend an estimated $18 billion annually for purchased feed, $7.4 billion for fertilizer, $4 billion for pesticides and $4 billion for seed. That adds up to national cost-of-production bill of at least $33 billion, without including other related input costs. Any reduction in these costs would certainly improve farm income, and presumably, U. S. agricultural competitiveness. Moreover, the public health and environmental costs of using many of these farm inputs are also high, and could be decreased accordingly with a reduction in use.

Third, it should improve consumer faith in the agricultural

system and possibly reduce prices once all associated "pesticide costs" were reduced throughout the system. And fourth, it could provide a powerful basis for rejuvenating the land grant universities and agricultural experiment stations. In fact, it should be the USDA, the land grant universities, and the agricultural experiment stations that are charged with pursuing and achieving this national goal.

The idea here is not to exclude the private sector, but for the public sector to "prime and pump", to take leadership in, and absorb the risks of, a major reorientation in agricultural production research that potentially could have wide public benefit.

However, with such a program, the definition of biotechnology and what is pursued in the name of non-chemical and/or biological alternatives will be absolutely crucial. What is needed is a biotechnology that embraces "common sense biology" and "common sense genetics", yet eschews the genetic engineering of organisms simply for the sake of making new products. What this means, however, is that gene splicing and other biotechnology techniques should be used, within reason, but that what we already know should not be disregared or overlooked.

There is already a lot of good biology 'on the shelf", so to speak. And there is a lot of good, innovative work going on throughout the country by researchers at our land grant universities and agricultural experiment stations. For example, Donald Barnes, a plant breeder at USDA's research center at the University of Minnesota, has developed a new variety of alfalfa called Nitro that produces good livestock fodder and puts high amounts of nitrogen back into the soil. When plowed back into the soil, Nitro puts 94 lbs. of nitrogen into the ground compared with 59 lbs. for the top standard variety. Nitro represents the kind of genetic improvement in agriculture that can save growers money on the input side of their operation, and thus, increase their profitability. Extensive information is available about soil tilth, plant breeding, crop rotations, intercropping, multilining, insect adaptability, agricultural diversification, and other fields--all of which has application today. However, we should not become so enamored of biotechnology that we go out of our way for a high-tech solution when a common-sense alternative is right in front of us.

If biotechnology is used only to divide up the natural world into its most numerous product possibilities, rather than using it to improve our ability to work with the biological realm, further economic and environmental problems for the future will surely be created, and a truly golden opportunity for righting some of the mistakes of the past will be missed.

RECEIVED June 7, 1988

Chapter 31

Genetically Engineered Viral Insecticides

Practical Considerations

David W. Miller

Genetics Institute, Inc., 87 Cambridge Park Drive, Cambridge, MA 02140

Baculoviruses play a central role in the natural control of insect pest populations, chiefly Lepidoptera. This has sustained an interest in the commercial potential of these as larvicides in several pest control situations; however, performance drawbacks have limited their general usefulness. Modification of the viruses through genetic engineering is anticipated to greatly increase their effectiveness. How this may be accomplished will be discussed. The effect modifications may have on the well-established safety of these agents as well as their perceived safety will also be covered. Pertinent examples from our work and that of others are discussed.

The need to find effective alternatives to chemical insecticides has kept alive an interest in the use of naturally occurring insect pathogens as control agents. With the advent of genetic engineering technology, an opportunity has emerged for alleviating the commercial shortcomings of these pathogens and fostering the creation of a new generation of products. Our desire to enter this challenging field led us to select insect viruses, specifically a baculovirus, as our model system. Their suitability to the techniques of genetic engineering and their well-chronicled safety and natural efficacy make them

0097–6156/88/0379–0405$06.00/0

an obvious choice. Equally important in our commercial setting was the opportunity to develop a well defined research goal and to design and, as described here, successfully demonstrate a biological containment feature.

Background and Biology

Baculoviruses are a family of rod-shaped, DNA viruses found exclusively within the Arthropods and chiefly within the larvae, i.e. caterpillars and worms, of the economically important Lepidoptera (For reviews that cover most of the general points raised herein see: 1, 2, 3.). Their association with natural disease epidemics, or epizootics, suggests their use as natural insecticides. Nuclear polyhedrosis viruses (NPVs) are the baculovirus group that is receiving the most interest as targets for change by DNA engineering techniques. The nucleocapsids of these viruses exist in two distinct morphological forms: a plasma membrane-budded, non-occluded virus (NOV), and an intranuclearly occluded form in which many nucleocapsids are collected together in a protein near-crystal, the polyhedral inclusion body (PIB).

The natural infection cycle begins when a leaf-feeding caterpillar ingests a PIB. Alkaline conditions in the midgut lead to the dissolution of the PIB, releasing the nucleocapsids to fuse with the midgut cells and initiate the infection. Progeny virus produced in these initially infected cells appear to be exclusively NOVs, acquiring a membrane as they exit through the plasma membrane of the cell. It is this form of the nucleocapsid that spreads the infection throughout the larvae. Subsequently infected tissues produce additional NOV and also produce PIBs. At the time of death, the larvae is completely packed with PIBs.

The model NPV system is an isolate from the alfalfa looper, *Autographa californica* and, in conjunction with Lepidopteran cell culture, makes an excellent laboratory system. In cell culture, the NOV is the exclusive infectious form; PIBs play no role. After infection, the nucleocapsid makes its way to the nucleus, where replication begins. The initial steps in the viral replication cycle are performed by cellular factors, leading to a cascade of viral gene expression driving the subsequent replication steps. From about 12 - 18 hours post-infection, progeny nucleocapsids form in the nucleus. These bud through cellular membranes to yield extracellular NOVs which spread the infection to other cells in the culture.

In the stages post-18 hours, progeny nucleocapsids do not leave the nucleus. A late viral gene product, the protein polyhedrin, commences expression as gene expression related to the production of NOVs winds down. By ca 18 hours post-infection, polyhedrin begins to crystallize in the cell nucleus. The nucleocapsids present here are caught, or occluded, in the growing crystal, the PIB. There can be several hundred PIBs per nucleus. This process continues until the cell eventually dies, ca 72 hr post-infection. The PIBs produced in culture are not infectious to cultured cells. They are perfectly infectious to caterpillars.

Several features of baculovirus biology have led to an impressively rapid increase in our understanding of their replication at the molecular level. The nuclear polyhedrosis viruses grow in cell culture, and, as described above, carry out all known aspects of the replication cycle. Furthermore, baculoviruses have double stranded DNA genomes making them amenable to all nucleic acid technologies applied to mammalian viral systems. Finally, there are the pragmatic forces driving the use of the virus as a pesticide and as an expression vector for heterologous proteins of experimental or commercial interest.

As mentioned, NPVs direct the cell to produce large numbers of PIBs. The PIB may be as much as 95% polyhedrin. The level of synthesis needed to sustain this leads to fantastic amounts of polyhedrin being produced in late-stage infected cells. Estimates of the amount of polyhedrin protein exceed 25%. One implication of this large amount of material is that a very active promoter is driving the expression of the polyhedrin gene, and it is now clear that very large amounts of polyhedrin RNA are produced in these cells. In such a case, a genetic engineer would be inclined to take advantage of the promoter and, leveraging it with other presumed useful features of the system, employ it in an expression vector for heterologous genes inserted under the control of that promoter. This has been the case for the NPV polyhedrin promoter. The features of this system have been elaborated in recent review articles (4, 5, 6). The only role for polyhedrin appears to be PIB formation, which is irrelevant to the production of *in vitro* infective NOVs. Genes can therefore be inserted in place of the polyhedrin gene, eliminating its function. The resulting viral genome will initially direct the production of NOVs then, with the time course of polyhedrin protein gene expression, go onto produce product as directed by the inserted DNA.

Experimentally, DNA is introduced into the viral chromosome as follows (described in detail with examples in 4, 6): To facilitate working with the polyhedrin gene, a small portion of the large viral chromosome, containing the gene and its flanking regions, are cloned into a bacterial plasmid. The polyhedrin coding regions can (but do not have to) be removed in the process of introducing useful restriction enzyme sites in the vicinity of the initiating ATG. Once constructed, foreign DNA can be introduced into the plasmid at one of these sites, placing it downstream of the presumptive promoter region. As an aside, the exact nature of the polyhedrin promoter and its ability to produce large amounts of RNA are only now being established (7 and references contained therein). It is a fact that genes expressed from such constructs produce far less protein than is produced by the natural polyhedrin gene. The reasons for this discrepancy have not been fully elucidated (see above references for discussion).

Once constructed, the plasmid sequences are introduced into the virus chromosome by co-transfection of the plasmid with naked wild type viral DNA. Recombination occurs at a low but useful frequency between these DNAs at their homologous regions, flanking the polyhedrin gene. The progeny virus from the co-transfection are plaqued, and the plaques are scored for type. Under a dissecting microscope, the recombinants can be selected from the wild types based on their distinctive plaque morphology due to a lack of PIBs. Once selected and a stock prepared, one has a virus which on infection of cells will direct the production of both additional virus and the product of the inserted gene. This system is now widely employed for the lab scale production of a variety of proteins from all varieties of organisms. Rarely does one encounter a gene which is not expressed in this system. Typically, 1 to 100μg per ml will be obtained. Maeda and coworkers have clearly demonstrated that the virus will direct foreign gene expression in intact caterpillars, as well (8 and references contained therein).

Baculoviruses as Insecticides

In our planning for the modification of an insect pathogen, we found that NPVs met many of the criteria in our mental check list: they have reasonable activity in their unaltered state, they have already received some commercial use and, as a result, have established a good safety record and regulatory history, they can be manipulated at the molecular level, and we could devise a realistic scenario for alteration, which we anticipate will increase their efficacy. Lastly, an unavoidable concern in this arena is obtaining permission for the experimental field application of such a product. A very definite plus in our selection of this group was that we could devise a biological containment scheme that would actually lead to a loss of the recombinant virus as it grew in its hosts and prevent its unchecked growth. This approach will be discussed in detail below.

The critical point in using the virus as an insecticide relative to its use as an expression vector is summarized as follows: A foreign gene, from any source, can be placed into the baculovirus chromosome so that upon infection of a caterpillar cell a gene product, perhaps never expressed, or never expressed at this time, or never expressed in this amount, is now expressed. One has the opportunity then to introduce novel gene products into a natural population of caterpillars via the virus and augment their control over and above that of the virus alone and perhaps in ways that cannot be supplanted by standard agricultural chemicals. This makes for some very exciting possibilities.

Baculoviruses have been commercially limited because of shortcomings relative to their chemical competition: Any one baculovirus is active against only a limited number of caterpillar species, which at one point in time was a feature thought to be an asset. However, in a control situation, there can be caterpillars present which are outside the host range, complicating the usage decision. Caterpillars which are susceptible take a number of days to die after eating a lethal dose, and later instars can be quite refractory to infection. Baculoviruses are rather environmentally unstable, as well. They are particularly sensitive to ultraviolet light, which decreases their effective lifetime on the leaf surface. To obtain commercial quantities of material, baculoviruses must be produced in larvae, which is perceived as a cumbersome retro-technology. This perception is in fact a misconception. While certainly not a standard way of producing an insecticide, it is a technically demanding but very workable process.

Before elaborating on the ways in which genetic engineering can alleviate some of these problems, it is worth restating some of the well-known safety characteristics of unaltered baculoviruses, all of which we anticipate will be maintained with an engineered virus. Baculo — meaning rod-shaped — viruses have no known counterparts in vertebrate viruses, yet they are widely distributed in nature. During epizootics in susceptible Lepidoptera, for example the gypsy moth, they are present in unbelievably high numbers and densities with no adverse environmental effects observed other than to the host larvae.

The animal safety testing of baculoviruses has been very extensive. Simply summarized, there have been no unexpected adverse effects on any vertebrate and no effects on any invertebrate which is not part of the host range (for review see 9). Baculoviruses have no phytotoxicity.

Several naturally-occurring baculoviruses have now been registered in the United States for use as insecticides on forestry and food crops. The current regulatory framework will subject the first engineered baculovirus pesticide to a high level of scrutiny. Questions of toxicity aside (as this will be handled on a case-by-case basis), an issue of general and prime concern will be the genetic stability of altered viruses. Going hand-in-hand with a rod-shaped nucleocapsid is the ability of the virus to accommodate gains or losses of DNA by simply expanding or shrinking the particle. For example, in natural populations, baculoviruses seem to be prone to acquiring host DNA sequences, which in several cases examined appear to be transposable elements (for review see 10). With this in mind, the following questions are of relevance to the field testing of genetically altered viruses: To what extent do introduced sequences change the overall dynamics of the genome and are the introduced sequences, in particular, more likely to be involved in untoward events? Having dealt extensively with engineered viruses both *in vivo* and *in vitro*, we have observed no unexpected instability of the genome over either the entire chromosome or the inserted gene. Neither have we detected any unusual recombination events involving engineered viruses. An additional safety feature of the engineered product we are developing is a biological containment feature which will minimize the spread, by growth, of the product from the application site, and in this sense, make the altered virus more "safe" than the wild type.

There are a number of technical means under consideration for alleviating the shortcomings of baculoviruses as insecticides. Of the several problems, perhaps the most readily approachable through engineering is improving the speed of kill. Shortening this to as little as one day would make a radical difference in the commercial utility and acceptance of these agents. It is currently assumed that infected larvae die because they are overwhelmed by the course of the infection, and it takes several days for this to occur. However, employing the virus as an expression vector to deliver some deleterious gene product into the larvae could greatly increase the speed of kill. It should be noted that kill as used here really refers to a disabling or inactivation of the larvae such that feeding ceases and, ideally, a farmer can visually assess the control he is achieving.

Most genes chosen for introduction will code for a protein that falls into any of several differing classes which might fulfill these criteria. This concept has been elaborated in 11, 12. There are a number of proteinaceous toxins produced in living systems. Arthropods produce a great variety; however, these are generally quite poorly characterized (for reviews see 13, 14). Perhaps the best known and most characterized caterpillar-specific toxin is that produced by the several varieties of *Bacillus thuringiensis*. Additionally, there are a number of enzymes, such as proteases and chitinases, as well as enzymatic inhibitors which might be examined. Exciting to contemplate are insect neurohormones. Their expression at high levels could lead to provocative phenomenon within the larvae and perhaps within the pest population. In thinking through a plan for modifying the virus by gene insertion, it must be kept in mind that the ideal selection would not have activity much outside the target range. Furthermore, it should be compatible with producing the virus product commercially in larvae. This loose grouping of candidate proteins will hereafter be referred to as toxins.

It is ironic that the initial focus on baculoviruses as insecticides stemmed from their high degree of specificity for a limited number of Lepidopteran larvae and consequent lessened environmental concern. This has in part been their undoing and mechanisms are now being contemplated for enlarging the host range. As adopted in baculovirology, the concept of host range treats a permissive infection as one in which the larvae is fully susceptible to the virus, leading in the case of NPVs to the production of PIBs and ending in the death of the larvae. Within this, it is now appreciated that there are also a wide range and probably high frequency of semi-permissive infections in which some aspect of the viral lifecycle (as measured against some reference) is curtailed. There are also non-permissive infections.

There are two approaches to improving the host range of baculoviruses. There is the confrontational approach of clearly delineating in some particular case of virus to host where the host range restriction(s) lies. This is likely to be a considerable undertaking for even just one case. The problem with this approach as it relates to developing a better pesticide is that the answer is likely to be case specific; that is, the deficiency in a virus' interaction with a particular host will be unique and knowing the answer will not tell us how to make that particular virus interact better with the next desirable target that is not part of its host range.

A better approach would be to actually ignore the specifics of host range determination and instead develop a method that allows an end-run around viral-host interaction deficiencies and permits the virus to (if perhaps not carry out a full replication cycle) act as an expression vector and deliver to the host a toxin gene that will be expressed and dispatch the pest. The idea here is that the correct promoter driving the expression of a toxin gene during a weak or inapparent infection or perhaps no replication at all would nonetheless allow sufficient expression of the gene to achieve the desired result.

An example of how this might work has been elegantly demonstrated by Lois Miller's laboratory (15, 16). By employing the technology described above, Carbonell et.al. constructed an *Autographa californica* NPV which contained two foreign genes, each under the control of a different promoter (15). The polyhedrin promoter was used to express a fusion of the N-terminal region of the polyhedrin protein to *E. coli* β-galactosidase. The virus also contained, adjacently located, the chloramphenicol acetyltransferase gene under the transcriptional control of the long terminal repeat (LTR) of the Rous sarcoma virus (RSV). The RSV LTR had previously been characterized as a promoter cassette which would function in a wide variety of cell types, including *Drosophila* and mammalian cell lines. It was of interest to determine if this arrangement would allow expression of either of these easily-assayed reporter genes when the virus was introduced into novel hosts.

When the virus was used to infect a known, fully permissive cell line from *Spodoptera frugiperda,* both genes were expressed. CAT activity was detected early in the infection at about six hours, a not surprising result since the RSV LTR contains an RNA polymerase II promoter and this enzymatic activity, pre-existing in the cell, should be amongst the first to act on the viral chromosome. Late in the infection, at the time that the polyhedrin promoter is known to become active, β-galactosidase activity was detected.

The NOV form of the virus was also used to infect a *Drosophila* cell line *in vitro*. Flies are not part of the host range of this particular baculovirus and no progeny virus were anticipated. The situation with regard to gene expression as driven by these two very different promoters was not known. As it turned out, starting at about 12 hours post-application, CAT activity was detected and continued to rise for a number of hours. Polyhedrin promoter activity, as reported by β–galactosidase, was never detected.

The authors were able to detect what they felt was a very low level of DNA replication in *Drosophila*. The important point is that the detection of CAT activity in *Drosophila* demonstrates that the virus is entering the cells and is being uncoated to free the genome for transcriptional activity. The presumptive reason for lack of β–galactosidase activity is the absence of cellular or viral products necessary for polyhedrin promoter activity.

A logical experimental extension of this work is also presented by these authors. The virus, as NOVs, is administered to mosquito adults by (of all things) enema. The authors were able to detect CAT activity in mosquito midgut. This illustrates the point raised above: It may be possible to effectively increase the host range of the virus by employing it as a delivery vehicle for a toxin gene. Presumably, had the CAT gene been replaced by a gene whose product was toxic to mosquitoes, it would have "killed" the mosquito adults.

This kind of construct will be very useful in determining the safety of engineered viruses to non-target organisms. When the two-gene virus is applied to mouse cells, no activity is detected from either promoter (16). Since it is known that the RSV LTR is active in mouse cells, the lack of activity is thought to indicate that the virus is not being uncoated inside these cells. This very important observation had an immediately positive effect on our thoughts regarding the safety of baculoviral pesticides. It also provides some guidance in designing viral modifications. Given the potential for activity in non-Lepidopteran hosts, one would ideally choose a toxin gene that was as Lepidopteran-specific as possible and a mode of expressing this gene, which if at all possible, added another layer of safety. These options exist and in conjunction with our biological containment feature (described below) have allowed us to become comfortable with the idea of developing and field testing a genetically modified baculovirus insecticide.

Other opinions do enter into such a decision. The Environmental Protection Agency has jurisdiction over the field testing of pesticides. Their most recent policy statement indicates that all genetically modified organisms will be subject to review before any field testing begins (17). Modifications such as those implicit here, that is, the addition of non-viral DNA to the virus chromosome, will get the maximum level of scrutiny. Likely to be of particular interest are questions of environmental persistence, genomic stability and safety of the specific construct to non-target organisms.

What can be said about the ecology of baculoviruses? As with all viruses, they are obligate cellular parasites, relying on their hosts for replication machinery. This has some important implications for field monitoring and should serve to restrict interest to replication proficient virus and virus that has the opportunity to get into a host. (See 18 for a thorough discussion of these issues.). A factor such as sunlight will inactivate the great majority of applied virus over a short period of time. The inactivated virus, while perhaps detectable by some antigen-based assay, is of no interest in terms of safety. To assess the level of active virus will require direct assay in their host, caterpillars. These assays are very sensitive because live virus is very infective.

Many workers have noted the sensitivity of baculoviruses to the ultraviolet light component of sunlight (For a recent review of all aspects of virus environmental stability see 19.). Unformulated and unprotected virus applied to a crop loses much of its biological activity with rates approaching 50% a day. This is presumably due to genomic damage. Virus which survives this and makes its way to the soil has a much better chance. Baculoviruses, perhaps due to their polyhedrin covering, are very long-lived when sequestered in the soil, with a time frame measured in years. It is currently thought that this reservoir provides the inoculum for most of the naturally induced viral epizootics. The stability of the virus in soil is something that must be anticipated and accepted by both the pesticide developer and the regulatory agencies.

A genetically engineered virus applied to a field will still be detectable several years later.

As part of our development effort, we have conducted a survey of the stability of the viral genome. The clear summation of this work is that an engineered virus is no less stable than a wild-type virus, when examined both over the entire chromosome and when restricted to the altered portion. For example, we have constructs in which we have utilized various LTRs as promoters for β–galactosidase in place of the polyhedrin promoter and gene. In another case, we have left the polyhedrin promoter as is but replaced the gene with β-galactosidase. Neither of these virus types is capable of producing PIBs and must therefore be passaged in cell culture or in caterpillars by injection. This has been done and many progeny plaques have been screened for the parental ability to convert the indicator x-gal to its blue form. We have never observed a deficient plaque.

We have also tried to force the virus to undergo unusual recombination events. In this set of experiments, cultured cells were simultaneously infected with two virus types. One, wild type virus (PIB +, clear plaque) and, two, a recombinant in which the polyhedrin promoter is driving β–galactosidase expression (PIB –, blue plaque). The progeny of this infection were plaqued and screened for any that were potential recombinants (PIB +, blue plaque). In fact, we have consistently found quite a high percentage of these. This work needs further development, but it is our feeling based on the analysis of a number of these plaques that they are not novel recombinants. Rather, we think they are either: genomic cointegrates (anticipated to be unstable), nucleocapsids with multiple independent genomes encapsidated, or nucleocapsids which have aggregated and consequently yield an impure plaque that initially appears to be pure. None of our analyzed plaques yielded a pure recombinant phenotype after several replaquings, arguing against stable novel recombinants but consistent with the above interpretations.

<u>Biological Containment</u> The safety history of non-engineered baculoviruses and the nature of the anticipated

modification work leads us to feel confident about the safety of our eventual product. Recent history has shown that there will be those who are not so sanguine. In an effort to increase their comfort level we have devised and tested a biological containment scheme for a modified virus. Our scheme takes into account the following: the engineered virus, having had its polyhedrin gene replaced by a toxin gene, will in general not be able to produce PIBs , PIBs are desirable for production, formulation, and application purposes, the virus is very stable once reservoired in the soil, and the virus requires an insect cell in which to grow.

To initiate the scheme, cells in culture are simultaneously infected with a wild type virus and the recombinant, polyhedrin deficient virus. Inside the cell nucleus, an infection is established by both virus types. The recombinant is as replication proficient as the wild type, except for its inability to produce PIBs. This function is carried out by the wild type. The PIBs that form in the nucleus occlude nucleocapsids of both the wild type *and* the recombinant virus, as they are indistinguishable at this point. We refer to such PIBs as *mixed-composition PIBs (MC-PIB)*.

The multitude of PIBs produced during this process are composed of both types of input nucleocapsids. These progeny PIBs can be fed to caterpillars, initiating an infection of both virus types. The nature of the growth dynamics between the virus types can be followed in two ways. The PIBs themselves can be dissolved and their nucleocapsid components analyzed by DNA restriction enzyme digestion and subsequent filter hybridization analysis. This allows an assessment of the proportions of wild type to recombinant virus in the MC-PIB. Also, the free NOVs present in an infected caterpillar can be collected and analyzed by plaquing in cell culture. This provides an assessment of the frequency of wild type to recombinant virus free in the host.

The combination of these two techniques allows one to follow the fate of a recombinant virus as it is passaged via the MC-PIB technique through succeeding caterpillars. We have now gone through the creation and tracking of several different MC-PIBs, and it is clear from this that the technique contributes to the solution of several problems. First, it allows the natural delivery of a recombinant, polyhedrin deficient virus to a pest population in the field. Second, it provides an economical means for mass producing a recombinant virus in agriculturally relevant amounts by allowing the production to proceed via caterpillars infected through their diet. Finally, it provides a means of biological containment.

It is our observation that if an MC-PIB of given ratio of wild type to recombinant virus is fed to a caterpillar, and the progeny PIBs are fed to another caterpillar and so on, each generation will see the frequency of recombinant virus decrease. At some point several passages hence, the recombinant virus has been diluted out by the more appropriate growth of the wild type virus. At this point, the recombinant virus has essentially grown itself out of existence and has been wholly replaced by the wild type in the PIBs.

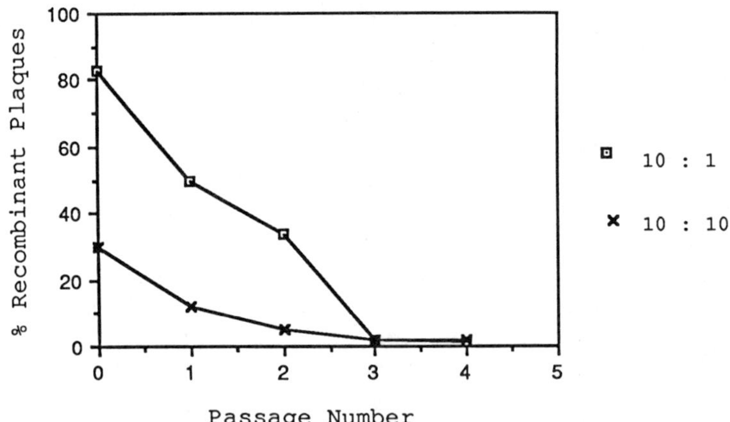

Figure 1. The ratio of recombinant to wild-type virus changes with each generation until only the wild type remains.

The number of passages the recombinant virus survives is influenced by the initial ratio of recombinant to wild type virus used to infect the tissue culture cells. This is illustrated in Figure 1. Cells in culture were infected with either a multiplicity of infection of 10:1 or 10:10, recombinant: wild type, plaque forming units.

Passage 0 is the percentage of recombinant plaques obtained when the progeny virus in the cell culture flask were plaqued. Passage 1 refers to the percentage obtained from the first caterpillars (*Heliothis virescens,* fourth instar) fed PIBs from the tissue culture cells. As can be seen, the recombinant persists at higher percentages for a longer period of time when the initial MC-PIB composition is skewed towards the recombinant.

In this example, the caterpillars were infected with the virus by placing 10^5 MC-PIBs on the surface of their diet. If this is raised to 10^6 MC-PIBs, the persistence of the recombinant is extended out at least one more generation. This sensitivity to inoculum levels is probably due to the requirement for a cell to be coincidentally infected with both a wild type and a recombinant virus to obtain an MC-PIB. The likelihood of this occurring should be sensitive to the initial number of viruses introduced into the caterpillars.

The MC-PIB phenomenon described provides biological containment for an engineered baculovirus. Baculoviruses require the packaging of a PIB to insure their survival in nature between infections. The provision of PIB-packaging for a PIB-deficient recombinant virus by coinfection with a wild type virus accomplishes this. This could have been counter to the safety issue, but fortunately in this case, the packaging mechanism is not foolproof and with succeeding infections, the recombinant finds its way less frequently into the PIB and is gradually lost. Whenever a MC-PIB harbored in the soil exits via the infection process, it will have contributed to a decrease in the proportion of recombinants. This will insure that the recombinant does not spread through the environment by unchecked growth after its application.

In this paper, we have discussed baculoviruses, insect pathogens which we feel have great potential as insect control agents. The shortcomings that have been their problem in the past may be eliminated by employing viral biology in the service of genetic engineering technologies. We have presented scenarios in which their effectiveness can be improved while their safety features are maintained and, in a sense, actually improved.

Acknowledgments

The author would like to thank Carol Ann Johnson, Pennina Safer, Peter LaPan, Lisa Hahn, and David Clemm for their invaluable contributions to the work described herein.

Literature Cited

1. Granados, R.R. and Federici, B.A. The Biology of Baculoviruses, vol 1 and 2; CRC Press: Boca Raton, FL, 1986.
2. Maramorosch, K. and Sherman, K.E. Viral Insecticides for Biological Control; Academic: Orlando, FL., 1985.
3. Doerfler, W. and Bohm, P., Eds. The Molecular Biology of Baculoviruses; Springer-Verlag: Berlin, 1986.
4. Miller, D.W. et. al. In Genetic Engineering, vol 8; Setlow, J.K. and Hollaender, A., Eds.; Plenum: New York, 1986; p. 277.
5. Doerfler, W. In The Molecular Biology of Baculoviruses; Doerfler, W. and Bohm, P., Eds.; Springer-Verlag: Berlin, 1986; p. 51.
6. Luckow, V.A. and Summers, M.D. Biotechnol. 1988, 6, 47-55.
7. Possee, R.D. and Howard, S.C. Nuc. Acids Res. 1987, 15, 10233-10248.
8. Miyajima, A. et. al. Gene 1987, 58, 273-281.
9. Groner, A. In The Biology of Baculoviruses, vol 1; Granados, R.R. and Federici, B.A., Eds.; CRC Press: Boca Raton, FL, 1986; Chapter 9.
10. Miller, L.K. In The Biology of Baculoviruses, vol 1; Granados, R.R. and Federici, B.A., Eds.; CRC Press: Boca Raton, FL, 1986; Chapter 11.
11. Kirschbaum, J.B. In Ann. Rev. Entomol.; Mittler, T.E., Ed.; Annual Reviews: Palo Alto, CA., 1985; p. 51.
12. Keeley, L.L. and Hayes, T.K. Insect Biochem. 1987, 17, 639-651.
13. Zlotkin, E. In Comprehensive Insect Physiology, Biochemistry and Pharmacology, vol 10; Kerkut, G.A. and Gilbert, L.I., Eds.; Pergamon: New York, 1985; Chapter 15.

14. Piek, T. In Comprehensive Insect Physiology, Biochemistry and Pharmacology, vol 11; Kerkut, G.A. and Gilbert, L.I., Eds.; Pergamon: New York, 1985; Chapter 14.
15. Carbonell, L.F. et.al. J. Virol. 1985, 56, 153-160.
16. Carbonell, L.F. and Miller, L.K. Appl. Environ. Microbiol. 1987, 53, 1412-1417.
17. Federal Register, June 26, 1986.
18. Miller, D.W. et.al. In Prospects for Physical and Biological Containment of Genetically Engineered Organisms; Gillett, J.W., Ed.; Ecosystems Research Center Report No. 114, Cornell Univ., 1987; Chapter 3.
19. Jacques, R.P. In Viral Insecticides for Biological Control; Maramorosch, K. and Sherman, K.E., Eds.; Academic: Orlando, FL., 1985; p. 285.

RECEIVED May 25, 1988

Chapter 32

Agricultural Biotechnology Research Guidelines

Considerations for Laboratory Experiments and Field Release

Richard M. Parry, Jr., and Jerome P. Miksche

Agricultural Research Service, U.S. Department of Agriculture, Beltsville, MD 20705

Biotechnology research supported by the U.S. Department of Agriculture must comply with principles and conditions contained in the NIH Research Guidelines. NIH is currently reviewing its existing Guidelines to include new appendices for large animal and greenhouse containment. USDA research projects are reviewed and approved according to the Recombinant DNA Advisory Committee (RAC) procedures. The biotechnology regulations being developed by other Federal Agencies will also effect the conduct of recombinant DNA research. A procedure to review modified organisms is presented which could be employed to predict environmental behavior. The two stage procedure used by APHIS to conduct environmental release studies of genetically engineered veterinary biologics, which may be applicable to releases of other organisms is also discussed.

The debate about biotechnology research and its products is indicative of an evolving system which can have broad impact on science, the economy and the public. As biotechnology projects begin to reach implementation stage, the public has become more involved in the ethical issues surrounding the transfer of DNA between different species which has led to questions about regulation of these new products. The establishment of a system to review research and to regulate new products is necessary to ensure that the benefits of biotechnology are to be realized by society. The public participation in the review of biotechnology work provides a unique opportunity for scientists to demonstrate that there is sufficient knowledge of relevant principles to evaluate projects and also that adequate precautions are being used to prevent any potential adverse effects to man and the environment.

The National Institutes of Health (NIH) have proposed amendments to their Recombinant DNA Research Guidelines to include containment procedures involving organisms of interest to agriculture. The amendments are based upon an extensive knowledge base developed over several decades of research with organisms that are potentially pathogenic to humans, plants and animals. The research procedures used for safely handling pathogens have benefited agriculture through the development of new methods to control pathogens that potentially threaten the United States food production system, and in some cases, have provided other countries with the tools to control debilitating diseases that ravage crops and animals.

Biotechnology Research Guidelines

The Federal Government sponsored a conference in the early 1970s to explore the risks and benefits of recombinant DNA research. This led to the establishment of the Recombinant DNA Advisory Committee (RAC) by the NIH and in 1976 NIH published Guidelines for Research Involving Recombinant DNA Molecules. The Guidelines were updated in 1986 (1). These Guidelines set forth procedures which scientists should follow when carrying out research with recombinant DNA materials used in biotechnology studies. The Guidelines describe the precautions which are necessary for handling laboratory organisms in containment, the role of the Institutional Biosafety Committee (IBC), and the oversight function carried out by the NIH RAC. The procedures recommended in the Guidelines to prevent adverse effects follow the principles developed for classifying and containment of pathogens which are described in the CDC/NIH booklet "Biosafety in Microbiological and Biomedical Laboratories" (2). These biosafety procedures have become an internationally accepted guide for handling pathogens in the laboratory and have proven to be effective for safe experimentation.

The NIH guidelines were originally developed to cover microbiological research carried out in a laboratory environment. As progress has been made in biotechnology techniques, many scientists saw the need to expand the research guidelines to include research on plants and animals. A Federal interagency working group was formed under the White House Cabinet Council on Natural Resources and the Environment to coordinate this process (3). The working group was charged to develop a framework to provide adequate guidelines for all biotechnology research as well as coordination of regulations being considered for new products. This proposal was finalized with the publication of a Federal Register document on June 26, 1986 (4) and the Biotechnology Science Coordinating Committee (BSCC) was established.

The June 26 document requested public comments on the Department of Agriculture's process for ammending the NIH research guidelines to cover research undertaken on agricultural biotechnology projects (see 4 pages 23367 through 23393). After considering the comments received, the USDA advised the BSCC that it would provide NIH information on containment of large animals and plants in greenhouses that could be included as annexes to the Recombinant DNA Research Guidelines. These proposed amendments to the NIH Guidelines were recently published in the Federal Register (5) and are to be finalized in early 1988.

The procedures for animal and plant containment currently being considered by the NIH/RAC follow the scientific principles that have been practiced for research on pathogens exotic to the U.S. They are designed to prevent escape of a genetically engineered organism and four biosafety containment levels are described for pathogens of plants and of large animals, depending on the pathogenicity of the microorganism. For example, containment at levels 1 or 2 would cover experiments where genetically modified plants or animals are being studied; higher levels of containment would be required for studies involving plants or animals which have been inoculated with genetically engineered pathogens that require high containment precautions.

Table I summarizes the practices necessary for four biosafety levels of containment of plants. Biosafety level 1 would be similar to a standard greenhouse, with the containment precautions becoming more strict at higher levels. The IBC would recommend the appropriate containment level for each experiment and will be responsible for assuring that the principal investigator adheres to the approved practices for that project.

Table I. Proposed Practices for Greenhouse Containment (X=mandatory; R=recommended)

Description	Biosafety Level			
	1	2	3	4
Practices Manual	X	X	X	X
Materials Rendered Inactive	X	X	X	X
Windows & Vents Screened	R	X	X	X
Only Persons Required May Enter		X	X	X
Protocol Adopted		X	X	X
Coats, Gowns, Smocks, Bootees		R	X	X
All Biological Material Autoclaved			X	X
Entry Through Two Sets of Doors			X	X
Windows Closed and Sealed			X	X
Air Through HEPA* Filters			X	X
Locked Access Doors			X	X
Double Doored Autoclave			R	X
Decontamination Showers				X
Central Sewage Decontamination				X

*High Efficiency Particulate Air (HEPA) Filter.

The containment procedures have been developed using the experience gained through operation of the special plant quarantine facility of the USDA's Agricultural Research Service (ARS) at Frederick, MD (6). Since plant pathogens exotic to the U.S. are studied there, the facility operates under the oversight of the Animal and Plant Health Inspection Service (APHIS). Thus the basic principles of the proposed greenhouse biosafety procedures have been tested under actual laboratory conditions and have been shown to prevent release of pathogenic organisms into the environment.

The development of research guidelines for procedures to conduct experiments with a genetically engineered organism released from containment are still being considered. The initial environmental release experiments have undergone intense scrutiny by regulatory agencies before granting approval. As scientists gain more experience with predicting the fate and behavior of genetically engineered organisms released into the environment, the approval process should become less onerous and time consuming. Clearly there is a need for procedures to safely conduct studies of modified organisms outside of containment, since society will not benefit from the new products unless they can be properly tested and then put to some practical use. The Council of the National Academy of Sciences convened a special committee to review the planned introduction into the environment of organisms genetically engineered using recombinant DNA techniques (7). The committee found there is no evidence that unique hazards exist either in the use of recombinant DNA techniques or in the transfer of genes between unrelated organisms. Furthermore the risks associated with the introduction of engineered organisms are the same as those associated with the introduction into the environment of unmodified organisms or organisms modified by other genetic techniques.

ARS assembled a group of experts to consider several aspects of biotechnology with relevance to research and product evaluation. One of the issues studied was the information needed to identify any potential risks in the release of a genetically modified organism into the environment. Table II presents a listing of the properties of a microorganism which should be known in order to be able to predict its behavior outside of containment and to avoid potential adverse environmental effects. By this procedure characteristics of the modified organism are compared to those of the "wild-type" organism in order to determine if the modified organism is of more concern than its naturally occurring parent. In this particular case, each characteristic was given a numerical value of 1, 2 or 3 according to whether the Observed Effect Level of the modified organism is potentially less, the same, or more than the wild-type, respectively. The value was multiplied by a weighting factor, since the biological significance of some of the characteristics obviously will be greater than others. The control potential of the organism is a desirable property and thus was given a negative score. A mutant strain of *Sclerotinia sclerotiorum* (a plant pathogen under consideration for weed control) in the example used was more responsive to decontamination cleanup than its natural parent organism and was given a higher negative score.

Using a comparison method to evaluate the safety of releasing genetically engineered organisms into the environment, such as the example presented here, one can identify areas of potential

Table II. EVALUATION OF A MODIFIED MICROORGANISM FOR ENVIRONMENTAL RELEASE

*Project Description: Evaluation of a weed biological control agent using a deletion mutant of the fungal pathogen *Sclerotinia*.

Characterization	Observed Effect Level** less 1	same 2	more 3	Weight	Score
Microorganism Classification					
(a) Pathogenicity		+		2X	4
(b) Host Range	+			1X	1
(c) Toxicity					
to animals		+		2X	4
to plants			+	2X	6
to workers		+		2X	4
(d) Characterization of genetic alteration		+		1X	2
(e) Mode of Action		+		1X	2
Biological/Ecological Profile					
(a) Longevity/Persistence	+			2X	2
(b) Mobility (Organismal, genetic transferability)		+		2X	4
(c) Control Potential Suicide gene, decontamination (negative score)			+	3X	-9
(d) Dose/Population needed to have effect			+	1X	3
		TOTAL SCORE			23

*Information on the organism and scoring adapted from a presentation by R. J. Cook and R. Linderman June 2, 1987, made at an Agricultural Research Service workshop on biotechnology held at Beltsville, MD.

**Observed Effect Level of the modified microorganism compared to the "wild-type" parent. The total score for the wild-type strain of *Sclerotinia sclerotiorum* is 26.

concern and bring to bear objective assessment procedures where none currently exist. This procedure could be employed by an Institutional Biosafety Committee or by a regulatory agency when evaluating requests to release organisms into the environment.

Biotechnology Regulations

The Federal Register document of June 26, 1986 (4) described regulatory proposals from the Environmental Protection Agency (EPA), the Food and Drug Administration (FDA) and the Department of Agriculture's Food Safety Inspection Service (FSIS) and APHIS. The proposed procedures for regulating biotechnology products were developed using existing statutory authority administered by these agencies and coordinated through the BSCC. The document generated considerable public comment. A common observation was that the regulations were overlapping and duplicative between agencies.

EPA has proposed regulating genetically engineered microorganisms under the Federal Insecticide, Fungicide and Rodenticide Act (FIFRA) for pesticide products and under the Toxic Substances Control Act (TSCA) for all other uses. The Administrator of EPA recognized that there is a lack of scientific consensus on the definition of pathogen, environmental release and greenhouse containment and appointed a Biotechnology Science Advisory Committee (BSAC) to establish a scientific consensus on the definition of these terms. Although the BSAC continues in its deliberations of these terms, it should be noted that the greenhouse containment provisions given in Table I closely parallel the BSAC current proposals. Thus, if the NIH Research Guidelines were amended to include greenhouse containment provisions, this might serve to adequately address the concerns of EPA.

APHIS is the first agency to publish final regulations on biotechnology products (8) using the authority it has under the Plant Pest Act. This rule has great significance for agriculture research in that a listing of plant pests subject to regulation is given and procedures are set forth to obtain permits for their release into the environment or for transporting them to new locations. The regulation stipulates that no person shall introduce an organism that has been altered or produced through genetic engineering, if derived from a plant pest, without first obtaining a permit from APHIS. Thus the APHIS rule regulates certain genetically engineered organisms and products that present plant pest risks, and does not regulate an article merely because of the process by which it was produced. The agency has adopted a flexible approach to regulating biotechnology, which will allow modification of its procedures as new scientific evidence becomes available on plant pests. It should also be noted that APHIS meets regularly with EPA to discuss petitions for release permits that are being reviewed, thereby avoiding duplication of effort by the two agencies.

APHIS has approved the release of several veterinary biologics using genetically engineered materials. The Veterinary Services Divsion has developed a two stage procedure to conduct field testing of the products prior to licensing as a biologic (9). After a biologic has been developed and tested in a containment laboratory, the company or institution will then submit a request for approval to APHIS to evaluate the material in a quarantined facility, which will allow larger numbers of animals to be exposed. The quarantine test will allow the development of critical information about the biologic, but takes precaution against uncontrolled environmental spread of the organism. These results are examined in an Environmental Assessment as required by the National Environmental Policy Act. If no significant impact is found from the test in the quarantined facility, approval is then given to conduct a restricted field trial involving larger numbers of animals at several locations. Should the material prove to be safe and effective in this expanded test, a final Environmental Assessment is prepared and a product license is issued. The two stage process for evaluation of genetically engineered materials prior to wide scale use will permit adequate safety testing and should be considered for other products of biotechnology.

Conclusions

Biotechnology offers significant opportunities to solve many otherwise intractable problems that confront agriculture but which may never be conquered using traditional approaches. Public concern about the environmental safety of using genetically engineered organisms has led to the development of research guidelines and regulations for handling and using the materials. The NIH Research Guidelines for handling recombinant DNA organisms have played an important role in allowing this research to move forward and have become mandatory for scientists receiving funding from the U.S. Government.

The NIH Research Guidelines are currently being amended to include containment procedures for both large animals and for higher plants in greenhouses. Various regulatory agencies are also developing protocols for the registration or licensing of new products. However, additional work is needed to develop a procedure for testing products outside of containment. Through application of biological principles currently used to handle pathogenic organisms, it is possible to conduct field release studies that will safeguard the environment. The development of data which allow for a comparison of the characteristics of the modified organism to the wild-type parent will permit a reviewing agency or committee to make a sound determination of the biological safety of a modified organism. Also, the two stage environmental release procedure used by APHIS to test biologics appears to be a rational process to allow field testing and virtually eliminate risk to man and the environment.

Literature Cited

1. Guidelines for Research Involving Recombinant DNA Molecules, U.S. Department of Health and Human Services, National

Institutes of Health, Federal Register, 51, 88, pp 16958-16985, May 7, 1986.

2. Biosafety in Microbiological and Biomedical Laboratories, U.S. Department of Health and Human Services, Pub. No. (CDC)84-8395, U.S. Government Printing Office: Washington, DC, 1984.
3. Proposal for a Coordinated Framework for Regulation of Biotechnology, Executive Office of the President, Office of Science and Technology Policy, Federal Register, 49, 252, pp 50856-50907, December 31, 1984.
4. Coordinated Framework for Regulation of Biotechnology; Announcement of Policy and Notice for Public Comment, U.S. Office of Science and Technology Policy, Federal Register, 51, 123, pp 23302-23393, June 26, 1986.
5. Recombinant DNA Research: Proposed Actions under Guidelines, U.S. Department of Health and Human Services, National Institutes of Health, Federal Register, 52, 154, pp 29800-29814, August 11, 1987.
6. Melching, J. S.; Broomfield, K.; Kingsolver, C. Plant Disease 1983, 67, 717-722.
7. Introduction of Recombinant DNA-Engineered Organisms into the Environment: Key Issues, Council of the National Academy of Sciences, National Academy Press, Washingto, D.C., 1987.
8. Plant Pests; Introduction of Genetically Engineered Organisms or Products; Final Rule, U.S. Department of Agriculture, Animal and Plant Health Inspection Service, 7 CFR Parts 330 and 340, Federal Register, 52, 115, pp 22892-22915, June 16, 1987.
9. Shibley, G. P.; Joseph, P.; Espeseth, D.; Gay, C. Presented at VII International Congress of Virology, 1987.

RECEIVED February 18, 1988

Chapter 33

Biorational Control of Crop Pests by Mating Disruption

Residue Analyses of Z-9-Dodecen-1-yl Acetate and Z-11-Tetradecen-1-yl Acetate in Grapes

Terry D. Spittler[1], Harrison C. Leichtweis[1], and Timothy J. Dennehy[2]

[1]Analytical Laboratories, New York State Agricultural Experiment Station, Cornell University, Geneva, NY 14456
[2]Department of Entomology, New York State Agricultural Experiment Station, Cornell University, Geneva, NY 14456–0462

Z-9-dodecen-1-yl acetate (Z-9-DDA) and Z-11-tetradecen-1-yl acetate(Z-11-TDA), primary components of grape berry moth pheromone, were applied in vineyards via tie-on dispensers for the entire growing season. Mature grapes (100G) were blended and extracted with acetone; following the addition of water, the analytes were extracted into hexane which, after evaporation, was adsorbed onto a Florisil Sep-pak. Elution was with 10% acetone-hexane. Chromatography (H-P Model #5890) was on a Supelcowax 10 capillary column, 30M x 0.25 mM I.D. x 0.25 μM coating. Temperature program: 80°C - 130°C @ 5°/min, 130°C - 200°C at 4°/min, hold 9 min. Detection by HP-MSD Model #5970B was in the selective ion mode at 166 M/E and 194 M/E for Z-9-DDA and Z-11-TDA, respectively. Corresponding retention times were 17.7 and 22.7 min. Sensitivity was <5 ppb for both materials at 80% recovery. Insect control by this biotechnical approach was good, and it eliminated the use of traditional chemical pesticides against grape berry moth.

As an alternative to the elimination of insect pests by the repeated application of chemical pesticides to crop acreage, much effort has been directed at developing technology to utilize biological substances inherent to the commodity or pest as safer, more environmentally tolerable, control agents. Macroorganisms have been introduced as predators and microorganisms are being used against a variety of destructive pests. In addition to these biological measures, investigations at the subcellular and molecular level have suggested other biotechnical approaches.

0097–6156/88/0379–0430$06.00/0

The role of insect sex attractants, or pheromones, has been studied for sometime; the corresponding entomological literature is extensive (1,2). Advances in analytical techniques and an increased understanding of the structural relationships between pheromone components has favored qualitative and quantitative elucidation of many insect specie´s pheromone composition and allowed for the synthetic reconstruction of these in quantities adequate for research and commercialization (3,4,5). The initial control application was, of course, a direct offshoot of one of the primary means used to verify their components and composition -- their use as trap attractants (6,7). However, since the pheromone source was a dispenser in the trap and only the volatile components dispersed over the study area in miniscule quantities, their legal status was that of a non-applied material. As such, it was considered that the commodities and the environment did not come in contact with the experimental materials and that they were safely confined to, and disposed with, the assessment/trapping device (8).

When it became apparent that pheromone application could confuse and disrupt the mating cycles of certain insect pests and thus biologically minimize the population of successive generations, the larger amounts of material and wider distribution necessary to do this caused the regulatory agencies to view these biorational control agents as chemical pesticides. While special consideration has been given to the amounts and type of toxicological data that may be required for registration, there is still a requirement that residue data for the components of the pheromone on the raw agricultural commodity be submitted in support of petitions for experimental use permits, pesticide registration, tolerances, and exemptions from tolerance (9).

Biotechnical control systems are beginning to make their appearance in the agrochemical marketplace. One potential commercial application of a pheromone insect control agent is that of the grape berry moth pheromone produced by Shin-Etsu Chemical Company and investigated in the field by Dennehy and Roelofs (10,11). Before petitions could be filed, it was necessary that an analytical method for the two components Z-9-dodecen-1-yl acetate and Z-11-tetradecen-1-yl acetate (Z-9-DDA and Z-11-TDA) in and on grapes be developed, and further, that residue determinations be conducted on samples from the cited field study.

Sample Production

Pheromone was applied in hollow, polyethylene tie-on dispensers placed at regular intervals throughout the vineyard at the rates and timing shown in Table I. Initially, each dispenser contained 88 mG of pheromone, a 10:1 mixture of Z-9-DDA and Z-11-TDA, which slowly diffused through the plastic walls and permeated the vineyard. Ten kG samples were harvested from each plot and transported to the laboratory for subsampling, spiking, and frozen storage.

Table I. Vineyards Selected for Treatment -- 1986

VINEYARD	SAMPLE#	RATE TIES/A	RATE (G/A)	APPLICATION DATES		HARVESTED
Hayward	4329-4332	400	(35.2)	5/15	--	10/1
Hayward	4333-4336	400	(35.2x2)	5/15 &	7/15	10/1
Hayward	4337-4340	0	(0)	--	--	10/1
Francis	4341-4344	400	(35.2)	5/15	--	10/1
Francis	4345-4348	200	(17.6)	5/15	--	10/1
Francis	4349-4352	0	(0)	--	--	10/1
DeGolier	4353-4356	800	(70.4x2)	5/15 &	7/15	10/1
DeGolier	4357-4360	400	(35.2)	5/15	--	10/1
DeGolier	4361-4364	0	(0)	--	--	10/1

Sample Preparation

Grape berries (100G) plus 5G Hyflo-Supercel and 100 mL of acetone were blended for two min. The resultant slurry was filtered through a sintered glass funnel, and the pad rinsed with acetone. To the filtrate in a separatory funnel was added 300 mL H_2O, 30 mL saturated NaCl and 100 mL n-hexane. After mixing and layer separation the hexane phase was retained, while the aqueous phase was extracted two more times with 100 mL portions of n-hexane. The combined organic portions were dried over sodium sulfate and evaporated under vacuum (35°C) to 5 mL. Waxy precipitates were centrifuged and a 1.0 mL aliquot was placed on a Florisil Sep-pak (Waters Assoc.). The Sep-pak was washed with 5.0 mL of hexane (discarded) then eluted with 2.0 mL of 10% acetone-hexane. Volume was reduced under dry N_2 to 1.0 mL.

Chromatography

Samples (1.0 µL) were injected on a Hewlett-Packard Model 5890 Capillary Gas Chromatograph utilizing a split-splitless injector at 245°C. Column was a Supelcowax 10, 30M x 0.25 mM I.D. x 0.25 µM coating with a He carrier velocity of 30 cm/sec. Temperature program: 80°C, hold 1.0 min; 80°C to 130°C @ 10°C/min; 130°C to 200°C @ 4°C/min, hold 9.0 min; 200°C to 250°C @ 30°C/min, hold/recycle. Transfer line to the detector was via a butt connector/ guard column maintained at 280°C.

Detection

Quantitation was with a Hewlett-Packard Model 5970B Mass Selective Detector run in the SIM (Selective Ion Mode) at the major unique M/E for each compound. Retention times of 17.7 min and 22.7 min were recorded for Z-9-DDA (M.W. 198) and Z-11-TDA (M.W. 254), respectively. Recoveries and sensitivities are found in Table II. Figure 1 illustrates the SIM response for standard materials, while Figures 2 and 3 are of a check material, and a check spiked with pheromone at the 5 ppb level. Figure 4 demonstrates the absence of

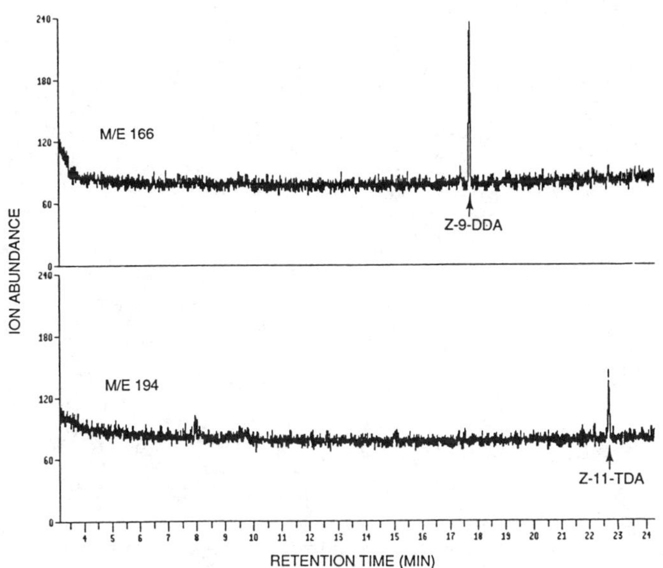

Figure 1. STANDARDS 0.1 μG/mL.

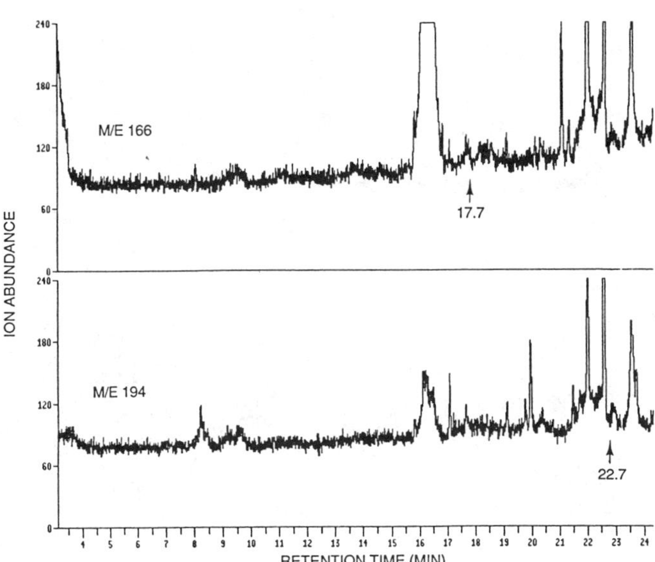

Figure 2. CHECK #4361.

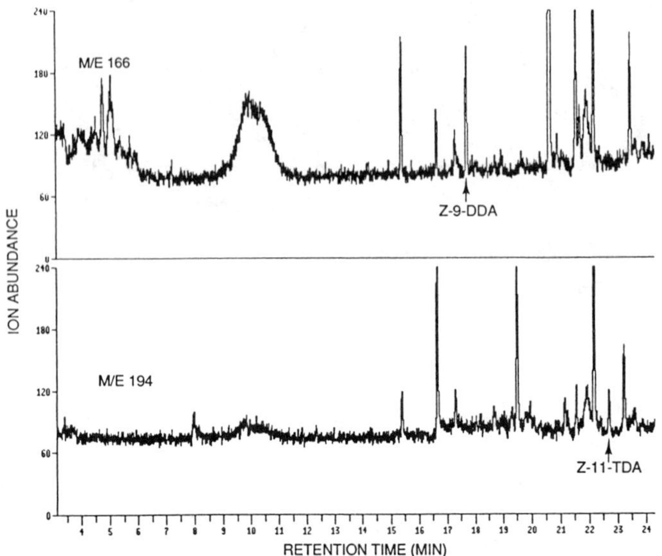

Figure 3. CHECK SPIKE 0.1 µG/20G ≅ 5 PPB.

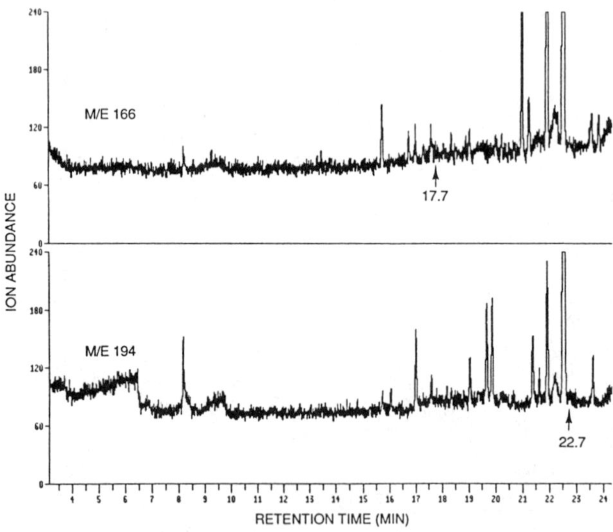

Figure 4. SAMPLE #4354 (800 TIES/A).

any detectable pheromone for a typical sample from the plots that received the heaviest application.

Table II. Recovery and Sensitivity

PPB	Z-9-DDA (M/E 166)		Z-11-TDA (M/E 194)	
10	85%		92%	
10	94%	$\overline{98\%}$	87%	$\overline{90\%}$
10	115%		92%	
5	80%		80%	
2	DETECTED BUT NOT QUANTITATED (2X BASELINE)			

Results

Grape berry moth control was excellent at all rates, with infestations well below the acceptable 2% damage threshold (11). A total of 36 samples was analyzed: none registered above the 5 ppb quantitation cutoff, or were detectable at the 2 ppb level. The GC-MSD methods described should be useful, with minor variations, for determining similar pheromone residues in other agricultural commodities. We were originally supplied a method developed by the manufacturer for Z-11-TDA determinations in tea but the extraction/cleanup was inadequate to handle the matrix obtained from blended grapes, and the GC sensitivity was limited by the flame ionization detector employed (Kinya Ogawa, private communication).

The absence of any detectable residues of Z-9-DDA and Z-11-TDA increases the data base for pheromone registrations and reinforces the argument that similar applications will leave to no detectable residues in other crops. It may eventually be recognized that exemptions from tolerance for pheromone use registrations are warranted. Because of the unique pheromone composition for each insect species, and because every combination would require a separate registration package for each commodity, the potential return on investment for many of these biorational systems is poor. Consequently, the adoption of an exemption policy would greatly increase the interest and feasibility of biotechnical insect control with pheromones by minimizing an expensive and time consuming step in the regulatory process.

Literature Cited

1. Karlson, P; Luscher, M. Nature. 1959, 183, 55-56.
2. Tamaki, Y. In Comprehensive Insect Physiology: Behavior and Pharmacology; Kerkut, G.A.; Gilbert, L.J., Eds; Pergammon Press, Oxford, 1979; 9, 145-183.
3. Tumlinson, J. H.; Heath, R. R. J. Chem. Ecol. 1976, 2, 87-99.

4. Tamaki, Y. In Chemical Ecology: Odour Communication in Animals; Ritter, F. J., Ed.; Elsevier, Amsterdam, 1979; 169-180.
5. Tamaki, Y. In Methods in Pesticide Science; Fukami, J.; Vesugi, Y.; Ishizuka, K.; Tomizawa, C., Eds.; Soft Science, Inc., Tokyo, 1981; p. 342-358.
6. Tamaki, Y.; Noguchi, H.; Sugie, H.; Kariya, A.; Arai, S.; Ohba, M.; Terada, T.; Suguro, T.; Mori, K. Jap. J. Appl. Ent. Zool. 1980, 24, 221-228.
7. Baker, R.; Herbert, A.; Howse, P.E.; Jones, O.Y.; Francke, W.; Reith, W. J.C.S. Chem. Comm. 1980, 52-53.
8. 40 CFR S162.5 (d)(2).
9. 40 CFR S158.165 (a-c).
10. Roelofs, W.L.; Tette, J.P.; Taschenberg, E.F.; Comeau, A. J. Insect Physiol. 1971, 17, 2235-2243.
11. Dennehy, T. J.; Roelofs, W. L. J. Econ. Ent. Submitted.

RECEIVED February 22, 1988

Chapter 34

Genetically Engineered Microbial Pesticides

Regulatory Program of the Environmental Protection Agency and a Scientific Risk Assessment Case History

Frederick S. Betz

U.S. Environmental Protection Agency, (TS–769C), 401 M Street, SW, Washington, DC 20460

Existing EPA regulations and guidelines for microbial pesticides are applicable to genetically engineered products. Additional review procedures and requirements have been developed to address questions about the nature and stability of the genetic alteration and whether the resulting microorganisms will have adverse effects on non-target organisms when applied in the environment. Two small scale field trials recently were initiated in California to evaluate the use of genetically engineered microbial pesticides to inhibit colonization of ice nucleating bacteria on plants. EPA scientists developed risk assessments which were the basis for the Agency's determination that the proposed field trials posed no foreseeable human health and environmental risks and could therefore proceed.

The Environmental Protection Agency (EPA) regulates pesticide products under the authority of the Federal Insecticide Fungicide and Rodenticide Act (FIFRA) (1). Section 6 of FIFRA requires that pesticides be registered by EPA before they may be sold or distributed, and section 5 gives the Agency authority to issue an experimental use permit (EUP) before an unregistered pesticide may be used on an experimental basis. EPA may grant a registration (2) or an EUP (3), provided that it has sufficient data and information about the pesticide product (4) to support a determination that benefits derived from the product will outweigh any risks that may be associated with its use. It is the responsibility of the applicant to provide the data to support this determination (5).

Greater than 99 percent of the registered pesticides contain chemicals as their active ingredients. The remainder contain bacteria, fungi, viruses or protozoans as their active components and are commonly referred to as microbial pesticides. Microbial-based products achieve their pesticidal activity by pathogenic or toxic modes of action or by indirect activity such as competitive displacement of the target pest. In order to fulfill their function safely, both microbial and chemical pesticides must kill, repel or otherwise reduce the destructive activity of their target pest while having no or minimal adverse effects on human health or the environment.

Microbial pesticides have been registered for use since the late 1940s. Yet today, there are still only 14 registered microbial active ingredients which are formulated into just a few hundred products. Several factors contribute to the limited growth of this group of products. Their spectrum of activity is usually narrow, their action in controlling the target pest is often slow, and their effective field life is generally short due to susceptibility to environmental factors such as temperature, humidity and ultraviolet radiation. As a result, microbial pesticides historically have not been commercially competitive with traditional chemical products even though experience to date has demonstrated no significant human health or environmental problems associated with their use.

This situation appears to be changing. Advances in molecular biology and genetic manipulation techniques over the past fifteen years have prompted a renewed interest in microbial pesticides. Genetic engineering techniques are being used to overcome some of the problems that have hindered the success of microbial pesticides. A few of these new products are now at the stage where they are ready for small scale field testing. This chapter summarizes EPA's current regulatory program for genetically engineered microbial pesticides and presents a case history for one of the experimental products the Agency has evaluated.

EPA Regulatory Program

The potential for human and environmental exposure to a pesticide product usually increases in a stepwise fashion as the product proceeds from discovery to commercial use. As pesticide exposure increases, so does the risk of adverse human health or environmental effects. Therefore, EPA's pesticide regulations are phased in as pesticide exposure, and potential risks, increase (Figure 1). For example, there is usually no EPA oversight of early product development activity in the laboratory, greenhouse or growth chamber since such activities can be expected to result in minimal environmental exposure. The next stage of product development usually entails small scale outdoor applications to evaluate product performance in the field. When naturally occurring indigenous microbial pesticides are tested in small scale situations, survival and dissemination are usually kept in check by natural environmental and biological control mechanisms. EPA does not generally regulate these limited and controlled microbial pesticide uses unless the treated crop is to be used for human consumption or domestic animal feed.

Large scale experimental applications of microbial pesticides are usually undertaken prior to registration in order to further evaluate product efficacy or to develop other data necesary to

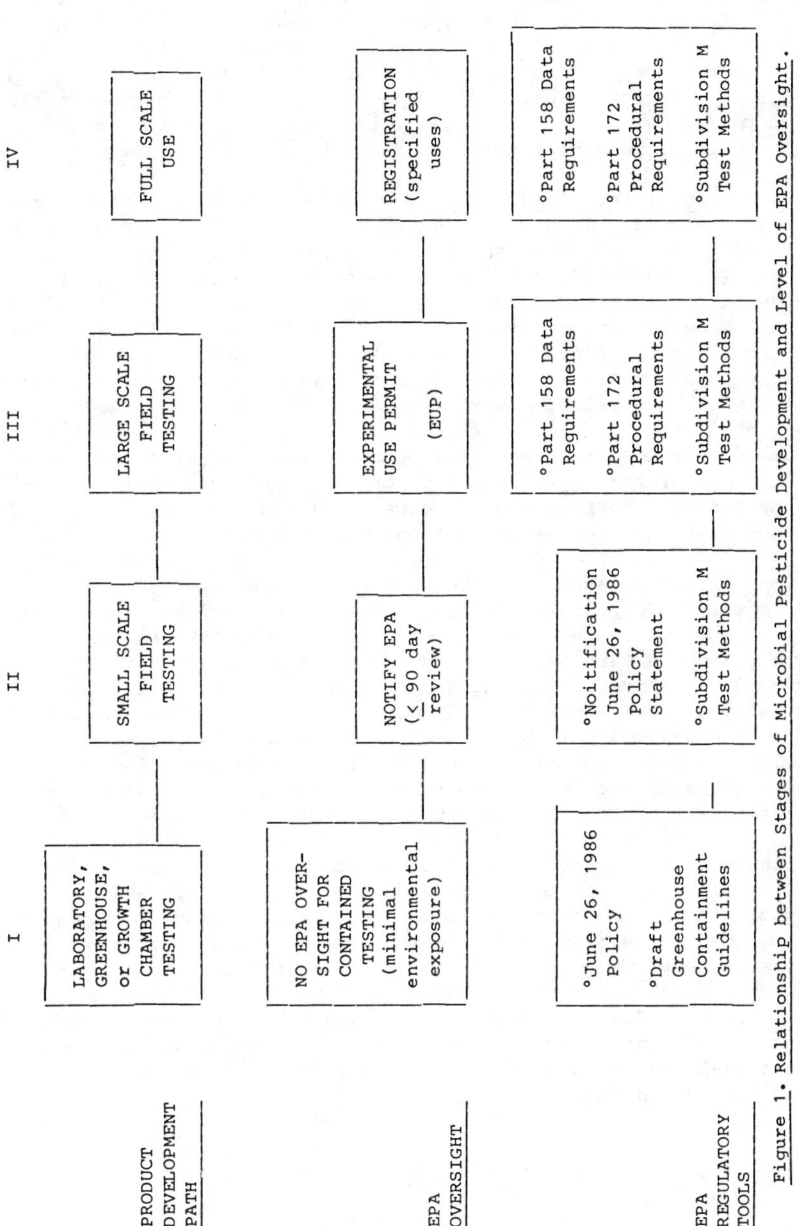

Figure 1. Relationship between Stages of Microbial Pesticide Development and Level of EPA Oversight.

support product registration. Such uses may pose greater risk concerns since they involve more widespread, repeated or intensive exposure of non-target organisms, including humans. As a result, EPA regulates all microbial and chemical pesticides when they are applied in a widespread manner either for experimental purposes or as commercial products. Generally, large scale outdoor experiments are subject to EPA oversight if the pesticide is to be applied on greater than 10 acres of land or on greater than one acre of water.

Genetic Manipulation. As discussed above, genetic manipulation of microbial pesticides is usually undertaken to improve the pesticidal qualities of the microorganism. Donor and/or recipient microorganisms well-adapted for survival in the environment are usually selected in order to maximize survival potential of the final product. Genetic manipulations may be accomplished by, for example, imparting a new toxin producing capability, increasing an existing toxin producing capability, expanding host range, and/or increasing microorganism survival or competitive capabilities. In considering the potential consequences of these kinds of manipulations, one can not take for granted that natural control mechanisms will prevent establishment of the engineered microorganism in a new environmental niche or that the toxin-producing genes would not transfer to organisms that would otherwise not possess such genes.

Interim Policy. Because of these concerns, EPA determined that it was necessary to scrutinize certain microbial pesticides before they were applied in the environment. As a result, EPA issued an interim policy in 1984 (6) requiring that the Agency be notified at least 90 days before conducting a small scale outdoor test with a genetically altered microbial pesticide. (The policy also required notification before small scale testing of nonindigenous microbial pesticides because such tests could involve the introduction of microorganisms into an environment where they did not exist before and therefore might not be subject to natural control mechanisms.) During this 90 day period, EPA would review the submission and determine whether an EUP would be required.

The policy also contained guidance on the kinds of data and information that should be provided to enable the Agency to evaluate the submitted request to conduct a small scale field test. This policy statement was reiterated in a proposed interagency policy statement issued December 31, 1984 (7) and was subsequently revised and set forth on June 26, 1986 as part of a final interagency policy statement (8). Under the June 26, 1986 policy statement, EPA reviews all genetically engineered microbial pesticides before they may be legally applied in the environment.

EPA Scientific Risk Assessment - Case History

Risk assessment for a microbial pesticide is the systematic scientific evaluation of hazard and exposure data, followed by the formulation of conclusions about the potential for human health or environmental risks as a result of using the pesticide. Such an assessment is based on the relationship between hazard, exposure and risk. Hazard is expressed in terms of an adverse effect such as death of a non-target organism due to a microbial toxin or other pathogenic action. Exposure of non-target organisms may be direct at the time of pesticide application, or may be indirect as a result of microorganism reproduction and dispersal. Before a non-target organism may be at risk, it must be exposed to the microorganism and the microorganism must have an adverse effect on the non-target organism (Hazard x Exposure = Risk).

In accordance with EPA's interim policy (6), the Advanced Genetic Sciences Company (9) and Dr. Stephen Lindow (10), University of California, Berkeley each notified EPA, in late 1984, of their intent to conduct small scale field trials with genetically engineered strains of _Pseudomonas syringae_ and/or _P. fluorescens_. Both groups sought to evaluate, under actual field conditions, the usefulness of the engineered strains as frost protection agents on crops such as strawberries and potatoes. Using recombinant DNA techniques, Advanced Genetic Sciences (AGS) and Dr. Lindow deleted a portion of a gene from their parental strains and with it, the ability to nucleate ice crystals at high subfreezing temperatures. The resulting non ice-nucleating (INA^-) deletion mutants were to be applied to crops at a time that would allow the INA^- bacteria to colonize the plant surface before significant populations of indigenous ice nucleating bacteria (INA^+) are established. If the tests were successful, the INA^- bacteria would inhibit colonization of INA^+ bacteria on the plant, thereby imparting protection against formation of frost. INA^- bacteria are considered to be a pesticide when used in this manner because they are applied for the purpose of controlling pest organisms (i.e., INA^+ bacteria that cause frost damage).

EPA conducted its initial review of the AGS and Lindow proposals under the 1984 notification policy (6) and determined that EUPs would be required. Both applicants subsequently submitted EUP applications that were reviewed and granted by EPA. As a first step in its review, Agency scientists with the relevant disciplinary expertise reviewed the submitted data and developed a risk assessment for each notification and EUP application. The EPA review process for these proposals was extensive (Figure 2) and included opportunities for the public to obtain copies of the EUP applications and to comment on the EPA risk assessment. EPA personnel also inspected the proposed test sites, inspected the applicants' books and records that supported their data, and participated in public meetings to discuss the Agency's analysis of the proposed field trials. In addition, the EPA risk assessment was subjected to rigorous peer review by scientists in other Federal Agencies and by a group of outside independent scientists convened by EPA as a subpanel of its FIFRA Scientific Advisory Panel (SAP).

Over 30 months elapsed between the time AGS and Lindow notified EPA and the tests were conducted. The EPA review process for the notifications and EUP applications consumed about six months, and the applicants spent four to six months developing additional data requested by EPA. The remaining 18 to 20 months were consumed by efforts to resolve legal challenges mounted against both EPA and the applicants in an effort to prevent the field trials. Table I

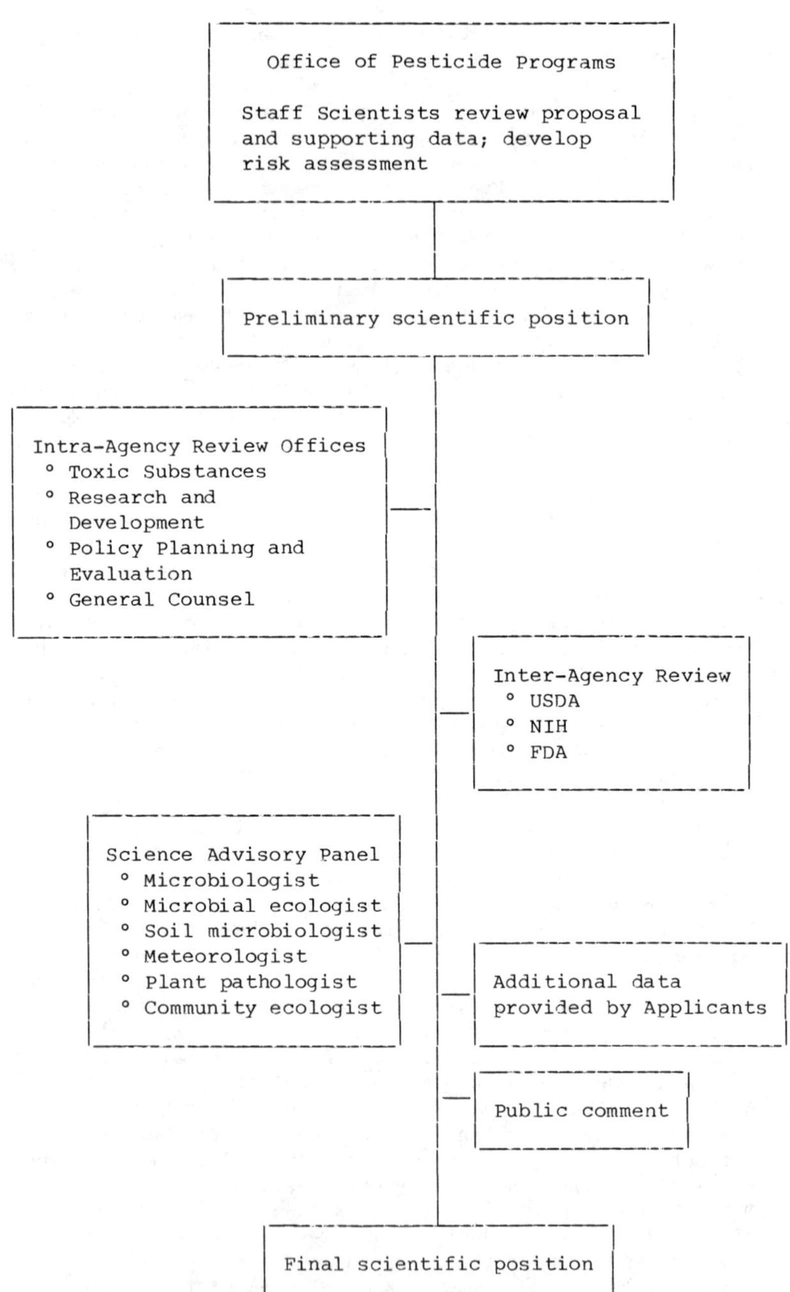

Figure 2 EPA Review Process for the Advanced Genetic Sciences and University of California (Dr. Lindow) Experimental Use Permit Application.

Table I

Chronological Summary of events in EPA review of Advanced Genetic Sciences' and Dr. Lindow's proposals to conduct field tests with genetically engineered microbial pesticides

Advanced Genetic Sciences

°November 1984	- EPA received Notification
°January 1985	- EPA completed preliminary risk assessment, FIFRA Scientific Advisory Panel (SAP) met to review submission and EPA risk assessment.
°February 1985	- EPA responded to AGS requesting additional information and an EUP.
°July 1985	- EPA received EUP applications.
°August 1985	- EPA announced receipt of EUPs in Federal Register and began 30 day comment period.
°September 1985	- Received public comments on EUP
°September/ October 1985	- Received SAP and intra-agency review of public comments.
°November 1985	- EPA granted EUPs.
°November 1985	- EPA sued by Foundation on Economic Trends.
°February 1986	- AGS' outdoor (rooftop) testing of INA⁻ bacteria disclosed.
°March 1986	- EPA audited AGS and suspended EUPs; plant pathogenicity testing to be repeated.
°July 1986	- EPA reviewed and approved AGS plant pathogenicity retest.
°December 1986	- AGS submitted proposed test sites.
°February 1987	- EPA approved test sites and reinstated EUPs.
°April 1987	- AGS sued and injunction requested in state court. Request denied; strawberry plot vandalized.
°April 24, 1987	- AGS applied INA⁻ bacteria at test site in Contra Costa County, California. EPA and State conducted monitoring.

Continued on next page

Table I--Continued

°June 1987	- Test concluded; AGS monitoring continued.
°August 1987	- AGS submitted summary of results of Spring 1987 testing.
°September 1987	- AGS submitted amended EUP application to conduct 2nd small test at same location in California.
°November 1987	- EPA approved AGS amended EUP.
°December 1987	- AGS began 2nd field test (fall and spring applications).
°February 1988	- 2nd application of INA^- bacteria made.
°December 1984	- EPA Received Notification.
°February 1985	- FIFRA SAP provided written comments.
°March 1985	- EPA responded to Dr. Lindow, requesting additonal information and an EUP.
°December 1985	- EPA received EUP application.
°May 1986	- EPA inspected test site, participated in public meeting, conducted books and record check.
°June 1986	- EPA granted EUPs - for multiple tests conducted over 3 year period.
°August 1986	- Dr. Lindow (U. California) and State sued by Foundation on Economic Trends et. al., experiment postponed pending development of further information.
°May 28, 1987	- Test plot was vandalized; spray appication of INA^- bacteria made; EPA conducted aerial monitoring.
°August 1987	- INA^- treated potato seed pieces planted for second test.
°September-December 1987	- Dr. Lindow submitted results of field testing.
°May 1988	- Further testing to begin.

summarizes the chronology of events that took place from the time the notification was received by the Agency until the actual field test took place.

EPA's risk assessment of the AGS and Lindow proposals consisted of a four step analysis: (1) identification of key risk issues, (2) development of a reasonable worst-case risk scenario, (3) application of data and information to address the key risk issues, and (4) formulation of conclusions with respect to likelihood of significant human health or environmental risks.

Risk Assessment. Identification of potential risks (step one) was based on an analysis of the taxonomic and ecological characteristics of the parental organisms, the functional changes in the microorganisms brought about by the genetic alteration, the mechanism of pesticidal action, and the nature and scope of the proposed field trials. Evaluation of these four areas was the basis for identification of potential hazards and mechanisms for exposure, which in turn were used to formulate the risk issues.

The parental strains (*P. syringae* and *P. fluorescens*), are plant surface colonizers (epiphytes) and all have the ability to nucleate ice crystals at high subfreezing temperatures. Based on experience and available information, none of the strains is from a species known to be pathogenic to humans or other mammals. However, certain strains of *P. syringae* are agriculturally important plant pathogens.

The engineered strains (INA^- deletion mutants) contained no new or additional genetic material. Phenotypically, they would resemble the naturally occurring non-ice nucleating bacteria (NON-INA) that are often the numerically dominant plant surface colonizers. No new function was imparted to the engineered strains. Instead, their pesticidal activity would be achieved by giving the engineered epiphytes the opportunity to colonize plant surfaces before indigenous INA^+ bacteria are able to do so. Frost protection would be achieved by using the INA^- deletion mutants as competitive inhibitors of the indigenous INA^+ strains. Since the INA^- deletion mutants were derived from natural INA^+ epiphytes, the engineered strains would be put back into the habitat for which they are adapted to survive, and thus should be ideal candidates for competitive exclusion of the pest strains (INA^+ bacteria). The proposed field trials would involve the outdoor foliar application of INA^- deletion mutants to blossoming strawberries and emerging potato plants. The application would be on small test plots (< 0.5 acres) using hand-held sprayers.

Potential mechanisms for pesticide dispersal and exposure of nontarget species would include direct contact of treated plants by animals (e.g., insects) entering the test plot with subsequent animal-borne dispersal of INA^- bacteria, aerial drift from the test plot at time of application, and post application movement of INA^- bacteria into the atmosphere with subsequent deposition in habitats that may favor bacterial colonization on plants, or movement into the upper atmosphere.

Based on the available and provided information on the parental strains, details concerning the genetic alteration, and parameters of the field trials, it was then possible to identify the specific issues that needed to be addressed in the risk assessment. These issues included non-target plant pathogenicity, possible alteration of geographic range and distribution of frost sensitive or frost tolerant plants and insects, and possible adverse effects on climatic patterns due to reduction of INA^+ bacteria in the atmosphere.

Risk Scenario. Although there was no indication that the latter two risk concerns were likely to be significant issues for the proposed test, EPA assessed the potential impacts in the unlikely event that one or more of these effects occurred. Therefore, in step two, EPA scientists addressed all of the identified issues in a systematic fashion. This was accomplished by developing a risk scenario and asking: what series of events must take place in order for these risks to materialize? It was determined that a sequence of three events must occur: (1) INA^- deletion mutants must be dispersed outside the test plot; (2) INA^- bacteria must colonize a wide variety of plant species over a large geographic region; and (3) INA^- bacteria must displace INA^+ bacteria on a wide variety of plants over a large geographic region. In other words, beginning from a small inoculation, the INA^- deletion mutants must be able to outcompete indigenous bacteria in habitats outside the test plot. If any one or more of these events did not occur, then no significant risks would result from the proposed use.

Addressing Risk Issues. Step three in EPA's evaluation was to apply the available data to address the risk issues and to determine whether the data were sufficient to negate the risk concerns. This effort focused on dissemination, colonization, competitiveness, and plant pathogenicity the INA^- deletion mutants as well as the field test parameters.

Dissemination of applied INA^- deletion mutants. In its assessment, EPA scientists assumed that, given the nature of the proposed experiments and the ecological characteristics of epiphytic bacteria such as those to be applied, the INA^- bacteria could not be contained within the boundaries of the test plot. However, field studies conducted with the INA^+ parental strains suggested that dispersal of the applied bacteria from the test site would probably be minimal.

Colonization of INA^- deletion mutants on non-target plants. AGS and Dr. Lindow conducted greenhouse and laboratory tests with a wide variety of agronomically and ecologically important plant species to evaluate the ability of the INA^- bacteria to colonize other species. These tests indicated moderate host specificity of the INA^- strains.

Competitiveness of INA^- deletion mutants. Greenhouse tests had been conducted to evaluate the competitiveness of the INA^- strains compared to their naturally occurring parental strains as well as other epiphytic strains. Results indicated that the INA^- strains are competitive equals with naturally occurring NON-INA and INA^+ strains. Application of equally low or high dosages of INA^+ and INA^- bacteria resulted in co-existance without either population predominating. When unequal dosages were applied, the microorganisms applied at the higher level always predominated while the other was present at low levels.

Plant Pathogenicity. Over 100 plant species had been tested using a variety of inoculation methods. The INA⁻ bacteria were not pathogenic for any plants tested. Results of *in vitro* biochemical and diagnostic tests also suggested that the INA⁻ bacteria lacked plant pathogenic potential.

Parameters of Field Tests. The scope of the test and the manner in which it was to be conducted and monitored was considered along with the supporting data on INA⁻ strains in order to evaluate potential environmental impacts. Factors considered included:

* small plot of land to be treated,
* single test to be conducted,
* INA⁻ bacteria to be applied by hand held sprayer to minimize drift,
* test plots to be surrounded by bare soil buffer zone
* INA⁻ populations to be monitored post application on plants, soil, insects and in air both within and outside of the test plot,
* Contingency plan - biocide to be applied if INA⁻ bacteria observed to colonize plants outside the treatment area.

Having evaluated all available data in the light of the actual field tests being proposed, EPA then completed the fourth and final step in its risk assessment; to develop conclusions concerning the likelihood of risks to human health and the environment. Collectively, the data provided by AGS and Dr. Lindow supported the following conclusions: (1) low numbers of INA⁻ deletion mutants applied on a small test plot (< 0.5 acres) will disseminate beyond the borders of the test plot; (2) the number of INA⁻ deletion mutants drifting on to plants outside the test plot (if any) will be much lower than the indigenous INA⁺ and NON-INA bacteria already present on these plants; (3) plants outside the test area will probably be inoculated, by insects, with mixed populations of INA⁻ and indigenous bacteria; (4) INA⁻ deletion mutants were not pathogenic to the plants tested; (5) INA⁻ deletion mutants have a restricted host range; they cannot survive at a detectable level on all non-target plants; and (6) INA⁻ deletion mutants will not have a competitive advantage over indigenous bacteria on plants outside the test area.

Considered in the context of the above described EPA risk scenario, these conclusions mean that limited dispersal and colonization of INA⁻ deletion mutants outside the test plots will occur. However, the extensive experimental data on competitiveness developed by AGS and Dr. Lindow demonstrate that INA⁻ deletion mutants disseminating from the test plots would not outcompete and displace INA⁺ bacteria. Thus, the risk scenario would not be realized.

Based on these findings, EPA scientists were then able to conclude that the proposed small scale applications of INA⁻ bacteria would pose no foreseeable risks to humans or the environment. Hypothetical adverse effects on precipitation patterns or on the survival and geographic range of plants and insects would not be realized.

Summary

(1) EPA's regulatory program and review mechanism for genetically engineered microbial pesticides is in place.

(2) Risk assessment for genetically engineered microbial pesticides is based on an evaluation of potential hazards and exposure of the pesticide.

(3) The assessment process consists of four steps:
  * identify key risk issues,
  * develop potential risk senario,
  * apply data and information to address key risk issues,
  * formulate conclusions with respect to likelihood of significant human health or environmental risks.

(4) Genetically engineered products are reviewed on a case-by-case basis to address any special questions that may arise. Most risk assessment considerations and methods for data development are similar to those established for naturally occurring microbial pesticide products.

(5) Individual small scale field test proposals have been reviewed. The Agency's regulatory decisions have withstood legal challenge and two field trials have been conducted.

(6) The Agency expects to refine and streamline its review process as additional data and experience are gained, and the public becomes more comfortable with the outdoor application of genetically engineered pesticides.

Literature Cited

1. Federal Insecticide Fungicide Rodenticide Act, 7 U.S.C. 136 et. seq., as amended.
2. Regulations for the Enforcement of the Federal Insecticide Fungicide and Rodenticide Act, Title 40 Code of Federal Regulations, Part 162.
3. Experimental Use Permits, Title 40 Code of Federal Regulations, Part 172.
4. Data Requirements for Pesticide Registration, Title 40 Code of Federal Regulations, Part 158.
5. Pesticide Assessment Guidelines - Subdivision M, National Technical Information Service, Springfield, VA. #PB83-153965, 1983.
6. Microbial Pesticides: Interim Policy on Small Scale Field Testing, FR 49 (202), 40659, 1984.
7. Proposal for a Coordinated Framework for Regulation of Biotechnology; Notice, FR 49 (252), 50882, 1984.

8. Coordinated Framework for Regulation of Biotechnology; Announcement of Policy and Notice for Public Comments, FR 51 (123), 23302, 1986.
9. Advanced Genetic Sciences Experimental Use Permit Application, 1985, OPP Docket No. 50645.
10. University of California Experimental Use Permit Application, 1985, OPP Docket No. 50651.

RECEIVED June 14, 1988

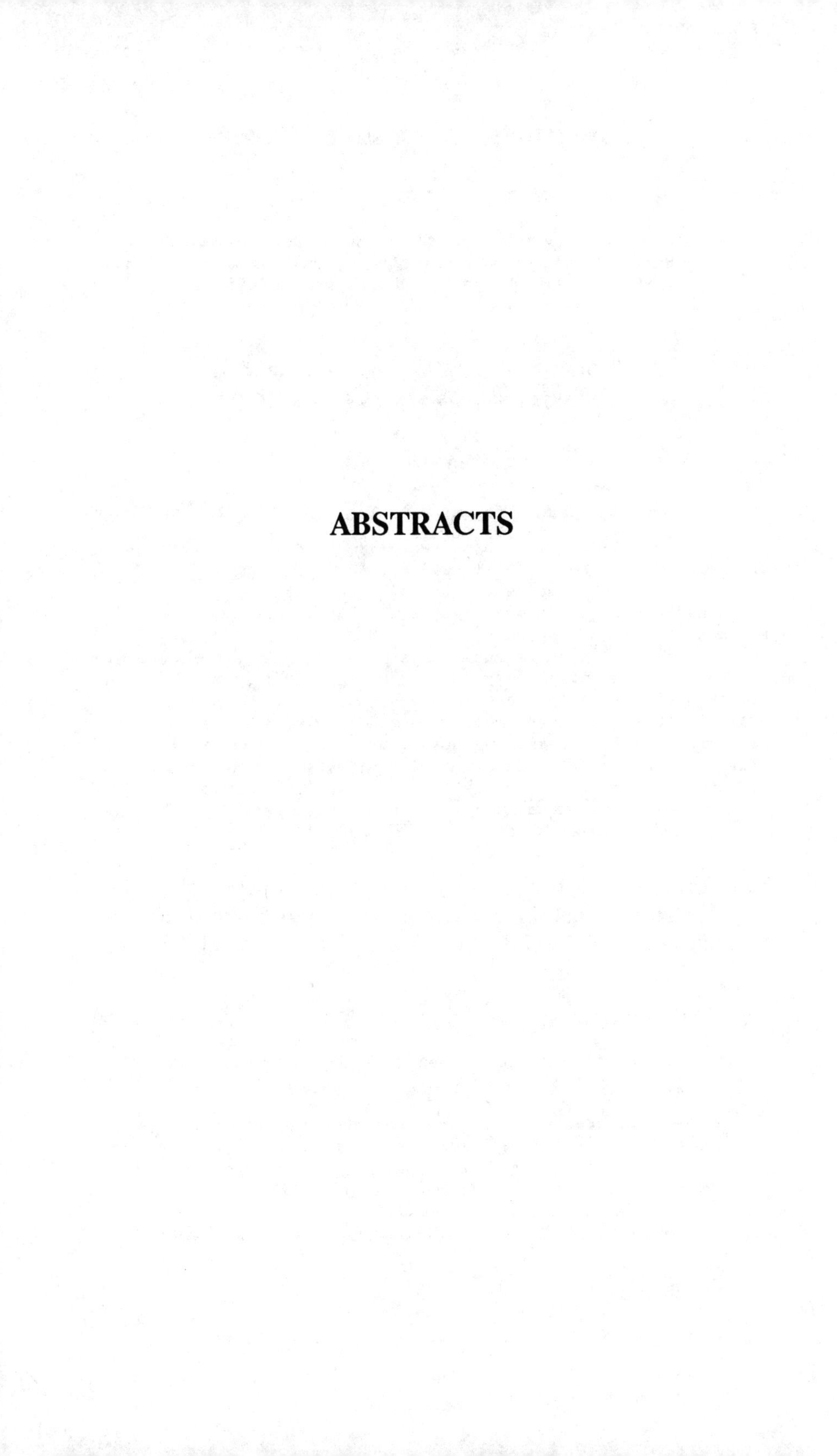

ABSTRACTS

Abstracts of Other Papers

Abstracts of papers that could not be included in this book but were pertinent to the symposium are included here. Copies of these presentations can be obtained by writing to the editors, if a complete set is desired, or by contacting the author for individual papers.

Applications of Restriction Fragment Length Polymorphisms to Crop Plants

Tim Helentjaris

Molecular Biology Group, NPI, Salt Lake City, UT 84108

Restriction fragment length polymorphisms (RFLPs) can be used as genetic markers, but with significant advantages over conventional marker systems. We have generated large numbers of these markers and in some cases, essentially saturated the genomes of some crop plants. Among their many applications, one of the more important is the analysis and dissection of complex traits into individual components. Using RFLP analysis, one can determine how many loci are involved in a complex trait, evaluate their relative importance and gene action, and in some cases relate them to known genes. This type of result can increase our general understanding of traits such as insect resistance, but also allow us to more readily manipulate them by conventional breeding strategies and potentially aid in cloning and genetic engineering of the major factors.

Role of the U.S. Environmental Protection Agency's Scientific Advisory Panel in Review of Proposed Environmental Application of Genetically Engineered Microbial Pesticides

James M. Tiedje

Department of Crop and Soil Sciences, Michigan State University, East Lansing, MI 48824

Having participated in all FIFRA reviews of genetically engineered microorganisms, and after genetic modifications included under the revised Fed. Reg. Guidelines, I will report on the role the committee has played, types of questions asked, and share insight into what I think are the most important scientific issues. The focus will be on scientific issues.

DNA Probe Technology To Track Engineered Microorganisms in Nature

James M. Tiedje, W. Holben, J. K. Jansson, and B. K. Chelm

Department of Crops and Sciences, Michigan State University, East Lansing, MI 48824

Methods have been develeoped to adapt DNA probe technology for use in detecting specific gene sequences in soil. Principle improvements were in methodology for general lysis of total soil bacterial populations in purification of DNA especially from humic acids, and in case of a most specific and sensitive vector system - M13. The npt II gene was inserted into *Brady rhizobrasin* and *Pseudomonas*, and this segment was used to specifically detect those modified strains. As few as 10^3 gene copies/g soil or 1 organism in 10 gene copies/g soil or 1 organism in 10^5-10^6 could be detected. Genuine digest of soil together with Southern blots allowed us to detect more than one organism in the same sample, and to detect genetic rearrangement of the engineered gene.

Immunological Approaches to the Study of Neuronal Development in Insect Embryos

P. H. Taghert

Department of Anatomy and Neurobiology, Washington University School of Medicine, St. Louis, MO 63110

Within the past several years, the application of immunological probes has advanced the studies of neurogenesis and neuronal differentiation. In particular, poly-specific and mono-specific antibodies have found use as histological tools to analyze the complex cellular events and interactions that take place as the nervous system is formed. This talk will review such studies and, in particular, will focus on the development of a simple neuroendocrine system in embryos of the moth, *Manduca sexta*. Serum and monoclonal antibodies to specific neuropeptides are used to establish the cell-specific patterns and onsets of neuropeptide expression; serum antibodies in conjunction with injected fluorescent dyes are used to study the morphology of the embryonic neurons; monoclonal antibodies are used to elucidate the roles of specific non-neural cells in helping to establish the elaborate terminal fields of the neuroendocrine neurons. Finally, the use of several of these antibodies to clone relevant genes will also be discussed.

Expression and Alteration of a *Bacillus thuringiensis* Endotoxin Gene in *Escherichia coli*

James Rusche, Cindy Lou Jellis, Kevin Farrell, John Farley, James Hodgdon, Helen Carson, and Daniel Witt

Repligen Corporation, One Kendall Square, Cambridge, MA 02139

A Lepidopteran specific δ-endotoxin gene was cloned from Bacillus thuringiensis var. kurstaki HD-1 and engineered to produce the active domain of the toxic protein. The engineered gene was expressed in E. coli and the recombinant cells were found to be toxic to a variety of Lepidopterae larvae. Two regions within the gene of about 150 nucleotides each were selected for directed mutagenesis. In vitro mutagenesis was carried out on these target fragments with the goal of randomly introducing about one amino acid change per mutant clone. Over 15,000 resultant clones were screened for increased activity and, of these, about 12 mutants were identified which were 5-50 X more toxic toward tobacco budworm larvae than the non-mutant recombinant parent. These mutants were sequenced to identify the point mutations responsible for the enhanced toxicity phenotype. Novel combinations of these point mutations were created and gave rise, in some cases, to further increases in activity. Site directed saturation mutagenesis was also carried out on selected codons and resulted in additional enhanced activity mutants.

Biotechnology Risk Assessment Research in the Environmental Protection Agency

A. W. Bourquin[1,3], R. Seidler[2], T. Barkay[1], F. Genthner[1], and S. Cuskey[1]

[1]U.S. Environmental Protection Agency Research Laboratories, Gulf Breeze, FL 32561
[2]U.S. Environmental Protection Agency Research Laboratories, Corvallis, OR 97333

Deliberate release of genetically engineered microorganisms poses potential risks associated with survival of released cells, the potential for gene exchange and the effects of released microbes on ecological processes. Efforts to improve cell and/or DNA detection in environmental samples focused on the use of DNA "detection cassettes" based on unique DNA sequences and on improvements in DNA extraction methods from soil and aquatic sediments (thus obviating the need for cell growth for detection). Biological control of released cell survival was attempted through the use of conditional lethal genetic determinants. The potential for plasmid gene transfer to the indigenous community was assessed in environmental samples. Self-transmissibility, plasmid incompatibility, group and cell densities, among other variables, were important determinants. Model systems, based on the addition of mercury-resistant bacteria to microcosm environments or on changes in the 16s riboromal RNA profiles from bacterial sub-populations were developed to assess potential effects from released genetically engineered microbes on ecological processes. These, and other techniques were employed in assessments of risks in a release of ice-nucleation-negative bacteria in California.

[3]Current address: ECOVA Corporation, 15555 NE 33rd Street, Redmond, WA 98052

Author Index

Affiliation Index

Subject Index

C

Q

R

Production by Joyce A. Jones
Indexing by Colleen P. Stamm
Jacket design by Lisa Damerst

Elements typeset by Hot Type Ltd., Washington, DC
Printed and bound by Maple Press, York, PA

Recent ACS Books

Biotechnology and Materials Science: Chemistry for the Future
Edited by Mary L. Good
160 pp; clothbound; ISBN 0–8412–1472–7

Chemical Demonstrations: A Sourcebook for Teachers
Volume 1, Second Edition by Lee R. Summerlin and James L. Ealy, Jr.
192 pp; spiral bound; ISBN 0–8412–1481–6
Volume 2, Second Edition by Lee R. Summerlin, Christie L. Borgford, and Julie B. Ealy
229 pp; spiral bound; ISBN 0–8412–1535–9

The Language of Biotechnology: A Dictionary of Terms
By John M. Walker and Michael Cox
ACS Professional Reference Book; 256 pp;
clothbound, ISBN 0–8412–1489–1; paperback, ISBN 0–8412–1490–5

Cancer: The Outlaw Cell, Second Edition
Edited by Richard E. LaFond
274 pp; clothbound, ISBN 0–8412–1419–0; paperback, ISBN 0–8412–1420–4

Chemical Structure Software for Personal Computers
Edited by Daniel E. Meyer, Wendy A. Warr, and Richard A. Love
ACS Professional Reference Book; 107 pp;
clothbound, ISBN 0–8412–1538–3; paperback, ISBN 0–8412–1539–1

Practical Statistics for the Physical Sciences
By Larry L. Havlicek
ACS Professional Reference Book; 198 pp; clothbound; ISBN 0–8412–1453–0

The Basics of Technical Communicating
By B. Edward Cain
ACS Professional Reference Book; 198 pp; clothbound; ISBN 0–8412–1451–4

The ACS Style Guide: A Manual for Authors and Editors
Edited by Janet S. Dodd
264 pp; clothbound; ISBN 0–8412–0917–0

Personal Computers for Scientists: A Byte at a Time
By Glenn I. Ouchi
276 pp; clothbound; ISBN 0–8412–1000–4

Chemistry and Crime: From Sherlock Holmes to Today's Courtroom
Edited by Samuel M. Gerber
135 pp; clothbound; ISBN 0–8412–0784–4

For further information and a free catalog of ACS books, contact:
American Chemical Society
Distribution Office, Department 225
1155 16th Street, NW, Washington, DC 20036
Telephone 800–227–5558